Quest

Other books by René Dubos

Pasteur: Free Lance of Science

Mirage of Health

Man Adapting

So Human an Animal

A God Within

Beast or Angel?

The Professor, the Institute, & DNA

The Wooing of Earth

Other books by Jean-Paul Escande

Les Médecins

Les Malades

Quest

Reflections on Medicine, Science, and Humanity

René Dubos &
Jean-Paul Escande

Translated by Patricia Ranum

Harcourt Brace Jovanovich
New York and London

Printed in the United States of America

Library of Congress Cataloging in Publication Data

Dubos, René Jules, 1901-
Quest: reflections on medicine, science, and
humanity.

Translation of Chercher.
1. Medicine—Philosophy. 2. Science—Philosophy.
I. Escande, J. P., joint author. II. Title.
[DNLM: 1. Medicine. 2. Science. 3. Bacteriology—
Biography. 4. Ecology—Biography. WZ100 D817]
R723.D7613 610'.1 79-3347
ISBN 0-15-175705-4

Set in Linotype Electra

First edition

B C D E

To Hélène and Jacques Dufresne

Contents

Preface

Dr. Escande mentions in his Introduction the circumstances under which he and I met for the first time, during a symposium on medicine organized in Montreal by Jacques Dufresne, professor of philosophy. The symposium was one of the activities of the Canadian group Critère, of which Jacques Dufresne and his wife, Hélène, are the "*animateurs*." The spirit of the Critère group is similar to that which presided over my discussions with Dr. Escande.

Every year, Critère focuses its activities on one or more topics of contemporary importance and examines them from multiple points of view—historical, philosophical, ethical, economic, and of course scientific. Directly or indirectly, I have collaborated with the members of Critère in discussions about medicine, urbanism, and regionalism. I believe with them that all contemporary problems, even those seemingly of a purely technical nature, must be evaluated in the light of broad humanistic as well as scientific considerations. This is what Dr. Escande and I tried to do in Paris, with the gracious and enlightened help of Dr. Claire Lugassy, in the course of conversations that ranged from medical research to the meaning of freedom.

Dr. Escande eventually converted our dialogue into a book which he entitled *Chercher*, a title that fits the rambling, exploratory nature of our conversations. The more pointed word *Quest* selected by the editors of the English translation may help to convey the searching mood in which Dr. Escande and I approached the problems of scientists, physicians, and humankind in the modern world. I wish to tell all those who have contributed to the writing, translating, and editing of this book how much I have appreciated reading my own thoughts so accurately expressed and interpreted through their efforts.

RENÉ DUBOS

Introduction

1939. For the first time a systematic method is worked out for the discovery of antibiotics and for their production on a large scale. Gramicidin, also known as tyrothricin, is the first antibiotic to be developed rationally and produced industrially. The door has been unlocked. Penicillin and streptomycin are yet to be used. Who took this decisive step? René Dubos, a Frenchman born at the turn of the century at Saint-Brice, a village near Sarcelles, not far north of Paris.

1955. The effects of isoniazid, a powerful medication for tuberculosis, are being tested. One group of patients is being given this drug along with bed rest; the others continue to go about their daily activities. The two groups get well at the same rate. Conclusion: tuberculous patients need no longer spend endless months in bed. Where was this study carried out? In René Dubos's department.

1972. Pollution is making the industrialized world shudder, for that world is finally discovering that one day it may well smother in the waste it has been producing. Anxiously, world leaders meet at Stockholm to examine the situation and plan a program of action. Who is selected to prepare a book outlining the nature of the problem? The English economist Barbara Ward and René Dubos, who for years has been sounding warnings, although he continues to have faith in mankind.

1978. Where will we find energy sources? And how should we

use the energy available? A new organization is created, the René Dubos Forum, which takes as one of its themes "Better life with less energy."

1980s. Television stations throughout the world will be showing programs several hours long devoted to a man whose ideas are well worth the vast sums that the production cost. The man is René Dubos.

So what do you imagine René Dubos is like? Pedantic, pontificating, sententious? Quite wrong! Vindictive, grumpy, spouting maledictions and prophecies of doom? Even more wrong! A hermit, inaccessible, walled up in his ivory tower? No, wrong again!

René Dubos is very simply a splendid man with a rare blend of sincerity, knowledge, pensiveness, humor, and esteem for his fellow men, a man whose thirst for truth causes him to expand his horizons daily. For Dubos's whole life has never deviated an inch from his overall view of the world; he has managed to integrate into his thoughts all the realities that individually make up the fabric of our everyday life in its humblest aspects as well as in its most ambitious achievements. The result is the René Dubos of today: an American Frenchman, very tall, smiling, attentive, and accessible, who says marvelous and simple things, who demonstrates that there are no big or little problems but rather a chain of realities that you can master only by looking at the links that bind them.

This great scholar to whom many of us surely owe our lives today—this sought-after man who has been asked to preach in the Cathedral Church of St. John the Divine in New York City, to preside over meetings of the most eminent scholars, and to teach students in the most sophisticated universities—this unusual man defies easy categorization. He is not only a very great biologist, a pioneer ecologist, and a specialist in the social sciences who is recognized the world over; he is also, very simply, a universal man.

Dubos is also a talented storyteller to whom one could listen for hours without ever tiring. His life takes on the contours of a Grimm fairy tale, with its shadowy corners and its brightly lighted areas. His is a very new and very original conception of life that leaves stereotypes, prefabricated ideas, and modish theories behind.

I was very lucky one day—I met René Dubos. The Canadian philosopher Jacques Dufresne had organized a meeting of physicians and nonphysicians at Orford, near Montreal. Dubos gave the opening address, and the audience, which had been looking forward to hearing him, was not disappointed. For an hour he evoked in simple terms, with precise examples, the problems of man's health. An hour was all he needed to electrify the room.

Some lecturers are like some playwrights: as the days go by, every trace of the words and sentences that exalted the soul gradually fades away. But Dubos's parables are not forgotten. Indeed, as time passes they take root in you, very gradually become a part of your own thoughts, and without ever causing you to contradict yourself, make you see the most important problems from a different viewpoint.

That is why, when Laurence Pernoud and Christian de Bartillat asked me who could present an original view of research and medicine, only one name came to mind: René Dubos. But would he accept? When a person lives surrounded by Nobel Prize winners of all sorts, when he is considered by Americans to be one of the seminal thinkers of his day, will he accept an invitation from a stranger to play the dangerous interview game for two whole weeks?

I sent Jean-Loup Chiflet to sound Dubos out. He knows the United States very well. How would he get in touch with Dubos? No problem. It's a quarter after eight in the morning? He must be awake—let's get the phone book. Du . . . Du . . . Dub . . . Dubos! There's only one Dubos in New York City! Let's try it. "*Allô*," said Chiflet, with the strongest possible French accent. "*Allô*," someone replied in a similar accent. "Professor Dubos?" "Speaking." The contact had been made. Next came a brief explanation of the purpose behind the call. Everything was going well, but the moment of truth was approaching. When could they meet? "Well," proposed Dubos, "come to my office at the Rockefeller University* at nine."

* The Rockefeller Institute for Medical Research, created in 1901, was transformed in 1965 into The Rockefeller University.

After three hours of friendly discussion the matter was all arranged. Because, as Dubos said, in any life sometimes one has to risk wasting a bit of time. So in September 1978 Dubos came to Paris to talk about what he knows and what he believes, to talk about science, medicine, and humanity. Yet above all he did not want to be presented as a visionary prophet, although he was willing to answer questions that had nothing to do with his specialization.

This book serves as a testimony, an analysis of the past, and a harbinger of our future.

JEAN-PAUL ESCANDE

Quest

I

René Dubos Conquers the West

The most striking thing about René Dubos's astonishing career is its diversity. It would be easy to refer to the five lives of René Dubos, who has been in turn an agronomist, a microbiologist, an authority on tuberculosis, a committed ecologist, and finally an eminent observer in the social sciences. People have often said of him, "He's been so lucky! What fantastic good fortune to change his field constantly yet always remain at the top!"

What if this interpretation were completely false? What if, on the contrary, there was but a single life, a single philosophy for Dubos? And what if it actually was all the same battle, but waged on different fronts?

If this second interpretation were correct, it would mean that we must not view Dubos's life as a marvelous fairy tale. It would mean that we must study his life and his work in order to find an original conception of the world about us, a conception from which each of us could profit.

And so, through the narrative of his life, through René Dubos's conquest of the West, we must seek what he actually announced at the outset: the profound unity underlying his multifaceted thought. How can one not be astonished when this seventy-nine-year-old man, recalling his earliest memories, immediately comes up with the preoccupation that has guided him throughout his life? That preoccupation is this question: How can we understand the relationships between one thing and another? Everything began at the Collège Chaptal with a book about the French fabulist Jean de La Fontaine.

DUBOS: I began my scientific training in Paris, at the Collège Chaptal, a public high school. My recollections of this period are indelible. I recall having read Hippolyte Taine's *La Fontaine et ses Fables*. He asserted that if La Fontaine had been born in the forests of Germany, the deserts of the Sudan, or a Mediterranean country, he wouldn't have written his fables in the same way. Indeed, Taine asserted, the form of the French countryside gave the fables their form. Strangely enough, I've never forgotten this. I reread Taine's book three or four years ago, and when all is said and done this point isn't as clearly stated as I remembered, but it's there all the same. So at a very young age I was already prepared to believe that things aren't easy to understand and that, if they are to be understood at all, they must be studied according to their relationships with other things rather than individually, in isolation.

Having graduated from the Collège Chaptal, I first studied at the Institut Agronomique in Paris, and then at the Ecole d'Agriculture Coloniale in hopes of going to Indochina. But as a boy I had suffered from severe rheumatic fever that left me with a serious heart lesion, so I wasn't permitted to go to Indochina. Bad luck must be good for something, for a few months afterward I was fortunate enough to get a job in Rome at the International Institute of Agriculture, an organization administered by the League of Nations, now called the Food and Agriculture Organization (FAO). For two years I worked in Rome as an editor, preparing articles on the agricultural sciences.

One day, in the French magazine *Science et Industrie* (which I suppose still exists), I by chance read an article by a Russian who was working at the Institut Pasteur at the time. His name was Sergei Winogradsky. Winogradsky, a microbiologist, had come to the Institut Pasteur after the Russian Revolution. He was already very famous. In fact, he'd been offered a post at a section of the Institut Pasteur located at Brie-Comte-Robert, not far from Paris. His article made a profound impression on me. He was studying microbes in the soil and asserted in this article that microbiologists were committing a fundamental error in studying microbes solely in the laboratory, in pure cultures, in artificial surroundings, since in nature microbes never live in pure cultures. Microbes whose natural habitat is the soil live there in association with other microbes, under completely unique conditions that of course are different from those in the laboratory. In addition, this article had broader implications. In a word, it meant that if you want to study life, you have to study it in its natural habitat and not proceed only as microbiologists do, by isolating microbes from their environment.

I was so impressed that I immediately decided to become a microbiologist and promptly took a course in microbiology at the University of Rome. Since I had no money, I did all sorts of jobs, especially working as a guide. I spoke Italian well and had learned English by dating English girls, so I became a guide in Rome. I had absolutely no specific idea about what I was going to be, but for a long time I'd been yearning to travel. I wanted to go to the United States. By taste, by inclination. Perhaps also because I was a bit afraid of Parisian society. I really don't know. In any case, I wanted to get away. Then fate stepped in: in Rome I met two people who later opened the first American doors for me.

The first of these two people was the American representative to the International Institute, Asher Hobson, a professor of agricultural economics at the University of Wisconsin at Madison. Each nation sent a delegation to Rome, presided over by an eminent figure. Almost all of them wore top hats to official dinners, but Hobson would appear for dinner without one. I used to meet him on the way to the office because we often arrived a bit earlier than the others. Our paths crossed a few times; he soon recognized me and greeted me with a casual wave, and I would reply with an equally casual gesture. That's how we became friends! Such a thing would have been impossible with the French representative. One day Hobson asked about my plans, and I replied that I wanted to go to the United States.

"Well, why don't you?" "Because I don't have enough money yet!" Then he said very cordially, "Go to the States. If you have any problems, let me know and I'll help you." And he added a remark that I'll never forget: "I know how to pick a winner." That's America in a nutshell. Of all the young people there, I alone had dared wave casually to an official delegate, so he had concluded that I must have some originality.

ESCANDE: So Hobson, the American representative, was your first fairy godmother. Who was the second person who watched over young René Dubos?

DUBOS: That episode in my life is just as incredible. I was on my way to America. In those days the crossing took eleven days. I was leaning on the rail of the steamship *Rochambeau*, watching the French coast recede into the distance. I wasn't homesick, but I really didn't know what I was going to do. I had just enough money for maybe three or four months. I was standing there, lost in thought, when I felt a tap on my shoulder. A strange man and woman. Well, not total strangers—I'd seen them somewhere before. But where? They knew exactly where. I'd been their guide in Rome. Now at this time anti-Semitism was really very strong in America, and this couple, the Waksmans, had been totally snubbed by all the other American passengers. So I became very close to them and found out that he was a professor of microbiology at Rutgers University in New Brunswick, New Jersey, not far from New York City! Toward the end of the crossing, Professor Waksman said, "Why not come to New Brunswick? We have a good faculty." So shortly after I set foot in New York—about five o'clock the very first afternoon—I left for New Brunswick to register as a student in microbiology. Of course I looked for work right away, in order to earn some money. That was very easy at the time, and I did all sorts of things, even taught German. But above all I specialized in microbiology and completed my doctorate. So now I had a Ph.D.

To tell the rest of the story, I must add that after World War II, in 1952, Waksman won the Nobel Prize for his discovery of streptomycin and in this way confirmed that a good deed is never performed in vain. He had helped me, but I repaid him amply, since it was thanks to my earlier work with antibiotics—I'll talk about this later—that he was able to undertake his own work.

Now that I had my doctorate, I tried to get a research grant. But I learned that I wasn't eligible, because I wasn't an American citizen. However, the secretary (a total stranger) who sent me this bad news added a brief P.S.: "Why not go see your compatriot, Dr. Alexis Carrel, at the Rockefeller Institute for Medical Research in New York City?" I went to see Dr. Carrel. He received me cordially, told me that he couldn't help me, but nonetheless invited me to lunch in the dining room at the Rockefeller Institute. He seated me beside a professor, a most friendly man, who asked me what my doctoral thesis was about. I replied, "My dissertation is a study of the microbes in the soil that break down the cellulose of plants." And this professor answered, "That's similar to something we're trying to do in the lab here." He took me to his office. He was head of the department of infectious respiratory diseases and was especially interested in lobar pneumonia caused by a microbe, the pneumococcus. "We've shown," he told me, "that the pneumococcus is able to resist the white blood cells that would otherwise destroy it, because it's encapsulated in a substance—polysaccharide—that resembles cellulose. This capsule protects it. If only we could find some way to destroy the capsule!" Then, with the audacity of a twenty-six-year-old, I asserted, "I'm convinced that a microbe exists in nature that will attack those polysaccharides, because if such a microbe didn't exist, polysaccharides would cover the earth, which they don't. So something must exist somewhere that destroys them." And we left it at that.

Shortly afterward I got a job out West, but I was barely settled when I received a telegram. I'd been awarded a fellowship by the Rockefeller Institute to work with that very man, Professor Oswald T. Avery. It was the great turning point in my life.

Immediately I returned to New York City and set to work. Two years later I discovered a microbe that broke down the polysaccharide of the pneumococcus. From this microbe I obtained a substance, an enzyme, and demonstrated that when this enzyme was injected into animals with advanced pneumococcal infections, the enzyme did indeed destroy the capsule surrounding the microbe, and so permitted the white blood cells to do away with the pneumococcus.

Then immediately afterward I made another discovery—from my point of view, the more important of the two. I discovered that the microbe that destroyed the capsule of the pneumococcus produced the destructive enzyme only when it *had* to—that is, when it had to use the polysaccharide. If, on the other hand, this microbe was grown in an ordinary culture medium, it grew very well but it didn't produce the enzyme.

This fact taught me a lot. It shows that we can develop latent qualities to the degree that we function. I'll give you a very simple example. All of us are born with muscles, but they develop well only if they are made to work. In the same way we're all born with mental capabilities, and these mental capabilities develop only to the degree that the brain functions. So ever since my earliest research, I've been convinced that while it's important to know the organism, it's equally important to know the condition under which it develops and functions, in order to reveal its full potential. In passing, I'd like to say that in my opinion the medicine of the future won't necessarily invent a mechanical substitute to replace a lost organ, but will teach the individual to function without that organ.

But let's get back to the conflict between microbes. About 1935 I asked myself whether I could find some microbes in nature that would attack and kill other microbes besides the pneumococcus—the staphylococcus, for example. I began to search, and in a short time I did indeed find a microbe in the soil that would kill the staphylococcus. From cultures of this soil microbe I extracted substances, which I called tyrothricin and gramicidin, that accounted for the killing effect. They were the first antibiotics produced commercially. They were used topically rather extensively and were quite successful in veterinary medicine. Indeed, they were very effective against mastitis in cows, which is caused by a streptococcus.

At that point I read that a few years earlier Alexander Fleming had also found an antibiotic—penicillin—to combat the staphylococcus. But he had abandoned his work with it because he didn't know how to prepare penicillin in a stable form. In 1939 I told Fleming that he should resume his work on penicillin, and I also talked with Howard Florey who, in collaboration with Ernst Chain, prepared a stable form of penicillin in the early 1940s with the resounding success that we all know about.

But even before many antibiotics were available, I realized that microbes would become resistant to them, and I issued a warning about the dangers resulting from this resistance. As early as 1942 I even gave a lecture in which I stated that one of the great problems of the future would be the resistance that microbes would develop to antibiotics.*

In the meantime something much more important to me occurred in my personal life. At the time—1939–40—Europe was at war. My first wife was French, born at Limoges. In 1939, at the beginning of the war, she came down with tuberculosis. I asked myself, "Why did she get tuberculosis, when we live as well as we do?" So I dug into her past and discovered that she'd had tuberculosis when she was about six or seven and had gotten well by herself, as most people do. Then why did she have a relapse? I theorized that the disasters in her family resulting from the war had caused her very great anguish. I'm convinced that this anguish reactivated the tuberculosis that to all appearances had been cured. She died three years later.

In my grief I said to myself, "Infectious diseases obviously are caused by microbes, but countless people are infected and don't develop the disease. It only becomes manifest if the infection occurs in a weakened organism. And there are all sorts

* Dubos's warning, published in the *Proceedings of the American Philosophical Society* (1942), reads: "There exists the possibility, therefore, that as a result of the widespread use of sulfonamides in therapy and especially for prophylaxis, there may develop in the population strains of pathogenic agents which have become resistant to these drugs. Although we have presented only a hypothetical possibility and although there is as yet no evidence of any real danger, the problem should not be ignored, and it is to be hoped that laboratories throughout the land will find it possible to maintain a permanent survey in order to follow the shift in susceptibility of the different pathogenic agents to the drugs in common use."

of factors in the natural environment that determine whether or not the disease will appear." Then I thought that it would be intellectually more interesting to observe the sick persons in their surroundings than to restrict myself to studying the microbial causes of disease. I changed the focus of my laboratory research and began to perfect methods by which we could demonstrate that an animal's susceptibility to an infection (first to the bacillus of tuberculosis, then to the staphylococcus) varied according to its environment. From then on I tried to recognize the factors in the environment that influence a person's resistance to infection. In fact, my second wife, Jean, and I subsequently wrote a book showing how certain conditions in the surroundings were responsible for the great tuberculosis epidemics, especially during the nineteenth century.

During that century huge segments of the population moved from the country to the city for the first time. The new living conditions of these people were so different—even though they weren't always worse—that they posed enormous health problems, and tuberculosis became prevalent. It appeared everywhere and influenced literature, art, and music. This influence has a simple explanation: the behavior of the tuberculosis victims themselves. The benign form of tuberculosis causes a fever. This fever doesn't usually stun the individual but instead may stimulate him. Alexandre Dumas the elder, and his son as well, stated that being slightly feverish was stimulating and definitely good for you. Then there's Honoré de Balzac's extraordinary remark, which goes more or less like this: "Being a bit feverish for a week or a month may keep you from sleeping, but it lets you write a great deal more!" We've forgotten that a person didn't necessarily die of tuberculosis. Most people went through a long period of stimulation that was viewed as desirable. The pale complexion of the individual with tuberculosis became stylish; the romantic heroine was tubercular. But sometimes it turned out badly for the sick, as was the case with my first wife.

So I began to work on the influences in the surroundings that make one susceptible to tuberculosis. Soon I moved on to the influences in the surroundings that make one susceptible to other infections. Finally I realized that each civilization, each type of culture, has its own particular illnesses, and that the influence of the surroundings upon health is a widespread phenomenon. Ever since then I've continually asked myself to what degree illness results from infection or other assaults upon the individual by something in his surroundings—indeed, to what degree it results from his way of life in general. This has become an obsession with me.

Then, instead of thinking solely about the effect of the surroundings upon disease, I became interested in the effect of the surroundings upon human development in general. I often give students the following example. Before World War II the Japanese were short people. Now that they live like Westerners, they've become tall. In fact, they even play basketball! Another one of my favorite examples comes from the Central European countries and concerns the physical development of the Israelis. If you visit a kibbutz in Israel, you'll see that in a single generation the Central European type has completely changed, and that the tall young men and women of the kibbutzim are totally different from their parents.

General human development is what I'm interested in. I'm concerned about the environment, but always as related to the human being. And I've become convinced that people should give up the notion that human beings are so fragile as to constantly need all kinds of medication or therapy. Medicine must learn to recognize the potentials of human nature and also learn to cultivate them in our own environment. This is a totally new kind of science, a science with great potentialities. It almost disagrees with established doctrine, even with the teachings of Claude Bernard, who held that the human body adapts to new conditions through the mechanism of homeostasis, that is, by returning to its exact state before the attack. As I see it—and my position can easily be proved—the body adapts, but often by changing.

ESCANDE: That's a revolutionary approach!

DUBOS: As a matter of fact, it changes everything. If you look at the history of biological sciences, without exception you'll find confirmation of this. You'll see that if one branch of the tree breaks, the tree doesn't heal by growing another branch but by reorganizing its entire structure. An animal that loses a leg as a result of some accident learns to walk by changing anatomically. A person who becomes blind readapts without regaining his sight. Some thirty years ago a very brilliant young man was working in my laboratory. He became totally blind because of a severe case of diabetes. This young man, who had a will of steel, reorganized his life so that he could perceive many things in the world about him that we too could perceive if we cultivated our other senses. He could go into a restaurant and tell you what was happening there, simply by listening. But you'd have heard nothing.

ESCANDE: What you're saying reminds me of a professor of medicine in Paris named Siguier. Siguier was an exceptional doctor. He'd become blind for the same reasons as your colleague. He was a diabetic and developed a pulmonary infection. The extraordinary thing is that he failed to diagnose it. Thinking he had lung cancer, he stopped treating his diabetes and became blind. Yet even after he lost his sight he presided over meetings of his department as if nothing had happened, and he recognized the voices of every one of his students, even if they said only one word. This included students who spent only six months with him. I remember that one day during a big meeting in his office, someone made a little noise in the back of the room. He said, "Aline, be quiet." He had indeed developed—superdeveloped—his listening capabilities.

DUBOS: Your story proves humanity's enormous and unused potential. I repeat, this has preoccupied me most of all during recent years. I've turned to history to find other examples—here's a striking one. Living the civilized lives that we do, we've lost our ability to live in nature and to perceive the world like the primitive men of the Stone Age, for whom this was a necessity. After two thousand years of history and civilization, some Europeans decided during the eighteenth and nineteenth centuries to go live in the American wilderness more or less like primitive man. They were called *coureurs de bois*. Instead of remaining on the coast and farming, these Europeans chose a life of hunting and gathering that was the equivalent of Stone Age life. Whether they lived in Canada or in the Rocky Mountains or the Sierras, within a few years many of them had regained these forgotten powers—how to track down an animal, how to live in the wilderness, as Stone Age man did.

The lesson is obvious. We still have these powers within us and can call on them if conditions require it. In like manner, Cro-Magnon man bore within him all of our intellectual capabilities. If he returned to earth today he might become an IBM executive! We still have at our disposal powers we thought had permanently atrophied because of our civilized life.

ESCANDE: So you don't believe in what has been called the degeneration of the race? Because today, if you'll permit me to say so, it's accepted as self-evident.

DUBOS: I don't believe that at all, absolutely not. I'll use myself again as an example, and I hope you'll forgive me for it. My damaged heart has given me a lot of trouble. It's still very much with me. But in spite of it I've progressively become used to functioning again, so that everyone around me is surprised at the things I can do. I always climb the four flights of stairs to my office, and I saw wood almost like a Canadian woodsman. And I always try not to write down a lot of things. This keeps my memory active, so that it holds a lot more than most people's.

ESCANDE: That calls to mind a famous example—General de Gaulle. When he went to London at age fifty, he read his speeches, even if they were only fifteen lines long. But at seventy-five he would present an hour-long press conference from memory and would pass out the text to reporters in advance. So the stereotype that memory declines with age is an idea you don't agree with at all?

DUBOS: Not at all! I'll have to digress again to discuss speeches. I'm constantly giving you personal examples, but these are experiences with which I'm very familiar. I have rather severe problems with my vision. For many years I've had a great deal of trouble focusing. So I can't read a speech. Still, when I give a lecture, I always take a written text along with me so that my talk will seem prepared, but I don't use it at all. I don't recite a memorized speech, but I improvise upon a theme, and everyone gets the impression that I'm reciting from memory. That's not true, though, and if I improvise in this way, it's because I had to adapt to my defective vision very early in life.

Everything I've just told you has specific consequences for me. I'm not merely telling you anecdotes to astound you. On the contrary, it would be a promising research project to discover methods for activating all sorts of potentialities latent within us, temporarily lying fallow because of the way we live. Moreover—and I repeat—we must understand that useful adaptation isn't merely homeostatic adaptation, which strives to return the subject to its previous state. Just the opposite: useful adaptation is adaptation that helps us compensate for irreversible inadequacies, both physical and mental. Medicine today believes that individuals must conform to a norm and that medicine's only goal is to bring them to that norm. But *I* say that the important issue is understanding and promoting adaptation. On the one hand adaptation of course strives to repair damage; but on the other hand it strives to exploit the organism's reserve potentialities. A very famous Englishman who has written about physiology said that the big difference between the living organism and inert matter is that if you put on shoes, they will wear out. And if you walk with bare feet, these will wear out, too. But instead of remaining worn out, your feet will develop thicker calluses. That's a very big difference. It's even a fundamental difference, and it's just what I'm talking about. We must learn how to develop thicker calluses from every point of view!

The medical world today can be criticized for not putting enough trust in human nature, for constantly trying to give mankind crutches when human nature has the capacity to compensate for deficiencies by adapting itself and functioning on a different level.

Incidentally, I'd like to mention one of the most famous scientists in the United States today, a man who's made the boldest statements about medical research. I'm talking about Lewis Thomas, a physician who has held all kinds of very important posts. He's currently in charge of cancer research. In a very successful book Thomas made the following point: when a sick person sees a physician, ninety percent of the time he'll get well all by himself if allowed to bide his time while taking two or three aspirins. This is a slight exaggeration, but it emphasizes the enormous responsive possibilities available to the human body. I want to specify at the outset that this absolutely does not call the physician's role into question, and I'll repeat this later. Doctors have a whole arsenal of useful weapons at their disposal. We mustn't ask them to use only the most spectacular ones.

II

Scientists

Man is born inventive. That is his main superiority over the rest of the animal world. But such superiority! It totally changes the power relationship, it definitively sets man apart from all other living things, it permits him to subjugate them, to shape the world according to his own desires. To such a degree, according to René Dubos, that man can have no choice but to be anthropocentric. He really is *the center of the world.*

And man's inventive power continually nags at him, pushes him on, urges him to investigate. Scientific research is so important for man, it is such a great need at this point, that he has tried to turn it into an institution. All the industrialized countries have made a mystique of it, have organized and programmed research, and have given subsidies to scientists in the hope they will achieve everything.

Health first of all—the eternal youth synonymous with love of life. So you're sure to be listened to if you make critical statements: "If more money were given to research, we'd make more progress. If all the money spent on the atomic bomb had gone into cancer research, we'd have reached a solution long ago!" True or false? Surely false. Why? Because research is not a machine that works automatically whenever you ask it to, as long as you supply it with the necessary funds. Discovery is not the inevitable end of research, even of research conducted to perfection; nor is discovery synonymous with progress, still less with useful progress. Science, as Dubos observed by quoting

Einstein, permits us to explain a great many things, but very often it is not useful!

Still, for societies like ours, research is absolutely indispensable. But what sort of research, and for what ends? On this, Dubos has precise ideas.

DUBOS: The most extraordinary thing about science in nineteenth-century Europe was the creation of research institutes in all the major hospitals. In them scholars carried out physiological studies, anatomical studies, pathological studies—in short, they studied everything. The eighteenth century, the Enlightenment, had given birth to an immense confidence, best represented in France by Condorcet and all the Encyclopedists. They thought that reason could explain everything, that methods and theories would be found for every discipline.

When I began to study that period, what struck me most was the phenomenal quantity of scientific knowledge that was accumulated over the course of the nineteenth century, with no practical results but with incredible enthusiasm. For example, just think that when Claude Bernard published his *Introduction à l'étude de la médecine expérimentale*, the book sold extremely well—throughout Europe and even in the United States! Only a year after Bernard had introduced the idea of the fixity of the internal environment, his works had been introduced into the United States in the *North Atlantic Review*. Yet at that time, from the practical point of view, it led nowhere. The fact that during the nineteenth century all sorts of organizations, all sorts of people, were willing to expend enormous effort in order to acquire scientific knowledge that seemed to have no practical application gives me great confidence in the human race. People talk, and it has become a cliché, about the "Pasteurian revolution." In reality, during Pasteur's lifetime the Pasteurian revolution had very limited applications.

ESCANDE: True, Pasteur was able to protect sheep from anthrax, but the smallpox vaccination antedated him by many years. As for the rabies vaccine, it was never definitively proved that Pasteur's treatment saved little Joseph's life!* The great practical applications came later.

DUBOS: Likewise, as I see it, the first major practical application of scientific medicine is the discovery of insulin and its use as a treatment for diabetes. In the 1920s people saw for the first time that science could really lead to extraordinary therapeutic procedures.

ESCANDE: Let's turn to the relationship or connection between research and discovery. It's a perplexing relationship. I'm increasingly aware that so-called basic research is often based on accidental rather than purely scientific discoveries. Someone notices empirically that a substance—aspirin or cortisone—has therapeutic powers, and he says, "If it's active, we must learn why." At that point science intervenes and attempts to verify and analyze why and how this substance can act. This runs a bit counter to what everyone imagines.

DUBOS: You're saying just what every historian of technology is in the process of writing. For a long time everyone emphasized that technology, the Industrial Revolution, came from science, notably from physics and chemistry. Today the historians of technology are stressing that the opposite occurred. Artisans invented the steam engine—Denis Papin in France and James Watt in England. Watt was looking for a way to pump water out of mines! Afterward, and only afterward, was Carnot able to formulate the laws of thermodynamics, which explained how the steam engine worked.

ESCANDE: All this leads to another question. In a research institute, shouldn't enthusiastic support be given to the purely theoretical scholars, the professors in ivory towers—in short, to the harebrained individuals who assert with conviction, "*I* have an idea"?

* Nine-year-old Joseph Meister was bitten by a rabid dog in July 1885. He was the first person to receive Pasteur's attenuated rabies virus. Joseph survived the attack.—Trans.

DUBOS: If you'll pardon me, I'd like to use myself as an example once again. When I first came into contact with the Rockefeller Institute, I knew absolutely nothing about medicine or medical microbiology, or even what a pneumococcus was. When Avery described the pneumococcus with "its capsule of polysaccharides that prevent phagocytosis," he was using words I didn't even understand. But when he said to me, "Unfortunately, we don't know of anything that can break it down, that can attack that capsule, except for chemicals that are too dangerous," I immediately replied, "If there wasn't some microbe in the earth that could break that substance down, it would have accumulated since the beginning of time and the world would be covered by it. So there must be something that breaks it down, and I think I could find it." Indeed, at that moment I had an idea, and that's why I was given the opportunity to work it out.

From then on I began to look for places in nature where there were polysaccharides similar to those of the pneumococcus. And it so happened that I found some in a marshy region rather close to New York City. I took some soil from that region, put it into contact with the polysaccharide produced by the pneumococcus —and you know the rest!

Obviously, when I tell the story this way, it seems extremely harebrained and not far removed from the eccentric scientist. It was a matter of intuition. Jacques Monod, whom I knew very well and who, unlike me, was a marvelous logician, used to tell me with charming and benevolent scorn, "You and André Lwoff, you're both intuitives!" Yes, I'm intuitive. Research must start with phenomenology, and I certainly am a phenomenologist. I've had numerous successes in the laboratory, and I've really demonstrated new things, but I did it as a phenomenologist, that is, as a man who looks at the problem as a whole but doesn't analyze its innermost details. But Monod and François Jacob made their extraordinary discoveries as analysts and logicians. In fact, they used some of my work as a point of departure. As I've told you, I discovered that the microbe producing the enzyme that broke down the capsule around the pneumococcus produced that enzyme only when it needed to. Monod drew his inspiration from the system that I brought into focus, but he was concerned with understanding the inner mechanism. As for me, I'd indeed suspected the genetic origin of all these phenomena. I'd written a very long chapter on the subject in one of my books, *The Bacterial Cell,* which was very influential. In it I gave examples of other microbes that put their genetic potentialities to work only when necessity prompted. One of these microbes produced a lactase, that is, an enzyme that destroyed the sugar lactose. But before it would do so, it had to be subjected to conditions where it had to use lactose. So I'd pointed out the phenomenology of the problem. I'd even tried to go deeper, but while I'd pointed to genetics, I lacked sufficient knowledge of genetics, sufficient knowledge of chemistry, and probably sufficient analytical capabilities. I'm almost certain that, even if I'd persevered, I couldn't have solved the problem. But Monod and Jacob applied their astonishing analytical capacities.

I don't in any way want to create the impression that I underestimate the importance of the Cartesian scientific approach, which is an eighteenth-century attitude. Monod was an extraordinary Cartesian. We were friends, true friends, and he'd always say to me, "I wish I could have lived in the eighteenth century." So when he said of Lwoff and me, "You're intuitives," he certainly meant it, but I suppose that at the same time he valued intuitives and thought they were needed to state problems.

Pasteur was also a great intuitive. When Professor Pasteur Vallery-Radot* talked about Pasteur, he would invariably say, "Pasteur was logical, Pasteur was supremely logical." Well, that's part truth and part fantasy! Pasteur wasn't one hundred percent logical. First of all, he was intuitive. Then, from the moment he was convinced that something *should* be a certain way, he became logical in his scientific work. But he always began—and he himself stated this with great precision—with extraordinarily intuitive insights.

I'd like to give you another example, that of a medical scientist who made essential discoveries that he was incapable of analyzing. During World War II this Harvard physician was surprised

* Louis Pasteur's grandson, an illustrious patron of medicine and member of the French Academy.

to note that, when a soldier was wounded, sometimes he didn't suffer at all at the moment of the injury, and sometimes not even for a long time afterward. This occurred when the soldier was certain that his wound would lead to a permanent discharge and that, as a result, the war had ended for him. In fact, the phenomenon proved to be far more general. When, in one way or another, an event occurred—even a painful event—and brought an agreeable change in a person's future, the person mobilized something within his body that prevented him from feeling pain. These facts were accepted as real, became part of medical thought, and were talked about a great deal, but for a long time they couldn't be explained scientifically. People argued about them. Then, very recently, someone using scientific methods discovered that under certain circumstances the brain secretes substances that make pain go away, substances that some people thought resembled morphine in their action. This led to a research program to try to isolate these enkephalins and endorphins—hormones produced by the brain that have been recently discovered and that we now know affect many aspects of human behavior.

Even more recently, at the Neurological Institute of Montreal, it's been shown that the secretion of these endorphins increases when acupuncture is done under the proper conditions. It has also been demonstrated that acupuncture doesn't work if the patient has previously received a substance that prevents morphine from working. So we've suddenly moved from Chinese empirical knowledge, dating back two or three thousand years and ignorant of scientific medicine, to precise scientific formulation.

ESCANDE: Instead of perceiving the overall phenomenon, you finally understand its innermost details, you understand what's going on at the molecular level?

DUBOS: Exactly. This is one more reason why we should pay attention to whatever phenomenology is available to us, thanks to which we can make progress. This phenomenology often depends upon the observations by individuals who know how to look at things and be surprised. And the explanation may not come for two thousand years!

I'd also like to mention that a major discovery can be made in a rather unscientific way. For example, during World War II, when war broke out in the Pacific, the United States organized programs to solve certain urgent health problems—malaria, for example. The areas supplying quinine were cut off almost immediately. So an enormous organization was created overnight to find some other substance that was active against malaria, for without it victory in the Pacific would be impossible. Research was begun empirically. Experimenters tested an incredible number of substances to see whether some might be effective against malaria. They tried every sort of substance, literally every sort. They organized programs to infect small animals and tested many products on them in an empirical way. There was nothing scientific about their approach. But although they tested many things, they nevertheless did it in a very organized way. For that sort of project the United States knew how to act on a broad scale and with real efficiency.

When the American pharmaceutical industry decided to produce large amounts of penicillin, it immediately began to isolate every possible strain of penicillium from everywhere to see whether it could find one better than Fleming's. It discovered one very soon and found that the initial strain used was probably one of the worst! This shows the sort of scientifically oriented disregard for established methods that gave America a certain kind of superiority over the rest of the world.

ESCANDE: You contrasted the nineteenth century, which investigated for the sake of discovery, with the twentieth century, which is in a position to use these discoveries. Don't you think that, since 1960, basic discoveries once again have rarely had an immediate practical application? When you observe medicine as practiced every day, don't you think we're behaving as if we thought discoveries can be applied safely and usefully to help the sick at once? As a physician, I've been struck by this very much. For example, look at immunology. I see immunosuppressors being used, even though we can't totally control their action.

The use of serum to destroy lymphocytes has almost totally stopped today, but I've seen patients being given—and not always with caution—BCG therapy, or an immunotherapy using the corynebacterium, which has not been proven to be absolutely safe. Don't you think this indicates that in a sense we've gone into a skid, that we've lost control and are claiming victory too soon? Without being accomplices, aren't scientists allowing discoveries still inapplicable to man to be used far too soon?

DUBOS: Once again I'll answer with an example. I was one of the first to show that the resistance of laboratory animals to all sorts of experimental infections could be increased by injections of endotoxins. First you inject the endotoxin, then you initiate an experimental infection, and you observe that the animal has become far more resistant to that experimental infection than the other animals which haven't received the endotoxin. No sooner had I published these results than people took this poorly understood phenomenon—one not even understood at all—and tried to apply it to patients immediately. And why did they do that? Because, since science (all sciences, but medical science in particular) is to a large extent dependent upon public funds, people think that the use of these funds must be justified by practical applications.

When you request grant money for medical research, you are always asked, "What use will it have?" You have to state a goal, so if cancer research is fashionable, you say that it will help cure cancer. In the days when tuberculosis still caused terror, they said that it would help cure tuberculosis. If I may say so, scientists have sold their souls for grants. And I've heard this attitude defended by colleagues whom I respect highly. They say, "Everyone in the know realizes that we're only saying this in order to get money, but we're not fooling the scientists themselves. Among ourselves we understand what we mean!" Everyone who asks for grants to work on a very fashionable subject—for example, recombinant DNA: that is, introducing fragments of DNA into a given genetic system—says, "This will help us understand how cancer develops, and in the future we'll be able to control it." Among themselves they don't necessarily believe it, but since they need money and since the money must be justified in the eyes of the legislators, they make fine promises. I clearly recall how, one day when I was testifying before a Senate committee, I heard the dean of a dental school ask for money to study the microbes that cause tooth decay. He said, "This will permit us to vaccinate against dental caries." I was horrified, because I was quite sure that at the time there was no chance of developing a vaccine against dental caries. And he knew it, too. But "vaccination" is a magic word, so he used it in order to get research funds.

ESCANDE: This makes you smile at first, or even wink in connivance. Scientists are conning the politicians, and we can understand why! For their own profit, however, the companies in turn later take up the promise that was made too hastily, and the supposedly proven property is put in the spotlight. At that point a real campaign is begun to intoxicate physicians, to make them believe that progress has really been made. From then on, no more brakes. As a result, among the most widely sold medications we often find many that don't live up to their promises, and many that have had new instructions put on their labels over the years. The medication most widely sold in France was initially used to combat high blood pressure, and then to oxygenate the brain cells. Maybe you can add oxygen to the brain cells of a rat that's perfectly healthy, but there are grounds to question whether, in a man whose arteries are as hard as plumbing pipes, the medication can really oxygenate the brain cells.

What worries me is that what most people take for science is really only a bunch of conniving winks that scientists exchange with one another. So should scientists allow themselves to "sell hope" to the public in this way? Instead, shouldn't they tell all the facts—I won't say all the truth—to the public?

DUBOS: My answer will be a bit more general than your question. In the United States both an antiscientific and an antimedical movement exist, or so it is said. Well, I for one don't believe that there's really an antiscientific or an antimedical movement; there's simply disillusionment in one segment of the population because too much has been promised. In all the sciences, and especially in medicine, so much has been promised that the public, or at any rate part of the public, is disillusioned. The promises simply can't be kept; but in order to create the impression that they *are* being kept, scientists try anything, make vague generalizations in hopes that they will calm the anxiety.

Instead of helping the public to realize that science makes only slow advances and that once a discovery has been made, the means by which that discovery can be applied must still be determined, those involved tend to cry victory at once and cite one of those fantasies of our day—statistics, which are made to show exactly what you want them to show. I believe that, by being too hasty in using something that isn't yet ready to be applied, they are creating a split in the public; as a result, a segment of the public is breaking away from the medical establishment. Here, of course, I'm talking about the United States, where over the past four or five years I've seen a large percentage of the population turn away from scientific medicine and seek help in Zen, Indian medicine, yoga, or transcendental meditation. But this is true of every wealthy country.

ESCANDE: Yes, it's noticeable in France, too, but here its manifestation is more paramedical, more "white-coat." The French, who are a bit more traditionalist, don't have a pioneering mentality, don't become devotees of Zen, but they consult other people who wear white coats: the herb doctor, the acupuncturer, the kinestherapist. They're fleeing medicine for other white coats. So official statistics—here they are again!—take pleasure in announcing, "For the first time there is a decrease in medical consumption." But if you look closely, that's wrong. There isn't a decrease in the appetite for care, but a decrease in medical consumption per se. People are simply going elsewhere, but it remains to be seen how useful this turning to the other white coats will be.

Now I'm going to ask you a rather blunt question about research. How many research laboratories throughout the world should be closed? I'm referring to the laboratories that spend fabulous sums for people who aren't well-qualified scientists, who use methods that aren't valid, and who work in facilities that aren't satisfactorily equipped. Take the pharmaceutical industry. In a research laboratory of a French pharmaceutical company, someone said to me one day, "It's amazing! Come see. We've done some work on cooperation between lymphocytes."* Under a microscope they showed me two lymphocytes side by side. And they said, "They're cooperating!" At that time I'd been studying very closely some scholarly works on that subject, and I knew the complicated technology required for a correct analysis. Well, these "scientists" were spending a great deal of money on this ridiculous "work," which appeared officially in their account books under the heading "research."

DUBOS: The important thing isn't just that this particular laboratory and all the others like it are spending a lot of money. As I see it, it's even more serious that afterward, in their attractive publicity folders, they can show photographs of equipment that looks extremely complex, scientific, and apparently usable only by eminent men, when actually it's all a sham. I don't want to name names, but there are so many! I'm continually observing that all the publicity folders distributed by the pharmaceutical companies include hard-to-believe photographs that have absolutely everyone hoodwinked. I'm a bit embarrassed to talk about this because three years ago, I believe it was, someone at a meeting in Paris asked to take my picture. Later I discovered that I'd been photographed because I was standing in front of an apparatus used to produce negative ions, and that this photo had appeared in several places, as if I were endorsing this marvelous scientific apparatus!

Having said that, I can assure you that in the United States most major institutions and most large businesses now have research laboratories employing first-rate scientists. So if something had to be criticized in the States, I believe it would be the

* Lymphocytes are blood cells that carry out protective immunological processes. Different types of lymphocytes cooperate with one another, especially to manufacture antibodies.

research conducted by very professional institutions. There are so many universities and research institutes, all of which have to find money—from the government and also from foundations—that scientists within the establishment have to play the very game you described for private businesses in France. Endless promises are made, and potential donors are shown mysterious apparatuses and are dazzled by all sorts of marvelous results.

But let's get back to your question. Should some research laboratories be closed and if so, which ones? Who'll be the one to decide? I've always considered this a very difficult decision, because I'm totally on the side of decentralized research. I want a lot of people to be able to do research, and I'd be very worried, for example, if I saw cancer research concentrated in one, two, or three enormous institutions—for the simple reason that such centralization would prevent the freedom of exploration so characteristic of the nineteenth century. So I wouldn't like to encourage the scattering of research, but on the other hand I know the dangers that you're referring to and I can't come up with much of a solution.

ESCANDE: Under the French system, in order to keep a research laboratory going you often have to ask fifty different institutions for a couple of thousand dollars each, and you have to make your request fifty separate times for the same project. When Roger Guillemin won the Nobel Prize, he was interviewed by the Parisian newspaper *Le Monde*, and he stated that in France doing research consisted of ringing thousands of doorbells in order to get money, while in the United States people said to him, "Here's the money you need for five years, and you have carte blanche. Do whatever you want with it, hire whomever you wish."

DUBOS: On that point he was exaggerating! He forgot to mention that he was connected with two institutions that have enormous amounts of money and prestige. I can assure you that if Guillemin had been in one of the small midwestern universities, he'd also have had to ring a lot of doorbells. At Harvard or in Texas you don't have to ring doorbells, but if you're in Indiana or Montana . . .

ESCANDE: Is there some advantage to giving only small amounts of money to scientists in a provincial French city like Clermont-Ferrand, or in the Midwest? In short, is there an advantage in scattering research funds around?

DUBOS: Personally, I make a special effort to pay frequent visits to the small universities and colleges of the Midwest. And I'm impressed by the number of young people who are very open-minded and who will certainly come up with the most original, the most unexpected ideas. When you go to Harvard or the Rockefeller University you find admirable people, but they've already started down one path and are not likely to leave it. So I'm rather glad when someone risks wasting a little money by giving it to a young professor at the University of Clermont-Ferrand, because there certainly are lots of young people there who haven't yet become part of the establishment's way of thinking and who can do something new.

Indeed, I'm summarizing Pasteur's career for you. If he'd been on the faculty of the Sorbonne in Paris when he first began his research, he might not have delved into those somewhat preposterous subjects—or rather, what seemed to be preposterous subjects—that he worked on at the beginning, when he was a professor at Lille, a newly created university. So I'd like to see our societies accept the idea that in small universities there are many very open-minded young men and women who possess a great deal of imagination. And I'd like these men and women to be free and to be told, "Don't try to imitate what's being done at the Institut Pasteur. If you think there's something more interesting to do, do it!" But I realize that this isn't easy to achieve!

ESCANDE: Let's turn the problem around for a moment and work our way back to my original question. I quite agree with you that each person who has ideas should be given a chance. But suppose it becomes obvious that he's not producing anything worthwhile. At what point should his money be cut off? In many countries, once the scientist has been granted funds, those funds are renewed for a very long time, even if he proves totally ineffectual. So at what point does it become obvious that someone isn't an effective scientist? Have you seen many people who had no business in research but who kept at it anyway?

DUBOS: Oh, yes, of course, and I see exactly what you mean. But how is that decision to be made? It's so very, very difficult. All the same, certain strategies could be changed. Here I'll turn to a subject on which André Cournand* and I don't quite see eye to eye. Cournand believes that the French should adopt the American system. That is, when a research project is presented, it is evaluated and judged by people in that field who belong to the research establishment. Now while these judges are very good at recognizing the quality of work being done on a classic subject, they often find it difficult to accept—or simply don't accept—something that doesn't fit into their established way of thinking. I'll give you a specific example that involves the polio vaccine. When research on a vaccine began, all the money went to projects trying to obtain a virus that had been killed. But John Enders, an immunologist who's a very good friend of mine, thought it would be much better to vaccinate with a so-called attenuated virus—a virus that hadn't been killed but simply attenuated. So he and two of his students who didn't even have their M.D.'s perfected a tissue culture in which they could grow the polio virus. This led to attenuated vaccines, and we know how successful they were. Well, they got almost no money at the beginning. Money poured in only after Enders and his students won the Nobel Prize! This work made it possible to develop another whole series of vaccines, although initially this approach was not very much in favor with the research establishment. The establishment had viewed Enders as a fellow who

* André Cournand, a French-born physician, has worked in the United States, where he won the Nobel Prize in 1956 for his research in cardiology.

was quite charming, agreed, but on the periphery of important research problems.

So I'm happy when life is made easier for all the John Enderses and for all who want to try a new way to do something. But I very much agree with you. There are a lot of people for whom the "new way" becomes a way of life and who eternally repeat what they've been doing—and not only repeat, but copy others.

ESCANDE: Then you believe that there are martyrs in science, that Galileos still exist in our day? People who say things very clearly, and to whom no one will listen because they contradict the dogmas of the moment?

DUBOS: Enders was an associate professor at Harvard and had a certain amount of personal wealth, so he could pursue his whims. But if he'd been a professor in a small midwestern university, I'll bet no one would have paid attention to him. Martyrs to science? People to whom science pays no attention? Yes, I'm convinced they exist. There are a lot of them.

ESCANDE: Because our institutional structures are too rigid, we risk being deprived of a certain number of discoveries?

DUBOS: I'm almost certain of it. For example, let's go back to the matter of resistance to infections. I'm convinced that the weather and the climate have an enormous influence upon the appearance and manifestations of infectious diseases. I'm convinced that infectious respiratory diseases occur mainly during certain periods of the year, not simply because those microbes are circulating at that particular time, but because at that time our bodies are different, are more receptive and more sensitive. Well, that's not an "established" view, it's not a modish way of thinking. I know a lot of people who'd like to work on these hypotheses, but they can't get the money. Another subject that would have been rejected less than thirty years ago, but for which grants are now beginning to be given, is the way in which our physiological activities change by cycles—daily cycles (the so-called circadian cycles), seasonal cycles, and annual cycles. And so, depending on the point that you've reached in the cycle, you react differently to an infection or medication.*

About ten years ago a very distinguished publisher of medical texts wanted to award a prize to celebrate the centennial of his firm. It was a rather big prize. I was one of the judges, along with three of my colleagues whom I not only esteem but admire. They presented their candidates and I presented mine. "I propose a candidate whom you haven't thought of," I said. His name is Halberg. He's a professor at Ohio State Medical School and he's done what I consider fine work on circadian cycles, seasonal cycles, and the very different ways in which test animals, and therefore humans, react to all sorts of insults, medications, and infections. He'd like to write a book on the subject, but since he's a professor in a state university, he doesn't have time. If he won the prize, he could write that book." Well, my three colleagues unanimously declared, "It's not a timely subject." The prize was awarded to two other men—a physician and a biochemist—who were very distinguished, and who of course were working on more academic subjects. That's the way it goes. The Halbergs of this world, who are working on subjects that, as I see it, may have enormous theoretical and practical importance for the future of medicine, have a lot of trouble finding their place in the sun. They are undoubtedly, if you will, martyrs.

In the nineteenth century, even if you didn't back a scientist's ideas, you at least permitted him to have ideas and express them. In fact, Claude Bernard's first propositions constituted a truly intellectual but unproven vision. The public and the scientists of today think that he started off with very profound scientific evidence. Well, when Bernard presented his theory about the fixity of the internal environment, it was really an intellectual vision, for the evidence that he presented to support his thesis was very weak. He gave only a few approximate values for sugar and other relatively simple components. Likewise, when

* See Alain Reinberg et al., *L'Homme malade du temps* (Paris: Stock, 1979), part of the French series published by Stock to which this volume by Dubos and Escande belongs.

Pasteur published his first ideas about the role of microbes it was in 1857, in an article about lactic acid fermentation. Honestly, if this article had been submitted to a journal today, it would surely have been rejected. They'd have said, "M. Pasteur is making assertions without proof and expressing ideas he also can't prove." In this short paper on lactic acid fermentation, dated 1857, he stated bluntly that he wouldn't be surprised if the same ideas were applicable to infectious and contagious diseases. Pasteur surely was criticized in his day by the establishment, but at least it let him think and write!

By the way, you could ask me another question about research: where should it be carried on? In existing institutes or in institutes not yet created? Let me give you an example. For several years I've been involved in environmental issues, and I've been asserting that people are wrong to present environmental problems solely from the viewpoint of air and water pollution. Of course I don't like air pollution, I don't like water pollution, and I'm very outspoken against them, even though it's been difficult to prove that air pollution is the *immediate cause* of as many illnesses as they say. But the real problem isn't starting a debate on this subject. I believe that other aspects of environmental problems are far more important. Of course we must create an agreeable environment, but when I say "agreeable" I mean more than people usually mean. We need an environment that satisfies all sorts of profound needs of the human organism, both visual and psychological needs. *This* is the important environment problem, and I say it over and over to anyone who'll listen. Because I'm known to have been an orthodox scientist, I was offered money to organize a group to work on improving the environment. They came and asked me, "Where would you like the money to go? To which university?" And I replied, "To none! If you give it to Yale, Harvard, or the Rockefeller University, everything that's said or done will inevitably be reformulated by establishment supporters within those universities. If you want to create something new, you must create it within a new institution. Something independent." That's how the René Dubos Forum was created, in the probably foolish hope that facilities can be developed outside existing organizations, so as not to be absorbed by them. As you may know, Franklin D. Roosevelt used to say, "If there's a problem, don't try to solve it within an existing organization or a long-established institution. Create a new institution, and then be courageous enough to disband that institution when the problem has been solved." To solve production problems during the war, or to solve financial problems, instead of turning the task over to existing government departments, he created new institutions that of course had their own powers but were supposed to solve only certain problems.

ESCANDE: What you're saying makes me think of a recent comment by Commander Jacques Cousteau after a general committee had been created in France to deal with marine problems. Cousteau said, "It's very simple. Either we'll choose people familiar with the problem and form a new group, and it wouldn't even bother me if M. Aymard Achille-Fould were named president of it. Or else we'll fit what we've just created back into the government framework and use people who've been working in the ministries on these very same problems, and then nothing will change." This amounts to the very same thing you were saying.

DUBOS: Exactly! But now I realize that the great difficulty doesn't lie in creating a new institution. When you're well-known, that's not so difficult. The big problem is knowing how to disband the institution when its mission has been accomplished. This is the very same problem we were discussing a while back about scientists. Indeed, Chairman Mao always referred to the perpetual revolution because he was well aware of the importance of these problems.

ESCANDE: Can a scientist remain "in top form" for a long period, or is his period of great productivity usually rather brief, like the career of a professional athlete?

DUBOS: I've known scientists who remained incredibly productive until the very end of their lives. They exist. Avery, with whom I worked early in my career, remained productive to the end. I can say that at the Rockefeller University about thirty percent of the scientists of my day remained active throughout their lives. Of course I have to say that I'm speaking only of the Rockefeller University, because what goes on there may not apply to the scientific world in general. Until very recently, and even today, our institution was made up exclusively of people who didn't want to work in industry, who didn't want to be administrators, and who had chosen the institute and all its difficulties. (Our salaries were a bit lower than those in universities, and the opportunities for promotion came less frequently.) So those who came there had a sort of monastic spirit and above all wanted to live in that sort of world. Well, at the Rockefeller Institute most scientists continued to do research almost to the end. It's my impression that today many more scientists abandon the laboratory earlier in life. I think one explanation for this is the great difficulty I encountered in my own life as a scientist—the difficulty of mastering techniques that are changing with extreme rapidity. Science per se and knowledge—I still possess them (well, almost!). But techniques! In my case it became obvious at least twenty years ago that I was no longer up-to-date. And I repeat, what has changed the most, and what continues to change most rapidly—even more than new scientific knowledge—is the tools we use. So for this reason among others (even though it's a bit dangerous to make generalizations of this sort), I'll say that many scientists—and I'm referring to those who've been highly successful—decide to do something else when they're past fifty or sixty. It's essential that they be given this opportunity, or else they risk drying up on the spot.

ESCANDE: I'd like to get back to a point you made in passing, concerning air pollution: that much of what people say is false. This raises a problem, the problem of the "scientifically proved" error. In medicine I'm very struck by the total changes observed today in a great number of fields that once seemed solidly backed up by definitive discoveries. Today we're shifting into reverse and declaring, "This is what we used to say, but it can't be said anymore!" Can we imagine a scientist who, having obtained results, would be humble enough to say, "It seems to me that it's true, but then again perhaps it's false"?

DUBOS: This reminds me of one of the most famous professors of medicine at Harvard. In fact, he was the dean. He used to say to his students, "Most of the things we're teaching you today won't be true ten years from now." So lots of people are aware of what you've just pointed out.

But I'd prefer not to give you a general answer. I'll reply with a specific example and return to the subject of pollution. I detest pollution for all kinds of reasons, and it's been clearly shown (see p. 85) that breathing "bad" air certainly makes people sick—eventually. But on the basis of general statistics, it's very hard to prove the harmful effects when they're just beginning, unless a person is in a place where it can be certain—and this is very rare—that a specific illness is directly brought about by a pollutant. So people must be wary of overhasty deductions, of what I'll call overly "sentimental" deductions. Indeed, I remember that in the France of my youth everyone had bronchitis during the winter. But when I return to Paris now, I scarcely notice any, so I could almost assert that our present kind of pollution prevents bronchitis! Of course I'm exaggerating, but I want to show you to what degree certain scientific "proofs" are open to criticism. People mustn't make generalizations too quickly on the basis of isolated evidence. Even so, a young scientist for whom I have complete respect demonstrated the effects of air pollution upon the respiratory process. He placed his apparatuses in places where there was likely to be a bit more sulfur dioxide and nitrogen oxide, and he performed every type of measurement. Finally he managed to demonstrate—not by distorting the facts but by choosing the situations—that in those places there was indeed a slight increase in the frequency of respiratory ailments. The same results can be obtained by studying the harmful effects of a large factory that spews out all sorts of fumes. If you measure the incidence of certain ailments upwind or downwind, you can indeed show that illnesses increase in the direction in which the prevailing winds blow.

I don't disagree with these observations, but I'm critical of the overhasty generalizations involved. People get very caught up in the idea of atmospheric pollution and talk of nothing else. They hold it responsible for everything and make totally ridiculous statements seemingly based upon scientific proof. In our day we magnify things and often make hasty generalizations on the basis of a certain number of isolated scientific measurements. Take saccharin, which is claimed to be carcinogenic—

ESCANDE: —Because a few poor rats, tucked away in some laboratory, were stuffed with saccharin and developed bladder cancer! So people have been influenced and extrapolate for man the dangers observed in the rat!

DUBOS: And at that point a law is passed without sufficient justification, and no one does any more research. Some fifty years ago an American humorist said, "Our great difficulties in life aren't caused by what we don't know, but by the things we know—and that aren't so." In point of fact, I believe the medical sciences are full of things that we think we know but that either aren't totally true or else are unimportant.

ESCANDE: Descartes used to say that you should never accept something as true without having proven that it's true. That's why he sat in his "stove," trying to filter through the sieve of his judgment everything that was accepted as true. And he finally realized that almost nothing was "true."

DUBOS: We must distrust hasty conclusions based on incomplete observations. Take an example dealing with tuberculosis. We know the death rate for tuberculosis a century ago: roughly five hundred deaths a year for every one hundred thousand people. But today the rate has fallen to two. In the end, what was responsible for this phenomenon? If you looked at the matter closely, you could defend various hypotheses. If you wanted to prove that discovering the guilty microbe caused the drop in the disease rate, you could show that after 1890 the death rate fell from five hundred to two hundred. If you wanted to prove that vaccination with BCG was of prime importance, you could demonstrate that after BCG came into use, the death rate fell to forty. If you wanted to prove the role of specific antibiotics, you could demonstrate that after the discovery of streptomycin, the death rate dropped to five.

But the other truth is that the death rate for tuberculosis had begun to decline shortly after 1845—even though no therapeutic program had been developed—and that between 1845 and 1890 this decline was just as regular as after 1890. These data are very widely accepted today; social changes occurring during the nineteenth century made people less susceptible to tuberculosis, even when they came into contact with the bacillus. It's essential to know this, for until a few decades ago ninety percent of all people in Europe and America still showed positive tuberculin reactions and had therefore been infected; they had reacted by developing a very mild infection and a positive reaction to the skin tests without suffering any serious effects. It must be pointed out, however, that during the 1950s the introduction of isoniazid made it really possible to speed up the decline in the death rate. But by this time tuberculosis killed far fewer people than in the past.

At this point the tuberculosis sanitariums were closed, for isoniazid not only cured the patient, but also let him be treated while going about his daily tasks. It was no longer necessary to remain in bed for months on end. This was demonstrated in our department, using a very small group of patients, or rather two groups. Some received isoniazid and stayed in bed; the others received the drug and continued to work. Well, the second group got well just as quickly and completely as the first one. So sanitarium treatment was no longer necessary.

Another example of an unexplained decline involves heart disease. As you know, in all the industrialized nations everyone was talking about an "epidemic" of heart ailments and about the steady rise in the coronary death rate, which seemed an inevitable consequence of modern life. But a few years ago it was observed that in both the United States and Europe the "epidemic" had reached its peak and was on the wane. Since then the death rate has been declining and continues to do so at an accelerated pace. Explanations have been pouring in: some people tell you, "It's because the intake of cholesterol has been reduced," while others say, "It's because people are doing more physical exercise," and still others say, "It's because of the early-detection services that were set up." All these factors probably played a part, but no one really knows.

ESCANDE: This makes me think of the famous remark by Jean Cocteau: "Since these mysteries are beyond our comprehension, let's pretend we created them." Certain publications frequently give the impression that everyone is saying, "Such and such a phenomenon has been observed, and I'm the only one able to explain it to you."

DUBOS: Yes, we must be prudent and learn to realize that a multitude of very poorly analyzed influences can affect one and the same phenomenon. Here's another illustration, one even more picturesque. You probably know that New York City experienced a gigantic blackout in 1968 that lasted many hours. Well, nine months later they noticed that the number of births increased significantly, and a lot was written about the phenomenon. I pointed out that the blackout had occurred during the full moon and that, as it happens, certain publications have claimed that primates, especially monkeys, are noticeably more active sexually at that time. So the full moon may have played a role!

ESCANDE: But you're raising a fearful question here! The work of scientists, of people working in laboratories in general, consists of decreasing the number of variables, and hence the number of variables to measure during the course of their experiments, so that the experiment can be reproduced exactly every time. The fewer the variables to be verified, the more solid the research. Once something has been demonstrated during an experiment with a very small number of variables, isn't it dangerous to extrapolate to other circumstances, other conditions? Let's get back to the example of the rat's brain cells that receive more oxygen thanks to certain medications, which is true for a normal rat, especially when he's swimming.* Do we have the right to conclude immediately that a man walking in the street or peacefully dozing in his armchair would have more oxygen in his brain if he took the same medication? Yet many people make this hasty generalization.

DUBOS: We must know how to use our knowledge wisely. As I see it, science can't avoid the method of work that you've mentioned, which consists of reducing the number of variables. This is called reductionism. Actually, it provides knowledge of a limited scope, but this scientific knowledge of limited scope can be put to very good use. If we really understand it, this information can serve as a guide to an improved perception of very complex subjects. To illustrate this rather vague statement, I'd like to tell you about two paintings that seem to me to reveal the different uses to which science can be put.

One is the very familiar portrait of René Descartes by Frans Hals. This portrait reveals his rigidity of thought, his desire to doubt everything, to find exact knowledge, to retain only the strict truth whenever possible. It's really a thing of great beauty.

The other is the engraved portrait of a physician by Rembrandt, which I consider by far the finest portrait of a physician in existence, although it's relatively unknown. I'd like to describe it for you, to help you understand my conception of the role of science in the practice of medicine. The physician portrayed by Rembrandt is making a house call. It's the portrait of a physician who really lived in Amsterdam and apparently was very learned. He's stopped at the foot of the stairs and is glancing around. You feel that he's in the process of perceiving the total situation of his patient.

Now I've always thought that the ideal physician, if I may say so, is a man who possesses very specific knowledge, but who knows that this very specific knowledge will never suffice for understanding the unique problem posed by the individual pa-

* The "swimming test" obliges the rat to swim for hours in order to trigger a certain number of changes. In actual fact, far more sophisticated experiments are used today in research on the brain.

tient. So he enriches his specific knowledge by attempting to perceive his patient's total personality, his family situation, and the way in which he lives. Then the physician asks himself to what degree he can apply what he has learned. All this in no way decreases my respect for the very exact sciences; above all, I in no way want to give the impression that I'm rejecting them. But I think the essential thing is to use them in a coherent way and surely not to use them to "prove" what can't be proved, or to "understand" what as yet can't be understood.

I'm under the impression that in medicine exact knowledge can best be used not only to perfect types of treatment but also to improve the understanding of the whole situation. I believe that people are confusing two very different things here: on the one hand, the practical use of science during direct treatments—which we make too much of, by the way—and on the other, the use of science in order to understand, to grasp each individual case—a use we underestimate far too much, doubtless as a result of overemphasizing the first use.

ESCANDE: To bear this out, I'd like your opinion about two promises made to the American people. President Kennedy promised that Americans would be the first to reach the moon. This promise was kept. The second promise, the victory over cancer, was dangled before their eyes by President Nixon. And here nothing, or almost nothing, has been achieved.

DUBOS: This fits in perfectly with what we're saying. I think scientists were certain that reaching the moon was a possibility, whereas they realized that the fight against cancer was impossible to plan with any certainty. Moreover, at present they're still unable to make precise predictions.

Jerome B. Wiesner, Kennedy's chief scientific adviser—formerly the president of the Massachusetts Institute of Technology and himself a very learned physicist—could make sure that the United States had the theoretical knowledge and the technological means to reach the moon. They knew how to provide enough energy and, by using available fuels, knew what speed had to be attained in order to break the attraction of the earth's gravity. They also had computers that could monitor the entire operation. So the bet was won in advance. All they had to do was obtain the necessary money.

But the situation was very different for cancer. The real scientists don't claim to know what causes cancer. There's no disputing it: they don't know much about its causes, they know little about its development, they know only the end result. Promising to solve the mystery of cancer is presently unjustifiable for lack of scientific knowledge. As a matter of fact, this was a political promise. Fortunately, that doesn't stop physicians from treating their patients.

ESCANDE: This reminds me of heart transplants. Everyone read the big headlines in the newspapers announcing the "discovery of the century." I don't understand why the most eminent scholars didn't immediately reply, "No, it's premature to rejoice."

DUBOS: Well, it's the same old story, and eminent scholars admit it to one another. All the big research institutes certainly wanted to profit from this windfall by requesting ten times more money to continue their work on immunology. At that time the big surgical departments in the United States asked for enormous sums—and got them—in order to create departments of immunology within their departments of surgery. So the departments of surgery became major research centers, and their prestige was no longer based solely on their surgical skills but on their great immunological research work as well.

But as far as the unjustified promises about cancer are concerned, such promises are really a form of dishonesty—and I used that word in one of my books—a form of dishonesty that's deemed justified because the goal is praiseworthy.

ESCANDE: As far as scientists are concerned, isn't it a bit of a shame that they keep the real facts to themselves and end up saying, "Oh, the public! We'll toss them a bone. We'll make them spin daydreams!" Or else: "Those poor legislators don't understand anything about our research, so we'd better tell them what they want to hear, whatever makes them happy." I'm surprised that my colleagues find it extremely curious, even inappropriate, when I occasionally say what's really going on. Furthermore, it causes me certain problems from time to time. Here's a recent experience of mine concerning a French law about donating organs.

In a newspaper article I wrote that this law was interesting because it was going to end the inevitable bartering that had begun to crop up in the intensive-care wards between those who wanted transplantable organs for their patients and those who wanted to continue keeping patients alive even if they were in an irreversible coma. The former would say, "You know he's dead," and the latter would reply, "No, he's not!" "Yes he is! And his kidneys could save other sick people!" Well, after this article appeared, I received four letters. One was from a scientist at the Centre National de Recherches Scientifiques, a government-sponsored French research foundation. He was a member of a religious order and he wrote, "I'm totally in agreement with you, and I'm sending you an article in which I say almost the same thing you do." A second letter came from a woman who said, "I had a kidney transplant. It's fantastic! You must certainly keep on saying the things you do." The third letter came from a man who wrote, "I'd like to donate a kidney. Do you know whom I should write to?" And the fourth was from two physicians who had signed it jointly and mailed it to the editor-in-chief of the newspaper. It read: "The gentleman who wrote that article had no right to say what he did. He's making things up. It's shameful!" So I wonder if telling the truth means betraying one's guild, betraying one's colleagues, not playing the game. I personally believe that scientists and physicians have everything to gain by telling what they know.

DUBOS: The director of a private hospital in the city of New York thinks so, too. He says the very same thing as you: the truth must be told, because that's the only way to regain the public's trust, which we've partially lost.

ESCANDE: In order to regain this lost trust, must scientists be asked to lift up their heads and look about them?

DUBOS: It's not that scientists or those doing basic research refuse to recognize the existence of problems such as climatology or the influence of the psychic upon the physical. It's not that they reject these problems. They simply think that the only subjects upon which they can work are those for which methods exist permitting measurable observations. And if this isn't the case, any work seems impossible to them. For example, take the effect of the foehn wind or the tramontane winds, an epidemiological phenomenon whose workings aren't understood. Although it's been asserted that all the observed effects result from the fact that the concentration of ions in the atmosphere is changed, phenomena of this type lend themselves to so many poorly defined observations, to so much intellectual cheating as a result of improper interpretation, that scientists don't want to get involved.

I believe that scientific philosophy must be reformed. We must admit that there are many things we can observe even though we can't understand their inner mechanisms, and that they ought to be studied anyway. We must accept being satisfied with general observations, until other scientific discoveries eventually permit us to zero in on their mechanisms. I consider it essential that we pursue further the study of relationships between phenomena. At the present time I'm convinced the medical world is depriving itself of much useful information because scientists tend to become involved only with things that can be analyzed precisely. For that matter, from the philosophical point of view, that's nothing but an illusion!

ESCANDE: In fact, the most resounding scientific discoveries are always made on the basis of a general phenomenon that seemed surprising. But it's a curious thing: people forget this later on, as if it were a point of honor to forget how the spark was kindled, as if it were almost something to be ashamed of.

DUBOS: It's a bit like that. I'd like to use Pasteur as an example again. About thirty years ago I wrote a biography of Pasteur that is still being read. When it was reprinted recently, I added a new introductory chapter but didn't change anything in the original text. This new chapter seemed indispensable, once I had reread Pasteur's writings from an ecological perspective. Everyone summarizes Pasteur with a maxim: "The microbe causes the illness. Look for the microbe, and you'll understand the illness." Of course Pasteur said that. But no one's noticed how all his life he stressed the fact that—besides the microbe, which is obviously important—there are all sorts of other factors involving both the climate and the patient himself, his past, his genetic structure—he didn't use the word "genetic" because it didn't exist yet—which are all essential to understanding the onset and course of the illness. At the end of his life Pasteur, whose initial work on infection had dealt with the diseases of silkworms, was even able to write: "If I were younger and had to resume my work on silkworms, I'd focus above all on the conditions under which they were raised: the living conditions of the worms—their food, and the temperature in the silkworm sheds and the air inside them—for I'm convinced that the worms could be made more resistant to illness if their living conditions were changed." And he even went so far as to write, "Moreover, I'm going to suggest to M. Duclos that he try to find substances that could be included in their diet to make the worms more resistant." Pasteur also wrote: "I'm convinced that when a wound becomes infected and festers, the course that the wound takes depends upon the patient's general condition and even his mental condition." All this can be found in Pasteur, and no one has paid any attention to it.

ESCANDE: Are Pasteur's successors, the Pasteurians, to blame?

DUBOS: That's just what I was about to say.

ESCANDE: Well, I'm going to ask you another question. Do you think it's dangerous to have successors or overly zealous students? In your own case, what ideas are you afraid will be attributed to you? Do they turn out to be the very ideas you've been fighting recently?

DUBOS: To a certain degree, yes, as far as the environment is concerned. I was one of the first to discuss environmental problems and the effects of the environment and pollution upon the onset of illnesses. And most of what I said has been adopted and exaggerated in an extraordinary way. For example, a few years ago most people were convinced that the air in New York City was deadly. But in reality New Yorkers are very healthy. There's no doubt that the initial statements that I made about the environment, which go back roughly twenty or twenty-five years, have been distorted to such a degree that I now must exaggerate in the opposite direction in order to compensate, and must say that in reality it's often difficult to prove that air pollution causes illnesses! I've often been very embarrassed by the irresponsibility of those who have become my followers, and by their assertions. There's less danger of this in the more exact sciences.

But I'd like to get back to Pasteur, who as I've said asserted all his life that all sorts of other factors besides microbes set the stage for an illness. He even conducted picturesque experiments to show this. I'll describe only one of them, which I believe is virtually unknown. Few people are aware that Pasteur was a professor at the Ecole des Beaux Arts in Paris, where he taught science as applied to the arts. Specifically, he taught the physiology of architecture, and to illustrate the importance of good room ventilation, he set up an astonishing experiment. He took a bird—a canary or a finch—and put it in a bell jar, in which the oxygen was not renewed and therefore gradually diminished. In order to survive, the bird adapted by decreasing its activity, by remaining almost immobile. The bird was taken out while still alive, then a new bird was put into the same glass container. Abruptly placed in an atmosphere low in oxygen, this bird didn't have time to adapt, so it immediately began to move about and promptly fell dead.

He wanted to demonstrate by this experiment that a person can unconsciously adapt to unfavorable conditions, as long as they come upon him gradually. In Pasteur's view, it was very dangerous to adapt to bad ventilation. So he told the students they must think about providing adequate ventilation and not wait until people begin to suffer.

In short, Pasteur was a very modern man. He placed great importance upon all those complex effects of the surroundings that we think of as our discoveries. And no one talks about these ideas. Pasteur was completely betrayed, I now realize, by everyone who wrote about him. A few years ago, when the Institut Pasteur celebrated the fiftieth anniversary of his death, I chose that very theme for my address. It is all too easy to distort a great scholar's thought by selecting only a restricted part of it.

ESCANDE: Still, that's what always happens, especially today, with the media eager to spotlight sensational news (or at least the news that will attract the most attention). It's very surprising today—when you know someone well, when you're very familiar with his work—to see what's said about him or to read the words that journalists put into his mouth. And you're even more surprised to read what they say about *you!* I learned by bitter experience that if you write something, you'd better not try to crack a joke. If you dare to be humorous, spiteful people or big talkers—the two are often one and the same—immediately comment upon your words as if you'd been presenting a fundamental truth. But that's another story. It's enough to make Voltaire turn over in his grave.

DUBOS: By the way (it's not completely applicable to medicine, but it's related to what we're talking about), one of my colleagues, with whom I was recently discussing this very question, told me about Albert Einstein's pertinent reply when asked the following question: "Do you believe that science can study every subject?" He said, "Yes, I think it can. Only I don't think it would be very useful, for there are a lot of things that would be no better understood as a result." And he explained his statement by giving an example: "I'm convinced that if scientists or physicists were asked to study music, they'd take one of Beethoven's symphonies or some other piece and reformulate it in terms of air pressure and vibration, which would be no help at all in explaining the important thing—that the music expresses an artistic emotion." This seems very far removed from medicine, but I believe that it's closely related even so. It's possible to explain everything scientifically, without having grappled with the real problems.

ESCANDE: Indeed, the relationships between medicine and science aren't simple. Hippocrates drew little distinction between medicine and philosophy or the other social sciences. Then a complete separation occurred, at least in modern industrialized countries. But in many parts of Africa, dance, religion, and medicine remain interwoven. It becomes clear that the three are connected when, for example, you study the whirling dervishes. So isn't the modern world at fault for having isolated medicine from everything else to this extent and for having made it into a science? Even though science permits us to understand a certain number of things, aren't there times when it mustn't be used?

DUBOS: Science, I believe, is by necessity reductionist—that is, it separates each phenomenon in order to study it suitably. If you work in a laboratory, you're by necessity a reductionist, and being a reductionist means being a Cartesian. By the way, completely by chance I came upon this marvelous and surprising quotation from Descartes: "Joy is the most important thing for keeping one's health." Descartes's thought probably went far deeper than the words that have been put in his mouth, and while people are continually citing his philosophy, I believe that he too has been betrayed. Doubtless Descartes wouldn't recognize himself in the Cartesianism that we teach today.

To end our discussion on the subject of research, I'd like to say a few words about the way the Rockefeller Institute and its hospital were organized. When the Rockefeller Institute was created around 1901, there was hardly any scientific medicine in the United States. American physicians who wanted to learn something beyond routine general practice had to go to Europe.

One day when Rockefeller's fortune was already very large, the Baptist clergyman who advised him went to him and said: "Here's an extraordinary charitable undertaking to which you could devote your fortune: trying to help create a medical science in the United States." He convinced Rockefeller, and as a result the Rockefeller Institute was founded in 1901. At the outset it was only a research institute. Then around 1908 they decided to add a hospital. This began a great debate among the administrators and members of the staff about how the hospital would be organized. Certain physicians wanted physicians to direct everything, and for the laboratory work to use biologists, chemists, and biochemists as their subordinates. Dr. Rufus Cole, who had been selected as director, was a physician. Well, this man ended up defending the opposing point of view. He maintained that physicians and scientists should be on an equal footing. Each department, whether it did research or provided medical care, could have physicians working under scientists, or vice versa. He probably thought that most department heads would be physicians, because it would be easier to organize along those lines, but that it ought not to be a requirement.

Actually, an equilibrium was established very rapidly: half the personnel were physicians and the other half scientists. A miracle? Physicians immediately began to work with nonphysicians and encountered no problems at all. Better still, a chemist who was studying the kidney soon began to be extraordinarily knowledgeable about kidney diseases. And he became head of the department. This man, Donald D. Van Slyke, set up one of the most distinguished and most famous departments in the Rockefeller Hospital.

By contrast, the department in which I worked was headed by Dr. Avery, a physician. He'd begun his career as a practicing physician, but very quickly he became interested in laboratory work and soon lost interest in purely clinical problems. Another very famous professor, the first to use the electrocardiogram, also gave up practicing medicine very quickly. He soon became exclusively interested in the history and philosophy of medicine. And so, free from all constraints, given the freedom to choose our individual paths, each of us discovered the role that best suited him, without any jealousy and above all without any worries about rank.

I've often wondered why this way of organizing an institute was so successful. I must say, I still don't quite understand it! Because among us there were all kinds of temperaments, some of them diametrically opposed; but that almost never affected personal relationships. What was begun at the Rockefeller Institute very quickly served as an example throughout the United States. A most famous hospital in America is the Massachusetts General Hospital. Well, it wasn't long before chemists, biologists, and physiologists—some of them physicians and some not—began to rub elbows at Massachusetts General in perfect harmony. One of the most famous was Fritz Lipmann, since then a Nobel Prize winner, who had studied medicine in Germany, then came to the States and gave it up completely, to the point that he no longer even thought of himself as a physician. That profession no longer interested him, but he worked at the hospital and continually talked with his physician colleagues. This manner of blending talents has continued ever since. At the heart of it lies the extraordinary flexibility of American institutions. If only that flexibility can last! Nowadays, when I talk to my American colleagues, I say to them, "What worries me is that you're beginning to become organized like the Europeans!"

At the Rockefeller Institute the department heads, the individuals in whom we were going to put our trust, were not chosen according to their scientific specialization, but according to other qualities very hard to define: a certain originality, a certain spirit of initiative. We wanted to feel we could trust them to do something reasonable and interesting with the responsibilities they'd been given. That's still the case at the Rockefeller University and in other institutions, too.

For example, when Lewis Thomas was named president of the Sloan Kettering Cancer Institute, he'd never worked on cancer before. But people simply realized that he was a fellow with original ideas! In the major American scientific institutes, men—or women—are chosen for reasons that have nothing to do with their formal training. The same is true for industry. No doubt this is why American industry adapts so rapidly to changing situations.

In most American institutions the door is, in the true sense of the word, always open. Really always open. When I was Avery's assistant, his door was always open. People came in and talked, asked a question, discussed every possible subject. I've always followed his example. I'm continually welcoming young people whom I don't know, though some of them have since become famous. I'd like to tell a story at this point. The great immunologist Baruj Bennaceraf, whom I never thought I'd meet in person, telephoned me one day and said, "I see that we're both on the program at the same meeting of immunologists. I'm driving. Can I take you along?" I was very surprised, but I accepted. During the drive I asked him, "What made you think of inviting me?" He replied, "Aha! You don't remember? Twenty years ago I had just about completed medical school and came to see you to ask if I could work in your lab. You said, 'I think the work being done in my lab isn't quite in line with your interests. It would be better to talk to Melvin Kabat.' " Bennaceraf had done so, had begun his career as an immunologist under Kabat, and had become one of the great immunologists of our day.

At our university we naturally receive a lot of young people from all over the world, and all of them are astonished at this extraordinary openness, this absence of formal interpersonal relations. Of course in France things aren't quite the same. In 1937 or 1938, when I was doing very productive research, I was introduced to the man who was then director of the Institut Pasteur. I was informed that if I wanted to work there, I'd be given important responsibilities. I recall wearing tennis shoes when I went to see the director. What was wrong with tennis shoes? Well, I didn't even get into his office. I was merely allowed to walk around the courtyard a few times, and that ended that!

ESCANDE: When the Rockefeller Institute was created, it was to promote scientific medicine in the United States. I believe that when Mr. Rockefeller created something, he had the entire nation in mind. What did he hope to get out of this new medicine? Nobel Prizes? More competent physicians? Or just improved medical practices in the United States, equal to those in European countries? What was the philosophy of the Rockefeller Institute?

DUBOS: The philosophy was very simple and can be summarized this way: "It makes little difference to us whether anyone ever makes a discovery having practical applications. What's important is that physicians and laboratory scientists learn to work together, learn to do things that will help the medical profession scientifically." And in a very explicit fashion the people promoting the experiment said, "What we want is for you to work on subjects that will illuminate medical thought scientifically."

At that moment they were above all thinking about the exact sciences, and they hired physicists, chemists, biologists, bacteriologists. Only much later (and so far it hasn't worked out) did they try to add the social sciences. But our institute was so centered around the exact sciences that it has been difficult to incorporate the behavioral and social sciences into it. The behavioral sciences survived, but the social sciences have gone by the board.

ESCANDE: Regardless of the field in which a person wants to do research, how does he become an investigator? By putting himself in contact with scientists who've already completed their training? Or by going through a long period of theoretical training? And can he call himself a scientist at the end of his studies? Or does he become one by doing research?

DUBOS: After all my broad experience, I'm convinced that a person becomes a scientist by doing research and not by studying books. And my belief is strengthened when I see that many of my illustrious colleagues at the Rockefeller University have become famous in fields different from the ones they studied in school. For example, some of the most famous geneticists began their careers as organic chemists, while some of the most famous chemists began their careers as physicians. I began my career in agriculture and at an early age became relatively famous in medical microbiology. So I believe that the best thing a person can learn from his formal studies is how to work. After that he must do research, and that's what turns him into a scientist.

ESCANDE: And now for an indiscreet question. When someone is at the Rockefeller University, does he have the impression that he belongs to the elite? Does he feel privileged—superior to other people?

DUBOS: To a certain degree, yes. Less so nowadays. But for a long time we certainly were a guiding light. And at the same time there's no doubt that we also were a sort of Benedictine monastery. It so happens that today the university is located in the heart of New York City. When it was built, the property was on the outskirts of the city, but now the city has swallowed it up and has spread far beyond it. But we've kept the park grounds, which are surrounded by an iron fence, so that when you enter you feel as if you were in one of those great monasteries. In my day it really was like a monastery and I dare say we were envied by everyone. I don't think we are as much any longer, perhaps because the example we provided was so successful that our imitators have outdone us.

III

Physicians

Why physicians? Why was this professional group created? To care for the sick, of course. But how were the sick cared for in the past, since medical science is a very recent phenomenon? This is an immense question.

If physicians have existed as long as man has, it is surely because, since time immemorial, people have realized the virtues of certain manipulations and of certain plants for relieving pain and prolonging life. Contrary to what people are all too prone to believe today, physicians have never been entirely without resources.

But physicians have other powers than the ability automatically and mechanically to prescribe the appropriate medication for an illness. Physicians know how to recognize the illness and consequently to foresee its course, which is tantamount to predicting the future. The physician is a soothsayer functioning on a rational basis. But he goes beyond that basis and is also kin to the sorcerer. Indeed, the relationship between the physician and his patient brings improvements that to a great extent are still not understood. Because one party involved is the physician, who has knowledge, and because the patient wants to believe, something intangible is transmitted that hastens healing or permits the patient to endure his illness better.

This is why the attempts to make medicine an absolute vassal of the exact sciences are certainly on the wrong track. Today we must learn to reevaluate the specific possibilities open to scientific medicine and, once we have made this evaluation, to recognize that now, as well as a century ago, the physician's basic role can be summarized very well by Dr. Edward Trudeau's maxim, which René Dubos cites: "Guérir quelquefois, soulager souvent, consoler toujours"*—to sometimes cure, often help, always console. And for our age, Dubos adds, "whenever possible, prevent."*

DUBOS: First off, I want to say that I've no technical competence when I talk about the practice of medicine. I'm not a physician, and I don't want to pose as a specialist on the subject. But at least one thing can't be overlooked: the fact that since time immemorial there have been physicians who succeeded admirably in helping their patients. Since time immemorial. They were, therefore, far more effective than we today imagine them to have been, or else their medical procedures wouldn't have survived for so long. Over the past twenty years a historian specialist on the history of Chinese medicine has written a book and a whole series of articles about medical practices in ancient China. In particular she translated a book entitled *The Yellow Emperor Treatise of Medicine*, which is a great classic about medicine in China approximately three thousand years ago. It provides descriptions of very precise and very complex surgical and medical procedures. Joseph Needham, a famous English biologist who lived in China for some thirty years, studied the ways in which the Chinese have worked over the last three or four thousand years, and he discovered that Chinese physicians had learned to make preparations from urine samples taken from men and women at certain points in their lives and used them to treat certain ailments. And so, despite their lack of scientific methods, they realized the presence of certain hormones. Naturally they didn't call them hormones.

We also know, of course, from the writings of Hippocrates, how precise his diagnoses could be. So in every period there have been great physicians, and these physicians have obtained cures that must have depended—I'm sure of it—partly on the physician's personal influence, but also partly on methods that are unjustly underrated today.

By the way, I'd like to give you an example that just came to mind. As you may know, the first formula for a contraceptive is found in one of the earliest Egyptian papyruses. It involves a mixture of plant extracts that, when placed inside the human body, begins to ferment, produces lactic acid, and therefore has a contraceptive effect. So you see, for a long time people have known many practical things that have mostly been forgotten today, but that people in the past probably used far more skillfully than we think. There's no need to remind you that the Peruvian Indians realized that quinine could lower a fever. They didn't know that quinine was effective against malaria, but it worked. And I don't have to belabor the fact that digitalis had long been used in the form of plant infusions.

And the discovery of salicylates! That's an extraordinary story. Long ago an Englishman was convinced that, wherever some disease prevailed, God in his goodness had established something effective for its treatment. Well, in damp regions people tended to come down with rheumatism. The Englishman looked at the plants growing in these damp regions and noticed a lot of willow trees. Was *this* the antidote to humidity? He began to make extracts from the willow tree and demonstrated that when these extracts were given to a sick person, the pain decreased somewhat. Now, the Latin word for willow is *Salix*, so the extract from the willow tree was called salicin. From salicin came salicylates, from which the pharmaceutical industry and chemists made aspirin, which is acetylsalicylic acid!

ESCANDE: So you think that, since the scientific revolution in medicine took place, old medications have been reused under new names and new cultural and technical approaches have been employed, but that man hadn't been completely ignorant before that? Yet it's a very general belief that once upon a time there were no medications, but then science arrived and permitted us to perfect new molecules, so we've been able to save people's lives. But these treatments used in the past weren't known the world over. Hasn't industrial society made the spread of these techniques possible?

DUBOS: True, there was no worldwide distribution, although each local population had discovered what it needed most, and today people are once again becoming interested in these old medicines. Dr. Mahler, the director of the World Health Organization, and his collaborators decided it was necessary to try to rediscover the substances that the shamans, the physicians, of each ethnic group throughout the world used in the past and often still use today.

Nowadays—and ever since the arrival of the scientifically oriented society—everyone talks like this: "Yes, it's obvious that shamans, physicians, had an influence, but it was a result of their personalities." Yet in every country these shamans, these sorcerers, were in fact using substances, treatments, and ways of caring for the sick that had a remarkable pharmacological effect. Today teams are being sent to every part of the world to try to study all these drugs that have been extracted from plants or animals and that are used in traditional medicine. And so, yes, these ancient medicines have a very, very solid base. The same is true for surgical methods. Phenomenal surgery was performed before our day. Scientific medicine began to make great strides only in the early nineteenth century.

ESCANDE: What is your present view of medicine?

DUBOS: In the United States, groups of physicians—usually eight or ten of them, with different specializations—will propose a sort of prepaid insurance to families whereby, in the event of illness, the families pay nothing beyond that fee. This movement has become quite extensive over the past two decades, to the point that a national organization was created and currently lists roughly twenty thousand physicians. They've been so successful that last year they decided to create a foundation that they named after the physician who first organized this sort of group medical practice, and they asked me to give the inaugural address. The theme they suggested was "Medicine and the Spirit of Man."

I replied that this seemed very vague to me and that I'm not fond of vague titles. I said, "Suppose I get down to the basic problems, such as 'What do we mean by illness?' and 'What help can the physician bring to the sick?' " It was at this meeting that I expressed the thoughts inspired by the two portraits I described earlier—Rembrandt's portrait of the physician and Hals's portrait of Descartes, expressing two complementary approaches to knowledge. On the one hand there's the Cartesian approach, which breaks down every situation into the simplest possible components, in order to study these components one by one and define them as precisely as possible. But on the other hand there's the approach captured by Rembrandt's portrayal of the physician, very wise and very conscientious, who is aware of the complexity of problems and tries to fit an event into its context.

Incidentally, I'd like to tell you more about a recent "discovery" that I made. People talk as if Descartes was nothing but a logical reductionist. Well, last week—purely by chance, for I wasn't doing research on it—I ran across a book in New York, *Descartes par lui-même* (Descartes in his own words), part of a well-known French series. Here's what I found: Descartes wrote to Princess Elizabeth of Germany, telling her, "In truth, the most important thing for curing illnesses and maintaining health is good humor and joy." This proves that there's not such a gulf after all between Descartes and Rembrandt's physician. Quite the contrary. Rembrandt's physician simply developed his insights as a result of Descartes's ideas.

But I based the lecture I referred to upon the maxim of a physician named Trudeau: "To sometimes cure, often help, always console." Who was Trudeau? A French-speaking American physician who was born in Louisiana and studied in New York City. Just over a century ago he came down with tuberculosis. He thought that he was going to die, for his tuberculosis had reached such an advanced stage that he was sure he had no chance of pulling through. So he said to himself, "Well, I'm going to spend the few months I have left doing the things I like most. I'll go hunting in the Adirondacks." He went to those mountains, which aren't far from New York City, and hired a guide. There are big lakes in those mountains, and he'd set off in a boat with his guide and hunt. He spent each day that way, and instead of dying, he got better. Almost cured, he returned to New York, once again became immersed in the medical and social life of the city, and suffered a relapse. Again he was told that he was going to die. "If I have to die," he said, "I'll go off again." So he went back to the Adirondacks, resumed his hunting, and once again got better. This time he decided to set up a medical practice in the Adirondacks and also to create a small sanitarium (it wasn't called that at the time) to admit sick people from New York City, to see whether life in the Adirondacks would improve their tuberculosis. This was the first important sanitarium in the United States. Trudeau became the great expert of his day on tuberculosis. He was the first person in America to employ Koch's methods for cultivating the tuberculosis bacillus. He became a very important man, a personal friend of the chief organizer of the public health services. When he died a statue was erected and bore in French the maxim that he was so fond of using to describe the physician's powers: "*Guérir quelquefois, soulager souvent, consoler toujours*"—to sometimes cure, often help, always console. Is this maxim still pertinent today, or has medicine completely changed since then?

First of all, one might object that these few words make no mention of prevention. And with good reason: in 1870 smallpox vaccination was the only technique of preventive medicine that existed. While there are grounds for skepticism about the progress of curative medicine, one certainly can't deny that preventive medicine has become quite effective during the intervening century, not only against infectious diseases but against malnutrition and work-related illnesses. This really constitutes an extraordinary chapter in medical history, a chapter based upon close collaboration between the medical and the nonmedical world. For I want to point out that those great campaigns supporting preventive medicine at the turn of the century were at first chiefly organized by people who weren't physicians. The physicians who cooperated—and of course they did so at a very early date—were physicians with considerable social concern. As I see it, this is the most extraordinary example of close and absolutely essential collaboration between the medical and the nonmedical world. People can invent all the preventive methods they want, but if these methods aren't incorporated into a suitable social structure, there's no chance of their being effective. Not even vaccinations. So a whole world of preventive medicine exists that isn't sufficiently recognized and that did not exist in Trudeau's time.

Now, lets look first at *guérir quelquefois,* to sometimes cure. Today a lot of diseases still aren't curable. I've never made a list of them, but I suppose that truly dependable cures exist for very few diseases, in particular for some bacterial infections and a few diseases treated by surgery. The rest of the time the patient's condition is merely improved, or else he gets well all on his own. Cures are the result of concentrated scientific effort to understand the disease in question. Curing infectious diseases by means of antibiotics became possible only when it had been demonstrated that such diseases were caused by microbes.

Next we come to *soulager souvent,* to often help. Exactly what does that mean? I believe you can't answer that question without asking what you mean by "illness." I've always noticed that people consider themselves to be ill when they can't do what they want to do, when they can't become what they want to become. People naturally like to know what's wrong with their body, but above all they're concerned about being able to do the things they want to do.

I'll give you an example that I've given a thousand times already, it's made such an impression on me. Some twenty or thirty years ago I was acquainted with two young women in New York City. I'm going to give you a lot of seemingly irrelevant details, but you'll see that they're nonetheless indispensable. These two young women were just about the same size, both quite tall, very slender, and extremely pretty, and both were Jewish. One wanted to become a fashion model for a New York designer, and the other wanted to live the extremely rugged life on an Israeli kibbutz. Obviously, the ideal of "health" for these two young women was completely different. If you want to be a fashion model in New York, your overall health must be very different from that required to do work on a farm in the Negev. Each of these young women aspired to a very different kind of "health." The one who wanted to be a model was satisfied with her appearance; the other girl feared that she was physically unqualified. I lost touch with them, so I don't know what happened to them, but I do know that each achieved her desired career. After thinking about them I suddenly became convinced that each of us wants to do something with his life and needs his own form of health to do it. And physicians must understand this. So there's no general definition of "health."

Now I'm going to give you a second example, a very personal one, the example of Dr. Avery, to whom I've referred several times. He had a disease that could have been considered serious, a goiter, Graves Basedow's disease. The laboratory was his whole life. He lived only a ten-minute walk from the laboratory and hardly ever went anywhere else. He used to come to the laboratory, do his work, go back home, sleep, then return to work. He scarcely ate, and he certainly could never have walked a mile without stopping. Yet he thought that his health was good, for he had everything he needed to keep functioning in the laboratory. That was his only barometer. Well, he began to encounter serious difficulties in his laboratory. Nothing worked anymore and worse still, something happened that made all his work appear useless. Most of his effort had been focused on the production of serums and vaccines for pneumonia. But sulfa drugs appeared in 1937, and overnight his approach appeared useless, out-of-date. He felt that he'd wasted his entire life. Then his Graves disease literally erupted; he couldn't bear it any longer. He had to have an operation, and for two or three years he was a dispirited man; he thought it was the end. Then he pulled himself together, we started on other projects completely different from his previous work on pneumonia, and this work went very well. He began to come out of his depression. He lived until he was seventy-five, happy and with no further worries about his thyroid. So he provides another illustration of how impossible it is to define what people call "health," how difficult it is to separate health from the person's total life.

The third example that I'd like to give you may seem a bit ridiculous. Near our house in the country lived an old Bulgarian tenant farmer who took care of his house and who lived with two cows in a manner that most people would have found squalid. His way of life wasn't very hygienic, his food wasn't very good, but he felt he was the happiest man in the world. Yet the day the people who owned the farm gave up their two cows, this man's health began to worry him; he "became" sick.

I've said all this to explain to you that a person becomes aware of losing his health when he can no longer live the life he wishes. If these few examples are used for reflection, we can see that a totally new role can be envisaged for the physician—but in the end, it's his oldest role. The physician can be marvelously successful in his profession without always relying on the physiology of the illness, but simply using time-tested empirical methods on his patients. Let's take a common example: a woman with arthritis. She can be helped to live with her arthritis, doubtless by being given a few drugs, but also by being told how she should behave in her daily activities, in her everyday life. There's a whole aspect of medicine that consists of thinking about what's important in each person's life—how he can be helped to live with physical deficiencies, without even hoping for a cure; how to let him live out the rest of his life experiencing a certain amount of fulfillment. The medicine built upon these foundations becomes extremely humane. Of course it depends upon a very complete knowledge of what medications can and cannot do, but also upon a sort of sensitivity toward the patient that it's very important to foster.

ESCANDE: Here you're putting your finger on what seems to me the essential problem encountered in training physicians—a problem, however, that isn't clearly perceived. When I give a course on dermatology for students specializing in the subject, I talk about the treatment of chronic childhood eczema and I say: "When a mother brings in her eczemous child, the first thing you must do is understand the relationship between her and the child, and then the important thing is to explain to her that there's no miracle treatment but that the child can be helped. And the first thing to teach her is to grease the child's skin. This is of prime importance; the skin must be lubricated. Then, in order to relieve tension, I tell her, 'Use butter if you want, but keep his skin greasy. You must keep it greasy all day long.' " Then I give my students specific details. I elaborate on how they must explain this to the mother, how she must go about doing it, and I see that everyone is bored to death. From time to time I say, "When I tell you this, I get the feeling my arguments are bouncing off a wall." And it's true! I feel as if the lecture hall had become a solid wall confronting me! Why? Because they believe in a purely scientific medicine that's the exact opposite of the medicine you seem to propose. On the other hand, I must admit that if a student were ever questioned in his examinations about the treatment of eczema and were to repeat what I said, he'd flunk the exam. Because it doesn't sound authoritative.

This is the real problem in medical studies. Reject common sense in the name of so-called science! Yet deep down inside, almost every physician is on *your* wavelength and senses that the physician's role most often consists of calming people by helping them to live with their illness. In dermatology—as in any medical specialization—a certain number of infectious diseases can be cured—scabies, for example. But that's all—others can't be cured. If they are cured, it's often because they go away on their own. As Jean Bernard likes to point out, we must keep the spontaneous evolution of illnesses in mind.

DUBOS: I wouldn't hesitate to let you speak for me on this point, for I must say, Jean-Paul, that I've been asserting this in medical schools ever since a lecture I gave at Yale University, a very scientifically oriented school. And you should have seen the students! In contrast to what you've been telling me, the students there suddenly said to themselves, "Aha, that's what life is all about!" They were persuaded by what I was telling them.

ESCANDE: It seems to me that you've encountered what I often find: a great desire to do good shown by the people who are going to practice medicine. But their teachers rein them in, examinations warp them, and at that point these students are forced to behave as if they could "always cure."

Medical studies create the impression that you must consider yourself a bad physician if, when confronted with a given patient, you can't make a perfect diagnosis followed by the appropriate therapy and cure the patient one hundred percent of the time. When I first began to work with Professor Vallery-Radot, I was so convinced of this that I felt very upset when someone mentioned a patient who'd died of alcohol-related cirrhosis of the liver. I said to myself, "Good Lord, how strange that a patient could die in such a famous department of this hospital." Still, I was nineteen—I was no longer a kid. I waited, expecting someone to say at any moment, "Of course the patient was going to die, but soon the famous doctor arrived and he was saved." For me, Paris, with its big hospitals, was a place where people's lives were saved, where people were always cured. And when you work with students who are preparing for the entrance examinations for internships! They ask, "And what should we indicate as therapy?" And we always answer, "Above all, when you answer that question, don't go out on a limb!" But we should be teaching them this: "There are cases where therapy is dangerous. There are cases where it's very helpful, but in the majority of cases we actually aren't sure what to do, but we can help the patient in other ways." Instead of teaching them that, we prefer to say, "Above all, don't go out on a limb."

Let's get back to Dr. Trudeau's second item, *soulager souvent,* to often help.

DUBOS: The meaning of the word *soulager* is very complicated. I tried to translate it into English and at that point I realized its deep meaning. First of all I translated it "to help," but *soulager* also means to lift a burden. In reality, at a certain moment most patients—and I think I'm one of them—ask not only that the burden of their illness be lifted from them in the form of physical help, but also that they be relieved of responsibility for their illness. Admittedly, what I'm saying here—and I believe it very deeply—runs counter to a movement that's currently very popular in the United States, the so-called self-help groups. In these groups patients meet without a physician and try to cure themselves on their own. Well, the physician should be there. And what I'm saying also runs counter to the experiences of another American that have caused a big uproar, an enormous uproar. Here's the story.

Several years ago Norman Cousins, the editor of one of the major American magazines, one of the most intellectual of them all, the *Saturday Review*, became seriously ill. He was completely demoralized, for he'd been told in vague terms that he had a fatal disease. This intelligent man, who has extraordinary willpower, reasoned to himself, "If the physicians can't cure me, I'll cure myself." At that moment he was in one of the best hospitals of New York City, and it seemed that he was going to die. But he asked permission to leave the hospital and, with his own doctor's approval, took a room in one of the hotels in the city. There he decided to do three things. First of all, laugh. So he bought a whole lot of very funny movies and hired people to tell him stories, amusing stories, so that he laughed, he was continually roaring with laughter. Second, he asked that all medications be stopped, because he had the impression that every time he took medicine, he felt worse. So all medication was stopped. And third, he decided to stuff himself with vitamin C, for he'd read—he's a voracious reader—that vitamin C probably has effects far beyond what biologists and physicians understand. He took sizable quantities of vitamin C and increased the dose daily. Since he's quite interested in science, he wanted all the same to have factual observations made of what was happening to him. So all sorts of measurements, all sorts of examinations were carried out on him. And the facts speak for themselves: he soon began to get better. Indeed, six months later he was completely well. I'm by no means trying to persuade you that this treatment cured him, but he was cured. Norman Cousins told this story everywhere. It was such a topic of conversation that an editor asked him to write about it. He described it in great detail in the *New England Journal of Medicine,* which has a great reputation as an authoritative medical journal. The editors expected strong protests from the entire medical world. Of course there were protests, but the extraordinary thing is that Cousins received more than three thousand letters—from physicians—who told him, "You're completely right. You were wise to break with established medical practice." So then he took up his pen and wrote a second article, in which he analyzed and quoted the statements of the physicians who were on his side. First of all these physicians said that he'd been wise to stop all medication, that all medications are toxic. Then they explained to him that he must realize that something exists called the placebo effect, and that this effect was undoubtedly what cured him. So he wrote a third article on placebos. A publisher asked to reprint these three articles in book form. Since we're friends, Cousins asked me, "Would you have the courage to write an introduction to these three articles?" I replied, "I certainly won't discuss your illness, because I know nothing about it, but I'll be glad to say why I find your experience interesting. It demonstrates that every living organism without exception possesses mechanisms for automatic correction, and that physicians often don't put enough trust in these mechanisms. As I see it, medicine's future task is to try to put to maximum use the human organism's power to recuperate." And I took this opportunity to bring up the subject of centenarians.

Most stories about hundred-year-olds aren't true, but a few are. One was told to me by a famous American physician, the late Paul Dudley White, a leading cardiologist and Dwight Eisenhower's personal physician. He lived near Boston, in the same town as a man who'd never been sick. Every day this man would go for a long walk in the country. At the age of one hundred and two, he came down with a sort of flu. People wanted to protect him, to care for him, and his doctor told him—ordered him!—to stay in bed, at which point the man began to waste away at an astonishing rate. Paul Dudley White, who knew him, came to see him and said, "Why don't you start taking walks again?" Immediately, even though he was still sick, the old man went for a walk, and he lived to a hundred and seven.

Here's another story about a centenarian, Michel-Eugène Chevreul, the famous French chemist who worked on fats. He was awarded an honorary doctorate from Harvard at a very advanced age and when, at a hundred and two, he was asked, "How do you feel?" he used to reply, "I feel fine, but I'm getting a little tired of living"—he felt *une certaine lassitude de vivre*. He died at one hundred and three.

Let's get back to the hundred-and-two-year-old walker. The physician's great wisdom lay in telling him, "Well, you've the right to continue your life as you've lived it up to now. It's your habitual way of life, and it's your best guarantee of getting well." So *soulager* of course means to relieve the patient by giving him a drug at the right time, but it also means helping him by advising him how to live in a certain way, it means helping him by getting him to change certain living patterns, to adapt those patterns to his individual needs. It also means relieving him of the enormous anxiety that's a major factor in his illness. Indeed, the definition of the word *soulager* reveals a whole aspect of medical activity that we barely know how to teach.

Soulager is an essential part of the physician's activities, but this activity is scorned by the medical establishment because it can't be expressed statistically! Just imagine a woman who thinks she has breast cancer. She's terribly frightened. She goes to see the doctor. If he tells her that it's nothing at all, her enjoyment of life suddenly returns. This is an essential medical activity—and only physicians can carry out such an activity—but it never appears in official medical statistics. This is something important to think about: a medical activity that doesn't appear in official statistics.

ESCANDE: I'd like to give two examples that reinforce what you've just said, but that show that *soulager* is very dependent upon the patient's profound need. It's not the same for everyone.

One routine reason for consulting a dermatologist is the fear of malignant melanoma, a very serious form of cancer. Hand in hand with this expressed fear go two very different types of behavior that require two different approaches by the physician. I'll give you an example of the first behavior. One day a very famous Frenchman called me long distance from Greece and said, "I think I have a malignant melanoma. Can I come see you?" I replied, "Yes." I was on vacation, but I knew he was very anxious, so I said, "Come." The next day he called back to say, "I'm coming." He cut short his vacation in Greece, and I cut mine short (in the Paris suburbs!). When he entered I saw that he was very worried, and he said to me at once, "Of course you promise to tell me the truth?" "I'll tell you the truth," I replied. "It's a promise." He undressed and showed me something that indeed might cause anxiety, for it had bled. I told him, "It's nothing. It's a fatty wart. Have a good vacation. I won't remove it now, because it would leave a scar." His relief was immediate. His problem was over. He had been legitimately afraid for a very specific reason. He hurried off. His wife was waiting for him downstairs, though I didn't know it; doubtless he wanted to reassure her at once. This is an example that, as you say, isn't included in cancer statistics, of a medical act that immediately relieved considerable anguish, was able to *soulager*.

But there's a second kind of patient. In his case you also examine the lesions carefully and here too you often say, "No, it's not cancer." But this time the patient doesn't get up, he doesn't rush off, he doesn't budge. He stays there and says, "Hmm! It's not cancer, but it could have been, couldn't it?" He's no longer frightened, but he really wasn't so frightened when he came. This type of patient has different problems. People like this who are terrified of cancer shift the difficulties they encounter in their daily lives to cancer; their actual problem lies elsewhere.

If you want to *soulager*, I think you must at all costs distinguish between the question and the worry, which are completely different. The stated question and the true worry aren't superimposed. If you want to *soulager*, the essential thing is to answer the call for aid, the cry for help, even if initially it's stated in a deceptive way. Having said this I'll add, so as not to sound like a demagogue, that this is something the public doesn't understand. It often makes the mistake of firmly believing that medicine is an exact science. This attitude, which crops up everywhere, explains many of the mistakes made by the medical profession. Why don't physicians dare to tell someone like your centenarian with the flu, "Get up and go take a walk"? They don't dare because they're afraid they might be sued if the gentleman went out into the street and stepped into an open manhole. Lawsuits against established physicians, like the pure sciences drummed into the heads of medical students from the very first day of their studies, are built upon the same foundations, that is, the belief in an exact medicine-science. At the present time the only recognized medicine is medicine that can be expressed in statistics. But as we've already said ten times and are about to repeat once more, the really interesting medicine is medicine that can't be expressed statistically!

DUBOS: Men always adapt, learn to change. They carry errors to extremes, but eventually they break free from them. Four or five years ago an enormous private medical foundation was created in the United States, second in importance only to the Ford Foundation, with an endowment, if I recall correctly, of two billion dollars. One of my former associates was chosen as president of this medical foundation, the biggest in the world. This man was in succession the youngest professor of medicine in America and the youngest dean of medicine, so you can see his abilities. And he's also an excellent scientist. So as head of this foundation, he and his adviser were besieged by people who wanted to see this immense sum of money devoted to basic research. Well, they decided that the problems of so-called basic research would not be their concern. First of all, because such research is sufficiently well-organized; second, because that research is already generously funded; and last but not least, because basic research seems unlikely to solve some of the big problems facing medicine. To their mind, the big problem confronting medicine now is knowing how to use total medical knowledge to provide better care for the sick. So they decided not to give any money to basic research, which brought them under considerable attack. But they held firm. So you see that in the United States a great movement is aimed in the direction you advocate.

ESCANDE: We both think the same way. When summarizing the problems facing medicine, I always say that their solution involves deciding how money will be invested. A lot of money is spent on health. Of course even more could be spent. But suppose we began to put what has already been spent to better use! Alas, when you talk like that, you're accused of "making a power play," of "wanting to manage the crisis." Such objections are absolutely unfounded, but admittedly they get the desired effect every time. The industry that invests in health feels protected and continues to make more and more big equipment for physicians, to promote expensive treatment for the sick. The results of all of this haven't been well thought out, although they make the public feel a bit more secure. But the public certainly would show even more confidence in another trend, if that trend were explained. But that's another story!

Having discussed *guérir*, to cure, and *soulager*, to relieve, let's go on to the third item in Trudeau's maxim: *consoler toujours*, always to console.

DUBOS: I've never really understood what Trudeau meant. Everyone, almost everyone, thinks that if a tragedy occurs, the patient himself or his family should be consoled. And perhaps that's all that Trudeau meant. But I get the impression that, in the context of this maxim, the word means much more. The term *consoler* seems to me to embrace the whole attitude of a physician who, having a long experience of life, can say the few words that a person always needs when confronted by even the most tragic situation. Here I can't help thinking of the marvelous final sentence of Maupassant's story "Une Vie": "Life is never as good or as bad as one thinks." I may be attributing to physicians qualities that they don't always have, but for me this is what Trudeau's *consoler* means: the physician uses his experience to tell someone who's suffering, someone who's in difficulty, that everyone experiences a certain amount of suffering, that everyone goes through difficult moments, and that in the end most people pull through. It involves more than merely expressing sympathy; it means giving encouragement by appealing to the patient's good sense, by evoking the complexity of life: "Life is never as good or as bad as one thinks." This sentence, which I first read when I was quite young—around fifteen—has stayed in my mind, because I've had many ups and downs in my life, in health and otherwise. So for me, Trudeau's *consoler* goes beyond consoling the family because someone has just died, or consoling an individual because he can't get well. It means, if you will, teaching him how to become master of the situation again.

ESCANDE: Let's change the direction of our discussion a bit. Doesn't the continual attempt to treat patients with ever more effective weapons against disease contradict the fact that the organism must be allowed to adapt?

DUBOS: I'm continually asking myself that very question. In the end the answer is simple and the compromise is rather easy to state. If one day I were to develop subacute bacterial endocarditis, which is known to be fatal if untreated, there's no doubt about it: I'd want penicillin. In fact this happened to me a few years ago. But if I have a minor infection that I know my body has the capacity and strength to conquer on its own, well, I don't want penicillin.

Here we come to an essential point, the importance of physicians, which as I see it remains considerable, contrary to what Ivan Illich says. I want to remind you once more that I'm not a physician, so I'm not defending my guild! But all the same, it's still the physician's clinical judgment that determines whether or not medicine functions harmoniously. A physician, a professor in a New York medical school, J. S. Maxmen, recently wrote a book called *The Post Physician Era: Medicine in the 21st Century*, in which he states that in a very high percentage of cases a computer can make the diagnosis as well as a physician—not quite as well as a very good physician, but better than most; and also that it can not only make a diagnosis, but also prescribe the treatment. I can well believe it. Only I don't believe that a computer will ever acquire the discriminating judgment that makes the real medical decisions possible. Now I myself have a very good physician in New York City. If he were to discover that I had endocarditis, he'd send me to the hospital at once (which he did); but if I have a mild sore throat or a routine complaint, he says, "Go to bed. That will be medicine enough." Here I believe that the physician's appropriate behavior is based upon vast experience. Before seeing the course that the ailment takes, he knows that most symptoms involve no real problems, but he also knows how to spot those symptoms that sometimes, even though they resemble other symptoms, can reveal very serious problems. The physician will always be able to make this distinction better than a computer. This is why I'm convinced that the physician's role will remain what it should be—or what it was, in any event, when the physician used to play his role, which was to see the patient, listen to him, become familiar with his case history, learn about his general way of life, get to know his family circle, and then add the factor of his exclusively human judgment, so important in diagnosing and treating an illness.

IV

-And Humanity

Scientists do their work: they do research. Physicians practice their profession: they treat. And humanity, the designated beneficiary of so much effort, ingenuity, and devotion, what does it think about all this?

Does it automatically find happiness as a result? Today everything suggests that this is not the case. Medical services have not brought the public nirvana, and the public says as much. Many physicians seem surprised: "After all we've done for them . . . !" Some are angry: "Well, if they aren't happy . . ." And René Dubos asks himself: Suppose it were wrong to show our interest in the health of men and women by reducing them to nothing but flesh, blood, and bones? Suppose we could only correctly grapple with the problem of a man's health by studying him in relation to other men, to the place where he was born, to the landscape in which he lives, to the moves that are forced upon him? And suppose the "progress" brought by the industrial world sometimes has harmful effects upon him?

Before creating a health-care system all these questions must be answered, if such a system is to be properly oriented. This is essential, yet in the most pacifistic sense of the term it is literally a revolutionary concept, since it requires no more and no less than giving up the idea that man's state of health will improve if more is done for him. Tomorrow's truth could well be: the more we try to study, to resolve each problem by placing it in a general context, the greater the chances of success for the isolated individual. Dubos summarizes this in a maxim that is a program in itself: "Think globally, act locally."

DUBOS: The Industrial Revolution created very serious health problems. There was of course malnutrition, but I believe that the worst problem resulted chiefly from the fact that masses of people, enormous groups of peasants, were shifted to the cities and at first didn't know *how* to live in this new urban context. After one or two generations these people generally learned to live in the city, even if they were still poor. Note that the city is a totally different thing from a village that grew bigger. A city is born a city. It began as a sacred place and a trading center. The first cities weren't agricultural villages; they were a complex of temples and commercial activities. This phenomenon gives cities a quite special character. The truly important big cities of the world—Rome or Benares, for example—probably didn't begin as small agricultural villages.

But let's get back to the peasants' adaptation to the city. I was born in Sarcelles-Saint-Brice long ago. Saint-Brice has remained what it was, a little village in the greenbelt surrounding Paris, while Sarcelles became an enormous new city, the first of the large housing developments built in France after World War II. At the beginning, as you know, people suffered from just about every complaint in Sarcelles. Meanwhile everything remained normal in Saint-Brice, which is separated from Sarcelles only by the railroad tracks. But the situation in Sarcelles was alarming—to the point that a few years back people were talking about "sarcellitis." Three years ago, when I was in Paris lecturing to a medical society, I stated that there were enormous powers of adaptation in both the human organism and the social organism, powers to adapt to all sorts of circumstances. And I said in passing, "I'm willing to bet that 'sarcellitis' will disappear all on its own, because people will learn to live in these huge developments." At the end of my speech a woman rose and said, "Sir, I'm a physician in Sarcelles. What you predicted for the future has already happened. 'Sarcellitis' has essentially disappeared."

Since then I've naturally been on the lookout for other examples involving adaptation to new cities. Four or five years ago an entire segment of the population of the city of Boston was moved from the slums and installed in quarters that were much better equipped for sanitation. Yet disorders of all sorts began to appear at that point, ranging from physiological to social ones. Everyone exclaimed, "You see, it would have been better to leave the people in the slums." But I'm opposed to such a conclusion, which is far too hasty. I say just the opposite: we must act, but by helping adaptation, because problems always occur at first, even when the change is a favorable one. And examples are easy to find.

The first new city built in England is called Harlow. Initially, the entire population of Harlow experienced difficulties, both social and medical. All sorts of more or less clearly defined illnesses cropped up, and Harlow seemed to be laboring under an eternal curse. Then last year a long article on Harlow appeared, describing the calm and tranquillity of the new city where everyone, or almost everyone, had adapted admirably. Today the only serious problem at Harlow is that things are going so well that teen-agers of sixteen, seventeen, and eighteen are bored! To satisfy their need to do something, these young people go to London to help those who live in the slums. So at present the inhabitants of Harlow seem to have completely, or almost completely, adapted. In two generations the first people to settle there have grown old but now feel perfectly fine, while those in the prime of life are also happy, as are the little children. But the young people bored by this excessive tranquillity go to London to find interesting activities. Who would have predicted such a thing? Everyone thought that the opposite was inevitable.

So you see, the phenomenon of adaptation seems to be one of the problems that isn't adequately studied in all of its ramifications. Adaptation has beneficial aspects, the ones I've just discussed, that make us transform ourselves mentally, socially, and physically, in order to adapt to new, unusual, or abnormal living conditions. In my case, I adapted to the fact that I have a damaged heart, and I've learned to live with it very well. And I could give you all the examples you want of beneficial adaptation, both organic and social.

But the word "adaptation" also includes phenomena that I consider very dangerous. It isn't always a good thing to adapt. For example, there's no doubt that a person adapts to air pollution. He adapts to air pollution by increasing his production of bronchial secretions. True, for a time the bronchi are somewhat protected, but this is a short-range beneficial adaptation. If this "beneficial" protection continues for months and years, its consequences are chronic bronchitis and emphysema.

Here's another example of adaptation that has very undesirable consequences. A person can adapt to noise, but he does so by losing some of the acuteness of his hearing. Of course he suffers less from the noise, but this constitutes a kind of mutilation since he now fails to hear many other pleasant sounds as well. And here's an example of social adaptation. If it's necessary to adapt to a crowd, a person adapts by putting on blinders, by no longer perceiving the crowd, and so he loses a certain quality in his interpersonal life.

I believe that the problem of adaptation is one of the most important problems confronting us today. And the problem isn't being studied, for the simple reason that it's very hard to study it. It isn't being studied? I'm exaggerating, but it's difficult to "reduce" it to a few variables because it always involves an extremely complicated situation, which implies a very complex response by the organism. At first it seems that the data obtained can't fit into the measurements made by what people call basic science. Well, I think that hiding the problem because it doesn't fit into the scientific procedures of our day is a mistake. I believe that we should say to ourselves, "Well, here's a new kind of problem, and we have to develop a new scientific approach to it." Kenneth Boulding, a professor of economics, has written in one of his books (I quote from memory): "From a certain point of view, science is the art of simplifying problems in order to solve them; but this is usually done by eliminating all of the most important problems." I've often observed that this is true.

As far as medicine is concerned, life is one continual adaptation. There are all sorts of beneficial adaptations, but there are all sorts of adaptations whose consequences, often rather long-term ones, can be disastrous. And this is a fact of life that we must learn to recognize and talk about openly. I know that it interests a lot of people, since I'm constantly being asked to write on the subject. I must, however, admit that every time I write or talk about adaptation, I go beyond the limits of factual knowledge because no one—or almost no one—is working on this subject as a whole. Naturally, studies of adaptation are being made, but on very limited aspects of the phenomenon. For example, studies are being made of how people adapt when they go live in the Andes at an altitude of fifteen thousand feet, for in this very specific case controlled studies are possible. Undoubtedly the results are very interesting, but there are adaptations that are so much more complex, so much more important —such as the shift from one way of life to another. I don't believe anyone has bothered to study the various phenomena that occur when someone shifts from the world of the farmer, with all its difficulties, to the life of a city dweller, which involves other difficulties.

ESCANDE: Do you think that broader studies of this sort are discouraged because politicians are afraid they'll trigger a challenge to the established order? In my contacts with those holding political power—Frenchmen and occasionally foreigners—I've often been surprised how frightened they are of science, of its intrusions into any field that could have repercussions on politics. When people refer to the problems of adaptation in France, it's indeed a subject that political parties, both in and out of power, find very easy to use against the government in order to assert, for example, "You see what you've done to those people—those poor people mired down in intolerable conditions!" So these subjects are looked upon as detonators. For a year I worked with a newspaper that absolutely refused to let me discuss these problems. It refused, saying, "No, no, the contents are too political." "Political" in the original sense of the word. The newspaper was afraid that what I said would be used by politicians—in this instance for political politics and not for politics in the noble sense of the word—as a tool to influence elections. I believe that this is one of the important reasons behind such resistance; but in any case it takes at least several decades for an idea to get anywhere, and I wouldn't be surprised to find that a lot of people will be working on adaptation ten, fifteen, or twenty years from now.

Here's another subject I'd like to hear your ideas about. In 1967–68 a lot of studies were being made all over the world about how people adapt to changes in altitude. This was before the Olympic Games in Mexico City, which is located at an altitude of 7,415 feet. Since every country wanted to win gold medals, a lot of money was spent trying to learn how an athlete could be helped to adapt to that altitude. Then once the Olympics were over, the research virtually stopped. Aren't a lot of scientific fields studied only when they become profitable or usable?

DUBOS: Yes, of course. The submarine is another well-known example. Subs are a military necessity. But as we know, the crew has to live in a closed world for extended periods, and all manner of new physiological and psychological problems arise. So people began to study this problem very intensively, for obvious reasons. But no one thinks of looking at far broader problems. Still, since it's true that some living conditions are changing at an ever faster rate, it's inevitable that more and more difficulties of adaptation will appear. The only solution would be to perfect techniques to anticipate this kind of trouble. But this means creating a new science, a science that obviously is very different from the sort taught in laboratories today (which consists of isolating very restricted phenomena), for this new science would have to learn to predict the total organism's response to very complex situations. And I don't know of anyone who's working on that anywhere.

ESCANDE: So you propose a totally new science? One that's neither a social science nor an exact science? In sum, a science as yet unknown?

DUBOS: I think so. Such a science could be considered partly medical, but it can't be looked at solely from the medical point of view. It must also be viewed from the perspectives of technology, architecture, and social life—in sum, all the really important perspectives. And this understanding of the many variables that make adaptation possible represents a new preoccupation. Less than fifty years ago people held exactly the opposite opinion. I'll give you a specific example—it happened back in 1933. An international exposition was held to celebrate the one hundredth birthday of the city of Chicago. Its theme was "Science discovers, industry applies, man conforms"—that is, man conforms to the products of industry. And this is obviously what has been going on since the beginning of the Industrial Revolution, when new surroundings and new living conditions were created and the human being was expected to adapt to them. Now I believe that the great revolution, the great revolution yet to come—I hope it's already begun—no longer involves saying, "Man conforms to industry," that is, man adapts to new conditions, but instead involves reversing the proposition and saying, "How can industry be adapted to human needs?" And when I say industry, I mean all of modern technology.

Anyway, although it still isn't being applied, the idea has existed for a long time. When Pasteur was named professor at the Ecole des Beaux Arts in Paris, his inaugural address, which was subsequently published, already echoed these preoccupations admirably. He said: "Architects think about construction from a purely technical and aesthetic point of view. They try to design buildings that will last for a long time and that will be visually attractive, but the real problems lie elsewhere. A building is built to be lived in, so they should rethink their plans from the occupants' point of view." This is one of the reasons why he conducted the experiment with the birds and the bell jar that I referred to earlier, to show that organisms can obviously adapt to anything, even to such horrible living conditions as an inadequate supply of oxygen, but that these conditions have very harmful consequences for those who don't have time to adapt. Take the second bird, who died immediately when placed in the bell jar with very little oxygen, although the first bird, which had had time to adapt, survived. This experiment appears simplistic, but I think it reveals the daunting problems we face in our time. Obviously, we all adapt to poor living conditions, but the consequences—and they aren't always long-term ones—can be disastrous.

I'm not saying that physicians ought to be the only ones promoting this new vision of things. I'm saying that medical science ought to be the first to show interest in it. I'm well aware that physicians have a variety of occupations and can't concentrate exclusively on this problem, but the medical profession surely ought to accept the responsibility for being the moving force behind it.

ESCANDE: So once again the situation is both very simple and very complex. The vast majority of physicians want to return to a more general medical practice. I believe that medicine will soon break free from the bottleneck that has been strangling it ever since it unwisely decided it should reduce medicine to purely measurable factors. As everyone now realizes, in most cases day-to-day medical practice doesn't involve quantifiable data. Students complain they're being taught things that won't be of use, and physicians who've completed their training often say, "They didn't teach us the important things." So people are ready to go along with your ideas, I'm sure of that. But on the other hand, what you call the scientific medical establishment isn't ready—not by a long shot—to accept this change, and it can be expected to have a lot of trouble doing so.

DUBOS: I'll give you an example showing where the expected change can start. At the turn of the century infectious diseases were one of the chief medical preoccupations, but no medical school in the world was capable of putting Pasteur's and Koch's discoveries to use. Then, independent of the medical schools, many institutes were created to study infections. This was when the Institut Pasteur in Paris, the Koch Institute in Berlin, the Paul Ehrlich Institute in Frankfurt, the Lister Institute in London, the Kitasato Institute in Japan, and the Haffkine Institute in India were founded. Today the study of infectious diseases and microbes has been reintegrated into medical schools. So Franklin D. Roosevelt's theories are borne out: when a new problem comes up, existing institutions find it difficult to adapt quickly enough, therefore new institutions must be created.

ESCANDE: Should a René Dubos Center be created in Paris? Why not?

DUBOS: Two people—I won't give their names, but one is French and the other American—have told me, "We must institutionalize René Dubos."

ESCANDE: I agree!

DUBOS: I'm joking about this, for I've no desire to do such a thing. I've no talents as an organizer or administrator. But I do think my role is to state my belief that almost all the major medical problems are in the end problems involving a failure to adapt to the conditions confronting us. I repeat, no one has worked out the means by which a correct study can be made. In all Western countries we act on the hypothesis that, to understand these problems, we must study the phenomena at the subcellular level and not study the total organism in relationship to its surroundings. I find no indication anywhere that people are being trained to study adaptation. There's one exception: the armed services.

ESCANDE: You're familiar with military medicine?

DUBOS: I knew people who worked at a submarine base, the big base at New London, Connecticut. Two were Germans who, I believe, once worked on such problems in Germany and now work in the United States. But everything is secret. From time to time we come across little bits of published information which indicate they actually know about adaptation to life in a submarine. I had personal experiences of this sort long ago. When I was at Harvard during World War II, they were trying to develop methods to adapt the American army to tropical climates. For example, I recall they demonstrated that, contrary to what people thought, adaptation to tropical climates required nothing more than doing strenuous exercises daily for two periods of a half hour each in a humid setting resembling the tropics. This intensive half hour brought about transformations as rapidly as if the individual had spent the entire day in those surroundings. I remember that detail because it surprised and impressed me. I also recall how they stressed that a person adapted only to the degree that he *functioned* in those surroundings. This was very important. It wasn't simply a matter of putting the individual in a room with high temperature and high humidity; what counted was that the person be active once subjected to those conditions.

I suppose that every army in the world conducts similar experiments, but this field remains highly secret and the results aren't published. I often see certain colleagues of mine who I know for a fact belong to this kind of research team. They never say a word!

ESCANDE: The same thing happens regarding the adaptation of athletes. It's kept secret. About a year ago a physician specializing in athletic medicine told me that, in preparation for future Olympic Games that might be held in a very hot country (an African nation), athletes are beginning to train in surroundings where the temperature is 104 degrees Fahrenheit. One of the Eastern European countries is said to have built a stadium that's a virtual hothouse and to be training its athletes there. But this is kept totally secret, and what we should be learning about adaptation will once again remain a closed book!

DUBOS: Let's get back to false ways of adapting, if you don't mind. What interests me about chronic bronchitis is that very specific observations are being made on the subject. Ten years ago in England it was demonstrated that, by the age of three or four, children raised in surroundings with considerable air pollution showed spirometric abnormalities.* Thus at that early point it's

* Spirometric studies involve blowing into a device that measures the volume of air used during respiration.

possible to detect those who will suffer from bronchitis in the future.

During the very first phases of the Industrial Revolution almost everyone in the Anglo-Saxon countries, northern France, and northern Germany was exposed to smoke and fumes, yet people said, "Well, it doesn't prevent crops from maturing, the economy is still thriving, so there's no real problem." Indeed, they were right. The fumes didn't prevent the crops from growing and the economy from thriving, but almost all these people promptly began to suffer from chronic bronchitis. Of course they didn't die right away, but they died from it indirectly, at around sixty. The population had adapted, but by agreeing to suffer from chronic bronchitis.

ESCANDE: One day you commented, "In choosing nuclear energy the French government is gambling with the health of French citizens." So there's a big risk involved?

DUBOS: There's a risk, but there's no certainty of a series of catastrophes, despite the events at Three Mile Island. The United States seems to have taken a different course from France. The current government energy plan is based on coal. If it's accepted, and it will be, the use of coal in the United States will at least double or triple. Well, using coal means accepting a certain number of mine accidents, accepting that large numbers of miners will come down with black lung disease, and naturally accepting that all sorts of other poorly understood substances emitted into the atmosphere when coal is burned will pollute the air a bit more. So I'm not at all certain that the risks involved in nuclear energy are any greater than the certainty of all kinds of accidents resulting from the increased use of coal. Besides, I find it extraordinary to accept coal as a solution without clearly stating in the press the repercussions that such a solution will have, even if coal technology is improved, as I'm sure it will be. You simply can't burn coal without certain disorders occurring. Now these are problems that involve environmental and medical policy, and medical circles say little about them. Medical science should learn to think in terms of the total medical environment. It never does, never has. It only does so when catastrophic accidents occur in very specific situations.

ESCANDE: So beneath the surface of everything you've said lies the myth of Hygeia and Panacea, which you often refer to. What do Hygeia and Panacea symbolize?

DUBOS: It's quite simple. Aesculapius had two daughters, one called Hygeia and the other Panacea. Panacea was the daughter who knew every remedy that could be extracted from plants or from the earth to treat the sick. Hygeia was the goddess of Athens and taught that one must keep a healthy mind in a healthy body by living in a reasonable way. She asserted (and Plato repeated it in his dialogues) that if one lives reasonably, one often needs little medical treatment. There's a magnificent bust of Hygeia at the museum in Athens, but I've never seen a bust of Panacea. She's always portrayed with Aesculapius, while Hygeia is often shown alone. This is all the more curious because basically Panacea has always been considered the more important of the two sisters. Of course people talk about Hygeia, but in reality she doesn't have a lot of influence over the way each individual lives. And schools teaching hygiene commonly go downhill, they collapse or lose their prestige, while medical schools continually triumph.

About sixty years ago—closer to seventy now—at the beginning of this century, all sorts of special schools of hygiene were created in American universities. Well, they all failed. I'm exaggerating when I say that they failed, for they still exist; but the level of their students is often low. Most of their good students are statisticians. Yes, indeed, they do statistical studies showing the incidence of diseases! But alas, when they try to study how to live well, they don't get anywhere. Because this field simply doesn't fit into our conception of science, which is totally reductionist and balks at the concept that men live in an environment that must be looked at as a whole.

ESCANDE: Panacea actually reduces the steps to be taken to swallowing a pill!

DUBOS: But imagining how to create a city in which life would be good is Hygeia's art, which is far more complex. For a city in which life would be good implies that there's none of that noise to which we adapt by losing our hearing, none of those fumes to which we adapt by developing chronic bronchitis, none of those constructions that prevent the eye from venturing to the horizon. Yes, that's Hygeia's art, but it can also become a science, a science that tries to understand the basic requirements permitting human nature to express itself totally. Artists talk about it, but scientists aren't at all interested. Yet enjoying life is a very profound biological need.

ESCANDE: Does history have a meaning? An irrevocably traced direction?

DUBOS: The entire history of civilization is in fact a series of attempts to adapt socially to new situations, attempts in which the course taken depends upon free will. Free will that can't be proven, of course, but that I for one accept as existing, although I don't even try to define it. Free will that, having propelled history along a certain path, permits it to make an about-face and go off in another direction when necessary. I believe that man's entire history is a series of attempts, struggles, and adaptations in one field or another. Then, when man realizes that he's gone a bit too far or taken a wrong path, he turns around to go in a different direction.

ESCANDE: Is it possible, then, to look into the future and assert that we've become too attached to the automobile, so we'll do an about-face and build fewer superhighways and fewer cars?

DUBOS: I'm completely convinced of it. Let's turn to the past for another example, the history of the Gothic cathedrals, which were built higher with each passing decade. Indeed, every decade the builders of a new cathedral wanted theirs to be ten or fifteen feet higher than the most recently completed one. When Amiens cathedral was built, it was the highest to date, so the folks in nearby Beauvais wanted a cathedral a little bit higher. They built it, but it collapsed. They started again, it collapsed again. Then another architectural style made its appearance. As I see it, this symbolizes what's always gone on in our societies. We go to the point of absurdity, so to speak, then suddenly they say, "No, that's really too idiotic," and we make an about-face or start down another path. I believe this example can be applied to all sorts of fields: medicine, automobiles, architecture —any phenomenon that's evolving.

ESCANDE: Once we realize that we're on the wrong track, must we go all the way to the point of absurdity before we turn around? Can't we ever be wise enough to say, "Come now, we've gone off on the wrong track. We have to stop before it's too late"?

DUBOS: Here you've touched upon the real innovation of our times. For the first time in the history of humanity, we're beginning to try to foresee the consequences of our immediate actions, and perhaps we'll succeed in not going all the way to absurdity. There's no doubt that at present every modern society is asking, "What will happen if we continue to make so many automobiles? If we continue to prescribe so much medication? If we continue to want a doctor every time we sneeze?" These questions are being asked everywhere, and it's very interesting to study this movement, which is evident on every level. Let's look at the United States. The Americans are far behind Europe where the social organization of medicine is concerned. Not medicine itself, of course, but its social organization. We've been endlessly discussing socialized medicine for years. Every thoughtful segment of the population is beginning to wonder, "What on earth is going to happen if everyone with a mild headache goes to see the doctor? It'll cost a fortune!" So the government undoubtedly will convince Congress that there's an acceptable sum to be spent on health care, and that that sum shouldn't be exceeded for any reason. Once such a decision is made, hospitals and physicians will have to adjust by rejecting a certain number of requests for medical help.

ESCANDE: With the means now available, can't reaching the point of absurdity in certain fields nevertheless be far more serious, far more serious than it used to be?

DUBOS: Yes, if everyone believes that equality obliges everyone to own a Cadillac, we'll obviously encounter situations as absurd as they are catastrophic. Yet I must say that there's a general awareness of the middle- and long-range danger. We know we must discuss the problems that we'll face in the year 2000. And I for one am persuaded that the art—I say "art" because what I'm talking about as yet is barely a science—the art of anticipating future problems will constitute one of the great revolutions of our time.

ESCANDE: Can we change the subject a bit and turn to nature?

DUBOS: Certainly. Man labors under delusions about nature. I don't know of any place in the world where nature is the same as it was before man appeared. I was born at Saint-Brice and raised in the Ile de France, the region around Paris. The charm of the Ile de France is admired the world over. Before man appeared there, this region was rather flat, rather ordinary, and covered with forests and marshes. Only during neolithic times did man begin to cut down the forests, control the rivers, and create fields and pastures. People think that the landscapes created in this way are nature, but they're a "nature" that was entirely conceived by man.

What I'm saying about the Ile de France is totally applicable to every part of the world occupied by humankind. Everyone the world over thinks that San Francisco is the most agreeable American city. And when you talk to Americans or Europeans about San Francisco, they especially praise its two large parks, the Presidio and the Golden Gate Park. Well, the land on which the Presidio stands was nothing but shapeless dunes until 1881, when a decision was made to plant trees. I'm very familiar with all the studies that determined which trees could grow there and what sort of vegetation was suitable for this very special region. Now it has grown into one of the most pleasant urban forests in the United States.

Or, turning to Europe again, I'll use the Black Forest of Germany as an example. Everyone thinks the Black Forest is a wild "natural" forest. But it's a forest in which a variety of species of trees have been cut down systematically for over a thousand years, so that only those considered most useful are favored. And they might not survive if man didn't continually intervene. And yet, while forests are extremely important for man, to my knowledge no human being could live in the heart of a very dense forest. Wherever we find people who lived in forests, it is because they were forced to do so by invaders. And when they settled in the forest, they immediately made a clearing. This becomes obvious if you fly over the jungle of Brazil. Here and there in the heart of that immense jungle you see big clearings near rivers. That's the only place where you'll find humans. They never settle in the interior of the jungle without modifying it first. This isn't merely a description of personal tastes; it's an absolute necessity.

This summarizes man's whole life. Humans have transformed the earth in order to satisfy certain needs acquired as they evolved during the Stone Age, when they lived on the savannas of Africa or Asia. The savannas have trees, shrubs, and rivers, but also vast open spaces and therefore great topographic diversity. Within this landscape man can see far into the distance, can keep an eye on the trails passing through it, and can also withdraw into a shelter if the need arises.

Now think of the Vézère region or the Dordogne Valley of France. The most striking example of this is the Cro-Magnon cave. If you stand at the mouth of the Cro-Magnon cave, you have a vast vista before you. It was easy for man to withdraw into his cave and look out into the distance to watch for passing animals. In addition, the forest is there close at hand. So I think there's something very important here, something vital from a physiological point of view. Man needs a shelter, water, and a distant and varied view. Moreover, think of paintings representing happy outdoor scenes. Such scenes always take place in a clearing. They always show a thicket or woods, a river, and a vast expanse. So wherever man is, in all the countries of the world where he's settled, in one way or another he tries to re-create the savanna.

ESCANDE: As far as paintings are concerned, when fixing a price for a landscape today, the professional asks, "Is there any water?" If it includes a pond or a river, the painting is worth far more than the same painting without water. I've always wondered why, and now you've given me the answer. In the last analysis, the salesman or the buyer himself is seeking to regain those favorable environmental conditions that you've just described. And water is part of them.

DUBOS: To be sure. And this leads us back to medicine, for medicine should consist of giving man back, insofar as it is possible, the conditions under which certain very profound visual needs can be satisfied. When I was working on tuberculosis, I routinely visited a lot of sanitariums, and I realized to what degree all these needs had been met, perhaps unconsciously, with a view far into the distance, and with woods and water nearby. Two or three years ago I went to the Aegean island of Kos and spoke in the amphitheater there, where Hippocrates once taught. From it you can see woods on all sides, a vast open space, and a spring. I'd be ready to bet that all the great temples to Aesculapius that the sick flocked to had similar surroundings. In his dialogues Plato tells of the ancient temples, before the great deforestation of Greece. He always mentions a river, a spring, and an open space across which the eye could wander. Doubtless this sort of landscape corresponds physically to something very important that the ancient physicians had unconsciously recognized as playing a role in getting well. And I believe that, as regards medicine, we shall once again have to be concerned with these essential needs. Here we're dealing with the most profound things in man's makeup. I think the medicine of the future is exactly that: satisfying once again *these very profound demands of human nature.*

I'd like to give another example, if you don't mind. Everyone tells me, "You claim that there's a unity in the landscapes man has created. Yet look at the fundamental difference between parks like those designed in France by Le Nôtre during the late seventeenth century, and parks created by the English landscape architects in the eighteenth century!" Well, they're indeed very different, but it's only a cultural difference in the way the same fundamental thing is expressed. Le Nôtre's individuality lies in his wide, straight avenues with woods on each side, and his inevitable reflecting basins laid out geometrically, as at Versailles. If you walk through the large English parks you see broad lawns where your eye can wander and then focus upon a lake that, instead of being rectangular, is a bit whimsical in shape and surrounded by woods here and there. These gardens are very different, but when you come down to it, both cases satisfy your need to look into the distance, and also make you feel that at the right time you can withdraw into the woods to seek shelter. Yes, there are these profound, universal, eternal things in human nature, and each culture expresses them in its own way.

This way of reasoning applies to all kinds of fields. For example, I think psychoanalysis is the way our culture does what the shaman did, that is, tries to dredge up deep, unconscious things in human nature, things that in one way or another must be explained, must find expression. I believe that in this sense psychoanalysis is merely another way of doing the shaman's work.

ESCANDE: Having discussed nature, how about turning to the natural and the artificial? Must we view as a great danger all those synthetic, chemical, and artificial substances that are becoming widespread in such forms as dyes and stabilizers in foods for both animals and humans? I'm thinking in particular of the problem posed by a dye such as tartrazine, which has been found to be extremely dangerous for people allergic to aspirin. The story is worth telling in detail. First it was discovered that aspirin can trigger serious disorders—a veritable "aspirin sickness"—in certain individuals predisposed to it. When they take aspirin they experience fever, generalized fatigue, and above all severe asthma attacks. Then it was observed that a dye called tartrazine yellow, which is used frequently, especially in the coatings of medicines, could cause the very same symptoms in some of these people. What I find interesting is that, before these observations were made, tartrazine was considered the safest and most easily tolerated dye. We've every reason to believe that certain substances, certain additives, must also provoke serious illnesses that we can't as yet trace to their true cause. I for one believe that today many heart attacks, kidney failures, and strokes are undoubtedly caused by factors that medicine pays absolutely no attention to and that aren't even taken into consideration, although they indeed exist.

DUBOS: Obviously, this is a very difficult problem to study; so many new things are being introduced that we don't know where to begin. We don't have time to study everything. The only reply that I can give you is that we should never use any new substance unless it's been definitely proved to be useful or pleasant. Obviously, we can't stop changes in the world, but we can keep them from going to extremes. One striking example involves commercial baby foods. All sorts of additives are used—dyes, flavorings, sugar, salt—simply to please the mothers' taste buds. We know that infants shouldn't be trained to crave sugar simply because we want to please their mothers. The same is true for salt. I don't have very precise knowledge about this issue, but certain colleagues of mine say that the great craving for salt shown by certain people stems from their having been conditioned as children to like salt, because mothers won't buy unsalted baby foods. It would be better to educate these mothers than to attract them by procedures that lead to a harmful conditioning of their children. Thus habits become necessities; people believe them to be necessities, because they've been conditioned that way.

ESCANDE: The example of the mother is really very significant. The child doesn't need salt, but since the mother likes it, she thinks her child needs it. And we're doing little to give her the correct information. The same thing can be observed in other fields. Let's go back to my example about colored medicines. They are often colored to make the consumer want to take the medication, and when the laboratory of a pharmaceutical company decides to market a product, a great deal of attention is paid to the packaging and to the color and form of the medicine. One day, on a French television program, I said, "You want me to demonstrate that medicine is a consumer's item like anything else? That's easy. Why are medicines colored?" And the anchor man sitting beside me—a remarkable and very agreeable man, but who in theory was there to speak as my adversary—was totally dumbfounded and said, "But what you're saying isn't possible!" Yet he was immediately convinced. Three months later I received a letter from the national association of pharmaceutical industries: "Sir, we cannot allow your statement to go unanswered. If medicines are colored, it is to keep them from being confused with other medicines." This isn't a valid argument, however, for if you don't want to mix them up, you merely make sure you put them in different pill boxes. A patient doesn't often mix up his medicines, and if he does it involves those left over from a previous illness that should have been thrown away. Food additives are a way the grocer can exert pressure. When you said, "We should never use any new substance unless it's been definitely proved to be useful," you're in fact asking for a drastic restriction that would trigger angry reactions.

DUBOS: We have to be satisfied with this: if we can show that a certain substance prevents bread from turning moldy, we can run the risk of putting that substance into bread, because mold can be even more dangerous. The thing is therefore justified socially—on condition, of course, that we show that the substance has only a rather slight chance of being dangerous. In the same way, using substances that help preserve foods is justified to a certain point, even though these substances are dangerous. But I'm categorically opposed to using dyes simply to make things look different, because it's technically impossible to conduct adequate tests of the toxicity of every new substance. We'd have to be so highly organized! Besides, we don't even know how to conduct the tests. We don't know what they are—they haven't been invented yet.

But synthetic substances aren't the only dangerous ones. Natural substances can also be dangerous, sometimes quite dangerous. In New York City much has been made recently of the fact that a carcinogen can be extracted from broiled meat. Actually, toxic substances and carcinogens can be obtained from almost everything. A priori, therefore, this is a terrifying problem, but in reality we don't have to be too terrified. It's a terrifying problem in that we know very well that every substance used in our daily lives involves risks, dangers. For example, there's talk these days about cancers found in certain workers exposed to sawdust. Everything around us undoubtedly involves certain dangers, but in the end the human race has survived and even thrived despite them. So although it may be terrifying, I'm no more terrified by it than I have to be.

ESCANDE: Can we get back to the definition of "natural" and "not natural"? I was very impressed by an article in the *New England Journal of Medicine* in which a leading immunologist wrote: "People drum 'natural' and 'not natural' into our ears. But what is more 'not natural' than the molecule of an antibody? It is created to fight something foreign to the organism and adapts itself exactly to that foreign body. It is the most antinatural thing in the world." The same author added that excellent, very natural things are found in nature, things called hemlock and digitalis, the most active poisons we know but very useful under the proper conditions. Then he concluded, "I really would like to know whether this distinction between natural and not natural is justified, and whether we should reexamine the stereotypes that equate natural and good on the one hand and not natural and bad on the other."

DUBOS: From that point of view, I'm far less natural than most people, because I tell anyone who'll listen that humans stopped living a natural life more than ten thousand years ago, not fifty or one hundred years ago. The day when humans stopped living solely by hunting and gathering and began to construct houses or shelters, to cultivate wheat or rice, and even more to build fires for warmth, they began to leave the natural life behind. We've retained our power to return to the "natural" life, as was shown by the people of European descent who settled in the Rocky Mountains during the nineteenth century and rapidly readapted to the life of the Stone Age. We still have the capacity to live naturally, but today we never live in a natural state, because what we call the Temperate Zone is a region that's theoretically incompatible with human physiology. If it were "natural" here, well, the room where we're now sitting would be in the middle of dense trees and probably marshes. And we couldn't live there. So the very fact that we've been willing to live in France or America shows that we've abandoned the natural and therefore assume the responsibility for creating an artificial environment adapted to our fundamental needs, which haven't changed. So I believe we've been sentenced to live in artificial surroundings. Then what is natural and what not? What I call "natural" is what is adapted to our physiological makeup and our mental needs, which—unless someone transforms our genetic structure —will not change. The problem of the natural doesn't impress me as being very important.

ESCANDE: So it's the latest rage, an "advertising" argument?

DUBOS: Of course! There's nothing natural about our surroundings anymore. I repeat, a few thousand years ago the region where we're sitting right now was covered with forests and marshes. And as soon as humans, biologically adapted to live on the savannas, arrived, they began to create surroundings that were adapted to them. And so, because our prehistoric past taught us to use only plants that live in the sun, we've cut down the trees in order to plant things that live and grow in the sun. For almost without exception, all the plants we use are plants that would never grow under trees. One of the few plants that will grow even a bit under trees and that we use frequently is coffee, which requires shade. But certainly none of the plants consumed as our basic foods could grow under trees.

ESCANDE: Does this explain why we haven't learned how to grow most kinds of mushrooms? Mushrooms usually grow in the woods. And we don't know how to cultivate them, or hardly so.

DUBOS: You may be right. In any case, what we call going back to natural things means creating surroundings that resemble the savanna. It's an astonishing paradox. Wherever he goes, man transforms nature in order to adapt it to his needs, and then, once he's transformed nature, he calls it "natural." I saw an extraordinary example of this at the international conference on the human environment held in Stockholm in 1972. As you may know, I co-authored a book that served as the working paper for this conference. At the end of the conference Olof Palme, the Swedish prime minister, invited five or six of us to visit him at his house in the country and talk about environmental problems. A very kind Swedish lady picked me up at my hotel in Stockholm and drove me the fifty miles to the prime minister's home. During this little trip the lady pointed at the landscape and said very sadly, "As you see, in Sweden we're in the process of destroying nature." By that she meant that a large number of the farms that weren't very fertile had been abandoned. Swedish farmers now earn far more money by working in the Volvo or Saab automobile plants than they would if they continued to work these farms with their poor soil. Well, as soon as one of these farms is abandoned, brush grows, then the forest reclaims the land. In the past, that entire part of Sweden was an unbroken forest. So I replied, "You say that nature is being ruined, but that's not so—on the contrary! At this very moment nature is reclaiming her rights." When I reached the prime minister's house, I told him this story and he replied, "I'm quite aware of this. We Swedes have become accustomed—and I'm glad—to seeing part of the Swedish countryside without trees, to seeing our lakes surrounded by fine meadows, cultivated fields, and pastures with cows. And we're horrified at the thought that once again there will be trees everywhere. The Swedes are haunted by the idea, so we've undertaken studies to see how we can most economically preserve the countryside cleared by the farmers. We could consider granting subsidies to farmers (for their work is no longer economically profitable), or else—and this might be more economical—every year, or twice a year, we could use big machines to cut everything down."

ESCANDE: A few years ago, maybe five or six, the late French president, Georges Pompidou, said on television, "The problem confronting agriculture is a very simple one. We must choose between two solutions: we either hire foresters—civil servants paid by the month—or else we give subsidies to farmers so that they will keep the landscape as it is." And he said, or at least implied, "I, a man from the central mountains, prefer to subsidize the farmers rather than pay foresters." This echoes Prime Minister Palme's observation.

DUBOS: I've just completed a book—*The Wooing of Earth.* In it I state that, as agriculture becomes increasingly productive and as marginal farmland is abandoned, every society encounters serious difficulties in preserving the humanized countryside, which for man is a physiological need. I repeat: we need to see open landscapes. So in the forests around Paris, the area under the trees is kept cut down because people wouldn't go into the woods if the underbrush were allowed to grow. I'm convinced that a human set down in the midst of an impenetrable woods would suffer. I'm not even certain he'd survive. We have a psychological need to see around us. This is why we must preserve the humanized countryside. So I totally agree with President Pompidou. No matter how we do it, we must preserve the countryside, for the temperate regions would no longer be livable for the human race if we let nature, true primitive nature, reassert its rights. I'm continually telling my ecologist friends, "We couldn't live in France, in the United States, in Sweden, or in England if we let nature revert to what it was before the advent of man." This is why the word "natural" seems to me a very difficult word to use. I don't really know what "natural" means; there are so very few things that are *natural* and hence better. If you tried to eat natural wheat, you couldn't digest it. It must be transformed by a bit of fermentation and by cooking, after which it's no longer "natural," but it *is* edible. Like it or not, from the moment we learned to transform things according to functions we developed a hundred thousand years ago, we drove the "natural" out. Today the "natural" is something that we've adapted to our needs.

This brings me to some very perplexing philosophical problems. We can't keep from being anthropocentric. Indeed, an early chapter of *The Wooing of Earth* discusses "the absolute necessity to be anthropocentric." It isn't possible for us to view or judge things otherwise. It isn't possible for the human race to survive on earth if (to use an expression that I coined and that's brought me a lot of criticism) it doesn't achieve the humanization of the earth. In France, England, Germany, or America nature is the humanized earth.

ESCANDE: What did they criticize you for in that statement? For saying that man was more than just another species of animal?

DUBOS: That's it exactly!

ESCANDE: Still, man exists as individuals. He's different from an animal, different from an inanimate object. One day, as a challenge to a friend of mine who's very fond of Mao, I said, "The 'masses'—I don't understand what that word means. Individuals exist one by one, and the concept of 'mass' means nothing at all to me." It was during a dinner party at a medical convention in Switzerland. My friend looked at me and said, "The Eiffel Tower is made up of three hundred million pieces of metal, yet it's the Eiffel Tower. The Chinese people is made up of seven hundred million individuals, and it's a whole." I found his statement so extreme and unconvincing that I remained absolutely speechless for at least five seconds. Being quite unaccustomed to seeing me silent like that, he added, "No reply, eh?" To which I answered: "The truth is that I was wondering how anyone can say such an absurd thing, because, in point of fact, there's nothing in common between a man who thinks and is autonomous, and a piece of iron that can be welded as one chooses. Seven hundred million human beings aren't seven hundred million pieces of iron welded together."

DUBOS: In addition you could have told him that the Eiffel Tower is also, above all, a structure created in the human brain. The Eiffel Tower isn't merely little bits of iron. The Eiffel Tower might also be made of some other material. The Eiffel Tower is a form—a form and a structure. We don't say that a house is five hundred thousand bricks. Five hundred thousand bricks are five hundred thousand bricks. A house is a structure. So we're right back to the criticism I received—that I give man a special place in nature. I say that those who refuse to accept this evidence should logically give up living like human beings.

ESCANDE: You often say that, of all living beings, man is the only one who modifies the environment to his advantage. Doesn't a herd of animals permanently modify the environment to its advantage?

DUBOS: The most powerful animals in the world never leave the restricted ecological zone to which they've adapted. Even monkeys—who know how to use certain objects; who are even capable of making simple tools; who know how to take a porous object, saturate it with water, and carry it; who know how to use a branch to catch termites and eat them—even monkeys, monkeys of every species, never leave the small ecological zone to which they've totally adapted. The other day my wife mentioned the extraordinary example of that charming Australian animal, the koala, which can live only on a diet of the leaves of a single species of eucalyptus tree. Animals of whatever kind are specialized. They live only in the ecological surroundings to which they've adapted. But for the past ten thousand years, and perhaps before that, men have been transforming the land in order to adapt it to themselves. And that, I believe, constitutes a most fundamental difference setting the human race apart from all other species.

ESCANDE: And migratory animals?

DUBOS: Migratory animals are total prisoners. They always take the same paths, they return to exactly the same places, and if you move their dens or their stopping places just a short distance, they become totally lost. All this shows the extent to which I think the word "nature" is very dangerous, unless a person agrees that nature is something man has transformed.

ESCANDE: Can you be more specific about the great threats hovering over mankind?

DUBOS: In the initial pages of *The Wooing of Earth* I state that I'm going to deal only with the most positive aspects of man's transformation of the globe. But despite this warning, there are sure to be criticisms that I've neglected the tragic problems confronting us. So I begin the book with a few pages that really have nothing to do with the material that follows, and state that I'll not only list the major problems—the tragic problems that the world faces—but I'll also list them in order of importance and tell why I put them in that order.

The first problem, nuclear war, is all too obvious. If a nuclear war should occur one day, it will just about destroy our civilization, and unfortunately I don't see that we've made any real political progress to prevent it.

The second great tragedy may surprise you. It's the fact that our society is increasingly unable to provide people with a function that has a profound meaning for their lives. On the one hand this constitutes a condemnation of unemployment, but it also goes far beyond. When you see the enormous proportion of young people who can't begin to integrate themselves into active life as they had hoped to, or who'll never enter that life, you have to observe that both a desocialization and a dehumanization are likely to follow. And this, it seems to me, is truly the most tragic human problem confronting the industrialized nations.

ESCANDE: And pollution?

DUBOS: I put pollution at the end of the list. Not because it's unimportant, but because throughout the world people have already become quite aware of the problem and have achieved quite a bit. Of course pollution is very important, but we've begun to take the necessary steps.

Third on my list is the destruction of the tropical forests. Man is destroying these forests with phenomenal rapidity the whole world over! Here again no one is doing anything, no one knows what to do to prevent it. The tropical forests are located in poor countries, which of course are using them to obtain money by selling the produce of these forests to the capitalist nations. If we let things go on as they are, the disastrous climatic and ecological consequences will affect the entire world.

There are also the problems of overpopulation, environmental degradation, and depletion of natural resources. I put the problem of energy sources in sixth place. I'm not very worried about finding raw materials and energy. Our societies can learn to produce energy and find resources, so the real problem is the excessive use of energy, which I want to turn to for a moment. I'm concerned about energy problems within the context of the René Dubos Forum, which organizes symposiums with a lot of eminent people from a great variety of backgrounds.

The first symposium was a discussion of how much energy we can produce from vegetation, the so-called *biomass*. To discuss this question we invited not only scientists but also engineers and businessmen, who asked, "How much energy can we obtain from the biomass? What will it cost?" We also had ecologists, who presented the dangers to the earth if we use this biomass too extensively. We even had humanists—some of whom, incidentally, turned out to be less humanistic than the scientists. And we reflected on what would happen to our conception of nature if we decided to create what might be called "energy plantations."

I didn't state the theme of the second symposium in question form. Instead, I made an assertion: "Live better with less energy." First of all we have to envisage the risks inherent in energy production. Whatever energy source is used, be it coal, nuclear power, the biomass, or solar energy, it can provoke atmospheric disturbances once a certain production level is exceeded. Thus enormous and very poorly defined problems confront us. Even if we decided to use only solar energy, which has the reputation of being a "clean" energy, we'd have to concentrate its production in certain regions and therefore increase the amount of heat generated in those regions, which might cause all manner of trouble.

That's one aspect of the problem. Another aspect to be considered is the relationship between energy use and physical health. For example, we all know only too well the consequences of the excessive use of food. But the numerous ways in which energy is used excessively in overdeveloped countries pose a multitude of other problems. Today, when people play golf, they ride around in little carts. While this may sound humorous, I'm citing only one example among a hundred thousand that ultimately all have profound biological repercussions. Because we're in the process of creating a society in which energy is used excessively, we're preventing the existentialization of our physical potentials. The overconsumption of energy in our societies favors a trend to replace the direct experience of things with an indirect experience. We prefer to watch a tennis match on television rather than play tennis ourselves. We prefer to see springtime or fall in pictures rather than go out into the countryside in all kinds of weather. The greatest danger in television, as I see it, isn't that it turns children into criminals because they've watched violent films, but that it prevents them from acquiring a direct experience of life. They watch the world passively, instead of experiencing it actively. This is a major factor in the dangers inherent in the excessive use of energy. Of course I can't prove this statement. It's still a hypothesis, only a hypothesis. Even so, it's already based upon solid foundations!

Architecture is another facet of the same problem. If we permit ourselves to consume enormous amounts of energy to heat and to cool, then we don't have to adapt architecture to our natural surroundings. In the past, one of the chief reasons why architecture was interesting—especially the architecture of modest houses—was their extraordinary diversity. Houses weren't constructed of the same materials in the South as in the North, or in the snowy mountains as in the regions where snow never falls.

ESCANDE: So according to your explanation, energy is to blame for having made modern architecture so monotonous?

DUBOS: Absolutely. And it goes deeper than that. Wherever a lot of land is available for construction, energy also permits us to build the little houses of those new "towns," houses that have space, a lot of space, between them. On the one hand it's all very nice, but on the other hand it destroys both the social structure and the human relations indirectly created when we strolled along Main Street or in the public square. The use of energy also has repercussions on the fertility of the soil. Energy permits us to make quantities of nitrogen fertilizers. As a result, the plants having the potential to use nitrogen from the atmosphere no longer do so if they are given large amounts of fertilizer. They no longer make the effort.

So by bringing in outside energy we prevent all kinds of adaptive phenomena: the adaptation of the body, the adaptation of the senses to perceive things, the adaptation of the land, the adaptation of plants to take nitrogen from the air. We introduce massive laziness. I repeat, I'd like all these problems to be studied in terms of their interrelationships, rather than one by one. Instead of saying, "Let's make energy and use the energy," we should think about the fact that when we use energy excessively, we stop doing a great number of things necessary to our well-being. Since I've no doubt that we'll manage to produce all the energy we need (all sorts of techniques exist that will enable us to do so), the major problem will be the pertinent use of energy.

I'll end this with a musical example. In the past, people used to sing, although they weren't musicians. Today almost nobody sings—they turn on the radio! But living means putting all your potentials to work. From the moment you replace your "self" with external elements, you live less, you live less intensely and therefore less well.

ESCANDE: You have a whole system of thought that stresses the relationships between human beings, the relationships between social groups, and the relationships between social groups and the land as far as the environment is concerned. I'd like to know whether your system of thought is planet-wide or whether it applies solely to the industrialized nations.

DUBOS: I've chiefly focused on the industrialized nations, that's true. But naturally I see the problem on a global scale. For the conference on the environment held at Stockholm, Barbara Ward, a famous English economist, was asked to write a book that would serve as a working paper for the entire conference. She's an extremely brilliant economist. But when she finished the book, she wasn't satisfied. So I was asked to collaborate with her, and I wrote part of the book, *Only One Earth.* This English economist had lived in Africa for many years, but what she wrote applied to the earth in general. As for me, I discussed the rest. And everyone says that it's impossible to distinguish what she wrote from what I wrote. By a lucky bit of chance, our two writing styles fit together perfectly. Yet she and I made very different contributions. If you look at any one chapter, you'll see that some assertions are worldwide, but also that particular stress is placed on the fact that each of us belongs to his own country, possesses a certain culture and a certain approach toward life, and develops certain needs that differ from the needs of people from another country with a different culture. I'm the one who added those particular points to each chapter. They reveal one of my deepest concerns, and in synthesizing them I came up with a very simple formula. In fact, I used it as the title of a lecture I gave in New York City and of an article I wrote recently: "Think globally, act locally." So you see, experience has shown me that when you set out to solve a problem for the whole world at once, you get nowhere. You get involved in things so complicated, so woolly-minded, so vague, that nothing useful comes of it.

I'd like to cite this example. Students always want me to talk with them about the problems of pollution in a general way.

And I reply each time, "If you haven't already made it a practice to clean up your campus and the gutters on your own street, don't try to deal with pollution as a worldwide problem."

But to get back to the developing nations, I want to say first that if a disaster occurs somewhere in the world, I definitely want an effort to be made to help the endangered populations. But I'm against a policy of systematic aid as it's conducted today. Sending American wheat to Lima, Peru, is nonsense. The rural Peruvians leave their land, come to Lima, and end up abandoning their farms. I think the result is disastrous. It seems to me that the problems of the developing countries should be handled differently for each nation. Let each of these countries find the needed formula. We can, of course, help each of them to find the formula that best fits its needs.

Let me return to a problem of enormous importance: the tropical forest. This is certainly a global problem, because if the tropical forests are destroyed, all sorts of climatic and other ecological consequences will affect the entire world, consequences that can be rather accurately predicted. In addition, we shall witness the destruction of species that may be necessary for the equilibrium of our planet. We must dare to tell the people who are destroying their forests, "Stop! The consequences may be too serious." But on the other hand, we Europeans and Americans are in a poor position to tell the Brazilians or the Malaysians, "Don't destroy your forests," for one or two thousand years ago in the case of the Europeans and three hundred years ago for the Americans, we destroyed our own forests for the sake of wealth. We must provide these people with technological aid—and we're in a position to do so—that will help them reforest immediately, as they gradually exploit the forest. We're quite aware where the major problem concerning the tropical forest lies: as soon as the forest is cut down, the soil becomes totally worthless as a result of the climate and other factors, so that nothing will grow there anymore. So I see this as a possible role for the world community. It should help the developing countries in the tropical zones to use their forests intelligently and help them to replant the tropical forest as it's being cut down. But I'm opposed to the disastrous approach that simply says, "We'll help the sub-Saharan countries by sending them food or any other foreign-made item," without seeing that this really is no solution at all, since once we stop helping them they'll be just as impoverished as they were before and even more so. In fact, we've already stopped sending them things! So we must help the sub-Saharan peoples to organize their production methods and their grazing methods, so that a viable system can be established there.

ESCANDE: Help them not to need help anymore.

DUBOS: Help them never to need help again—exactly! But that's very difficult because our societies present themselves as models, and also because these developing countries are captivated by the mirage of the modern world and want to construct extremely sophisticated hospitals before they even set up elementary sanitation systems, or else want to construct atomic reactors or computers before they know how to use the tools adapted to their country. So both sides encounter extraordinary difficulties.

ESCANDE: Which also benefits multinational companies.

DUBOS: Which probably benefits multinational companies. The solution involves a revision of our concept of things, and I summarize what should be done by repeating, "We must think globally but act locally." This formula, I repeat, is applicable not only to developing countries but to the United States and other industrialized nations. I find it ludicrous to go to a large American university, where the students want to talk about a whole bunch of issues involving worldwide pollution, and then go out into the street where pollution is right before my eyes.

ESCANDE: They toss away their beer or Coke cans and exclaim, "Stop pollution!"

DUBOS: I regret to say that's just what happens.

V

René Dubos
-No Holds
Barred

"Because I was the most scrupulous of scientists for forty years, today I can turn to more general considerations," says René Dubos. In actual fact, these "more general considerations" usually stay within what he considers his field of expertise. "I'd never have agreed to work on a book about the philosophy of death. I haven't given enough thought to it."

Still, at the end of our sessions I was tempted to get his ideas about subjects on which he is not a specialist but that constitute the warp and woof of our daily preoccupations: the reasons behind success, elitism, feminism, high-flown conversation (bavardage), public opinion, and liberty.

I won't say these ideas are the heart of this book, but the honesty and frankness of his answers were compelling. No condemnations on principle, no thoughtless partisanship, but instead a personal and discriminating point of view that reveals whether Dubos is attracted, repelled, involved, or indifferent. A point of view that above all does not want to be taken as law. Dubos played the game I proposed only on the express condition that I state clearly that he does not consider himself a prophet, a guru, the holder of truth, or even a free man since liberty, he concludes, can't exist in the absolute.

DUBOS: You wanted to know whether I was afraid that people would distort my thoughts. Of course everything depends on the use to which those thoughts are put. Having said that, there's indeed one thing I'm afraid of. Perhaps it's not exactly a fear that my thoughts will be distorted. Rather, it's a danger that I've discovered as I near the end of my life.

I'd never taught formal courses—never. When I retired nine years ago, I was asked to teach in two universities. One was part of the New York State system, the other was part of the New York City system. I seemed to be a big success. But I was a "big success" in a way I found extremely dangerous. The students saw me, at the end of my life, working on very general problems and making observations about every discipline, be it social, medical, or scientific. Immediately they wanted to do the same thing, to come to grips from the start with problems on a worldwide scale, without being willing to work before thinking. Yet I'd tell them every day, "I want to stress that for forty years I was the most disciplined microbiologist possible, and not until I realized that I'd mastered that discipline could I permit myself to look at it from the outside." But they simply wouldn't accept my explanation. I believe that from then on I began to feel that I was a bad influence on them. Because my courses were going too well, I gave up teaching and from then on gave no more courses of that sort. This attitude may seem a marked contradiction to everything that I've been saying, but on the contrary it's right in line with my thoughts. When you believe that, above all, the most indispensable thing is to view problems in terms of their interrelationships rather than in terms of the elements they're made up of, it means that you must first be very familiar with these elements. If not, you're just jabbering. I know that if I'd continued to teach, my influence would have been a bad one —no, a dangerous one. Young men—more than young women, by the way—were attending lectures about any broad subject without having first mastered with any precision even a bit of that subject.

ESCANDE: In a medical consultation, the first thing you must do is talk with the patient, so that he'll tell you about himself and his symptoms. Next you must examine him with your eyes, your ears, and your hands. Then and only then can you begin to treat him, that is, try to cure, relieve, and console. Nowadays I'm appalled when I see many people unfamiliar with medicine emphasizing the importance of a psychological relationship with the sick person. That's extremely dangerous. I also believe that, when a person begins to study medicine, he definitely must be in contact with people who will enlighten him a bit, who will show him the limits of the field in which he'll be working, and who from the very first will share with him some of their experiences and thoughts.

DUBOS: I must say that, at the time I described, the students in American universities were totally free to choose the courses they took. There was no required curriculum in American universities. There were all kinds of courses, and a student signed up for the ones that appealed to him. So the students would come to my classes solely because I made them dream. In those days there was another big danger: they spent their time at the university taking courses in jabbering.

ESCANDE: I'd like to ask you about that word, "jabbering," that is, *bavardage*. It's a word that Jean Bernard also uses very, very often (and, like you, in the most pejorative way). Several times in our talks you've been critical of jabbering. But when discussing the physician's role, you also said that in relieving and consoling, the important thing was that the patient shift all or part of his anguish to the physician during their talks. I think I see where you draw the line between useful words and jabbering, but I'd like you to be more specific and tell me when words are mere talk in scientific fields, and even in medical ones, and when words are useful. I'd also like to ask you about the word "brilliant." On several occasions you've discussed eminent people—your students or friends—and you've said, "He's brilliant." I want to tell a story and see whether you share the opinion of Fred Siguier, the blind physician I mentioned earlier, about brilliant people.

One day an intern in Siguier's class was presenting a paper on

Pickwick's syndrome. He was a young intern and was very frightened, because when Siguier wasn't pleased he would destroy people in a diabolical way. At the end of the paper Siguier didn't say a word, he didn't applaud (you never applauded in classes). Finally he said, "My dear colleague, I've heard about you for a long time. I've been told you're brilliant. It's true, but then, all great talkers are brilliant." Then he added, "They told me you're solid and know your stuff. That's true, but all plodders are solid." The poor intern turned white. Then Siguier smiled and concluded, "In the end, the important thing is to be both solid and brilliant at the same time, and that's what you are!" I agreed completely. Well, I too think that all big talkers are brilliant, in the way that Siguier used the term, and that all big talkers jabber. Do you agree?

DUBOS: In my case, as I've already told you, I always seem to be improvising, and everyone thinks I give improvised lectures. I really always write them at least six months in advance. In preparation I read all sorts of publications, even those that don't seem too pertinent but touch upon one aspect or another of a problem I want to talk about. So I believe that a teacher or lecturer must have acquired and assimilated the knowledge he needs, until it has become so much a part of him that he can permit himself a bit of whimsy—"brilliance," if you will—when speaking. But he musn't simply have accumulated the knowledge on the paper before him. He must have assimilated that knowledge until it has become a part of him.

When I was doing a lot of work on infectious diseases, microbiology was completely a part of me. And when I talked about it, I needed only to think about how I would communicate. I would put myself on the listeners' wavelength. If I was addressing a group of science students, I talked from a very technical point of view; if I was before a group of general practitioners, I talked in a more humanistic way; and with the general public I talked on yet a different level. But this is because I never had to think about my subject. It was inside me.

Well, I believe that jabbering is a totally different procedure, in fact the contrary one. Jabbering means thinking you have the right to have ideas about everything without having spent days and nights not only accumulating the facts but assimilating them. If, when preparing a lecture, a person is content to assemble facts and list them one after another on a piece of paper, the audience—even if it's dazzled—won't profit from that accumulation of absolutely unassimilated words that the orator has put together. I can assure you of that. Indeed, whatever the subject, the difference between assimilating things and simply collecting data is considerable.

ESCANDE: I'd like to discuss the role of women in the world.

DUBOS: I don't know much about the subject. But here's what I'd like to learn from women, and I think that they themselves don't really think about this point enough. My question is this: What is the cause of what might be called the "crisis of the responsible woman" during the nineteenth century? Before the nineteenth century, women actually played a very important role. As far as I know, during the seventeenth and eighteenth centuries women fulfilled essential roles, and not only in an indirect way. There were women in banks, in government, and in all sorts of positions of authority. It seems to me that in the late nineteenth century a very false climate of opinion appeared —a climate that originated in part in England and that excluded women from the world and its realities. For example, when the famous Johns Hopkins Medical School was founded, women were admitted from the first, because this was an absolute requirement stipulated by the wealthy woman whose bequest had helped make the school possible. At the beginning there were four or five women students per year. I later knew one who had been in the first or one of the first graduating classes and who became a very famous histologist and the first woman member of the American Academy of Sciences. I knew her very well because she worked at the Rockefeller Institute. Her name was Florence Sabin. She would tell me how, at the end of the nineteenth century and even during the early twentieth, it was possible for a woman to play a significant role in American society. Later it became increasingly difficult, and growing stress was placed upon the woman's role as involving children, church, and the kitchen—the old saw later taken up by Hitler in *Kinder Kirche Küche*.

As far as I can judge, women have functioned as well as men in every field they've entered. One exception all the same: very few women have been composers, which is a very curious phenomenon. There are a lot of female mathematicians, and women who are eminent in the various sciences and literature, of course. But there are very few in music, so far as I know. I've always wondered why. A lot of women have been performers, but almost none have been composers.

But getting back to women's self-effacement since the nineteenth century, I think that the feminist movement should try to find an explanation for this relatively recent situation. On the basis of my personal experiences with all the young women who came to work as technicians in my laboratory, I can state that a great number of them were far more effective than their male colleagues. Yet until just a few years ago, very few women laboratory technicians, however well qualified, were given a chance to move up the professional hierarchy. I never noticed the least intellectual difference between men and women. A physical difference, yes, that's obvious, and it may be the initial cause of the misunderstanding. It's certain that once the human species appeared on earth, specializations were spontaneously created as a result of the need to divide up the work. Here's a very simple example. Since men can run faster than women, spontaneously, out of necessity, men turned out to be better adapted to the hunt. Examples of this sort appeared on every level and automatically resulted in a division of labor. But these spontaneous phenomena had all sorts of social consequences that gradually led to the establishment of habits. And these habits spread to many other activities beyond those in which sexual differences actually came into play. In this way, perhaps, we can explain the "crisis" involving woman's position. But that's really not my field!

ESCANDE: What about the problem of elitism? Aren't the major questions of the day, the questions that preoccupy you, perceived solely by a small number of distinguished people who make up the intellectual hierarchy in our countries? Or is the public in general also preoccupied by the same problems? Does a man like you, who rubs elbows with the most intelligent and powerful men of our day, think that an osmosis occurs between what the public needs and the ideas being discussed at Harvard, Yale, or the Rockefeller University and in the government? Doesn't the elite decide for the public what the public needs?

DUBOS: Obviously, the danger can seem enormous. But I don't believe it's as serious as one might think, and as an example I'd like to tell you about my experiences as a visiting professor at an engineering and architectural school in one of the big state universities. It happened last spring in Seattle, Washington. Almost seventy-five percent of the students in these state universities come directly, as I did, from the humblest social groups. So they really represent a limited section of the population. Well, it was clear that these students not only were completely open-minded about what I think are the most serious problems of our day, but that they also personally brought these problems up for discussion. I don't want to be over-optimistic on this score, but there's no doubt that twenty-year-olds are aware of present-day concerns, often in a far more pertinent manner than adults. During the past five years a need for change has spontaneously been expressed by medical students, who are personally demanding courses that will better prepare them for day-to-day practice. So it isn't the established elite, but those who are *in the process of becoming* the elite, who are perceiving the critical issues and demanding that the universities do something about them. Obviously, once they actively enter the profession, their attitude may very well change, but new students will in turn perceive other needs and express them.

ESCANDE: So in the end, who makes the changes possible? Who pushes the established institutions off dead center? Does the machine we've already discussed go on irrevocably once it starts, until it reaches the brink of the absurd, and then, just before the point of no return, turn around and regain its equilibrium? Or are there social groups that can really make the course of history change? If so, do you think that the young will make the world change? Or do the ideas of a person's adolescent years triumph only once he's become part of the age group that holds power, that is, when he's fifty or sixty?

DUBOS: I believe, and I've often said so, that social transformations never occur through the initiative of existing governments. Not the government in Washington, nor those in the European capitals, nor anywhere else. They always occur within a community, within a group that's independent from the existing administrative units. This results in astonishing outbursts, because this sort of independent initiative develops under the influence of a few very strong personalities and also through the power intermediary formed by the press and television—the mass media, to use today's term. The vigor and rapidity with which these decisions and changes can occur, without government intervention, is truly phenomenal. I'll give you an example that's almost a caricature, yet perfectly true.

Three years ago two Harvard professors announced that they had scientific reasons to believe that the fluorocarbons used in numerous spray cans were destroying the ozone layer found in the upper atmosphere, which is indispensable to life on earth. I believe that this so-called destruction of the ozone layer has not yet been proved, but their statement was immediately taken up by the press and television, and the reverberations were astonishing. At that point people began to assert that ultraviolet rays were going to increase on the earth's surface, with all the dramatic consequences imaginable. Well, believe it or not, in less than two years the purchase of spray cans using fluorocarbons decreased by half, and a dozen firms had marketed other spray cans to replace those said to be so dangerous!

Another example concerns the war in Vietnam. In the space of three or four years—independently of government decisions, of course—the nation refused to participate in the war, and this phenomenon took on extraordinary dimensions. This shows, in my opinion, that in very complex societies—and all our societies are complex—it's illusory to hope that change will come from the established government units.

ESCANDE: Indeed, if they're established, it's in order not to change!

DUBOS: In order not to change! So a need must first make itself felt and evident among the ordinary people, and then the bureaucracy will heed this pressure for change that has come from outside.

ESCANDE: Jean-François Revel maintains that the revolution to come will start in America, and moreover that all new ideas come from America. I for one think that he's right. Besides, in a country like France we're only imitators (as they say in painting), imitators of the United States. This brings me to a question that I was especially eager to ask you. Must a young man or woman, bursting with ideas and enthusiasm, go to the United States today? And, to ask a very indiscreet question, if twenty-four-year-old René Dubos were going to leave for America today, would he find the same opportunity to adapt that you encountered?

DUBOS: Alas, I don't think so. Because when I left France, American institutions were really far more flexible than now. I keep telling my American friends how frightened I am to see all those constraints and all that organization characteristic of European societies creeping into the American social structure. For example, as far as medical research is concerned, I wonder if it would be as easy for me to enter the Rockefeller Institute for Medical Research today, lacking any training as a medical microbiologist as I did. Actually, I'm almost sure that I couldn't. American institutions are still more flexible than any in Europe, but this flexibility is continually decreasing. And yet, since there's so much discussion about this developing ossification that resembles the paralyzing framework found in Europe, America may be able to stop it in time. Countless books critical of the new situation are beginning to appear, some of them written by leading businessmen. For example, the president of Avis Rent A Car has given a fine description of how all the big companies are becoming ossified through their passion for overorganization. A lot of fuss is being made about these problems at the moment, but can this new trend toward excessive organization be stopped? We'll have to wait and hope. Everything may revert to what I encountered when I arrived.

ESCANDE: Let's move from René Dubos to the more general problem of the French in America. French immigation was very low compared to immigration by Germans, Italians, and of course the English and the Irish.

DUBOS: You know, in the seventeenth century America was almost a French country. Moreover, traces of the French remain, such as the French names given to certain regions of the United States, or in the names of towns such as Prairie du Chien, Wisconsin. But when other Europeans began to immigrate, the French scattered and were swallowed up completely in the new situation. By contrast, the English continued to preserve their social identity. So did the Italians, not only because of their distinctive religious practices (you know what Italian Catholicism is like), but also because the Anglo-Saxon population looked down on them. The same was true of the Irish, who were scorned by the English. All these groups remained unassimilated. But the French disappeared, having completely melted into the American surroundings. In addition, of course, emigration from France was never as extensive. Today, for all practical purposes the French don't form a distinct ethnic group in America. Yet I hear on all sides that the prestige of France remains high, although this prestige is hard to define. It's an intellectual prestige, independent of technological development—for which, I might add, the French are unfairly looked down upon, since the French contribution to recent technological developments is enormous. But the Americans view France as above all the center of high culture.

ESCANDE: Does France really exist for Americans, or is it only a little piece of Europe?

DUBOS: I think Americans find it very hard to believe that Europe will ever become a political unit.

ESCANDE: My next question is inspired by tourism as advertised in the United States. I recently heard someone say, "Americans are being 'sold' Europe in a week, with a day and a half for Paris!"

DUBOS: The word "Europe" is used because it's practical from a commercial point of view, but deep down what's being "sold" is the Eiffel Tower, Versailles, the Champs-Elysées—and that's France. Just as Schönbrunn Palace and the Prater are Austria, and so on. So it's not "Europe" that's being conjured up. And if they sell the Old World, it's because once you reach Europe it's easy to move about. But Europe as an entity doesn't attract Americans. I believe that Europe as an entity doesn't in fact exist. So everyone has a certain admiration for France, but this admiration goes hand in hand with an enormous amount of misunderstanding. The French are seen as very odd people. Now the Germans, the Austrians, and many others make a considerable effort to help American tourists, but it's obvious that the French haven't paid much attention to this sort of effort!

ESCANDE: I also wanted to talk with you about a problem involving the United States. A few years back a book published in France asked an important question: "Who governs in Washington?"* The banks? The President? The lobbies? I'd like to ask you not who but *what* governs in the United States: money or ideas? Is the intellectual movement the leader in the States, or in the end does everything do what money dictates?

* *Power in Washington* by Douglass Cater (New York: Random House, 1964).

DUBOS: Outside the United States, everyone says it's money, and a lot of Americans would say the same thing. But I don't think so. Because it so happens that I've been observing the phenomenal rapidity with which public opinion can influence very big political decisions. I've already mentioned Vietnam, but I can give you some other concrete examples in which I played a small part.

About ten years ago the directors of a very large brewery, a huge company with considerable capital, asked me to advise them during some studies on the effects of breweries on the environment. They intended to and did construct a very large brewery where the ecological conditions were very special. I didn't work with them very long, but all the same I got to know them well enough to ask the company president one day, "Who gave you the environment religion?" He replied, "Oh, it's very simple. I have two children, sixteen and eighteen, and each of the members of my board of directors has children between sixteen and twenty. All day long our children would ask us, 'What's the company doing for the environment?' " This is a very typical example. As far as the environment is concerned, the ideas generated among young people have achieved changes in the policies of big businesses far more profound than government directives could ever have achieved. Government directives only came later! I believe that phenomena of this sort are gradually occurring in almost every field. At the moment we see religious groups—Protestant, Catholic, Jewish—and student groups and citizen groups of all sorts putting pressure not only on banks but also on large universities not to invest their money in South Africa. And now these organizations are telling them, "Not only should you stop investing, but you must pull your money out of South African businesses!" Of course an official government position exists, but public opinion is far more powerful. And I believe its power is gradually increasing.

I'd like to give you one final example in which I personally played a part. It happened at Jamaica Bay, Long Island, a large bay on the outskirts of New York City and adjacent to Kennedy Airport. To expand the airport, which was overcrowded, they planned to extend the runways out over Jamaica Bay. Technologically the plan was extremely easy to carry out. When the decision was announced, a great commotion broke out in New York City. Numerous groups—not just ecology groups, but citizen groups as well—wanted to save Jamaica Bay. The governor of New York State therefore asked the Academy of Sciences to conduct an inquiry into the ecological consequences of extending the runways over the bay. At that point I made a public statement to the effect that the Academy of Sciences didn't have to spend two years on an inquiry, for people already knew how much the bird population would be upset, the climate modified, and so forth. This statement was immediately snatched up by one of the leading New York newspapers. This drew attention to my "speech," and then a group of people formed—I don't even know how it happened—and held a meeting at Jamaica Bay for all sorts of people. The organizers saw to it that one of the biggest television networks was present, and they asked me to speak. They managed to take my picture with my arms in the air, just as a large flock of birds was flying over one of my arms and a jet over the other. I seemed to be saying, "If you let jets come in, you'll have no more flocks of birds." The next morning my photo was on the front page of the *New York Times*. That's all I did, absolutely all. Well, the *New York Times* is naturally very influential, so the next morning the governor stated, "We don't need an inquiry by the Academy of Sciences after all, and no runways will be constructed over Jamaica Bay."

This completely true story is a good illustration of how spontaneous reform never comes from the established institutions. On the other hand, once public opinion has clearly and forcefully taken a position, everything becomes possible.

ESCANDE: Once a person understands that he mustn't rely on official bodies, must he withdraw from them? Must he choose splendid isolation?

DUBOS: It's quite possible to remain within an organization and on occasion oppose its activities. I for one haven't rejected organized medical research, but I've certainly taken positions mildly critical of it, and I continue to do so when asked. Of course this is a very widespread phenomenon. A lot of people have left the established system intellectually, although they remain a part of the university. But I think this is easy in the United States, because the universities and medical schools have great independence. Each of them forms a sort of entity, so that if a person doesn't feel at all comfortable where he is, it's almost certain that another university will be only too happy to snatch up someone who has original things to say.

ESCANDE: This explains how professors can move from one university to another. You're not part of the civil service system for life, as in France? You're professors with contracts?

DUBOS: Yes, and you can be tough if you wish—you can discuss salaries! This goes on in every university, even in the state-supported ones that officially should propose only those salaries authorized by the state legislatures. In actual fact, a few supplementary funds are always being given by citizens or by foundations, to permit a higher salary for someone whom the university is eager to attract or keep.

ESCANDE: So you transfer from one university to another just as football players shift from one team to another?

DUBOS: Yes, that's it exactly! Some people move every year. Lewis Thomas, whom I've already mentioned, was in turn professor of pediatrics, professor of pathological anatomy, professor of medicine, professor of I don't know what else, then dean, and suddenly (I must say he's an extremely intelligent fellow with a great deal of charm) president of the Sloan Kettering Institute for Cancer Research, even though he'd never previously done any work on cancer. My own story runs much the same, but on a smaller scale. I was at the Rockefeller Institute in 1942 when I was invited to Harvard, where two chairs were vacant simultaneously, a chair in tropical medicine and one in comparative pathology. Well, I'd never done any comparative pathology, and naturally I'd never done any tropical medicine. I remember having replied, "I'd accept with pleasure, but I've never seen a palm tree in my life!"

ESCANDE: They were offering you the chair in order to get René Dubos?

DUBOS: Apparently. Not because of my skill in the field, but simply so I'd be one of them. And when the president of the university, who was a very important person, answered my letter, he told me that the palm trees didn't matter, and that if he wanted me to come, it was because he knew that I could "pull rabbits out of a hat"!

ESCANDE: How about changing the subject? How should a person keep up-to-date? Should he try to talk to other people? Read? Listen?

DUBOS: I read about everything, a little every day. I'm a very fast reader. In fact I don't read, I look. I look at the page and I search for something that I'm especially interested in, or something I don't know, or a sentence that expresses an idea that I'm already very familiar with. Then, two or three days later, maybe a week later, I go back to that page and if I still like the sentence, I copy it down on a little piece of paper.

ESCANDE: What do you do with all your little pieces of paper?

DUBOS: I've never found a practical way to file them, so I function in a way that's all my own. I put all my papers in a box, unsorted, completely unsorted. About once a month I read them all and keep the ones that I feel really say something important. I pile these slips up in big boxes. Then, when I prepare a book or a lecture, I spend up to two weeks going through my mountains of paper. I look them all over and pull out everything that may be pertinent to my subject. Then, all by itself, the book develops, the chapters organize, the sentences are born. When I've finished the book, I put my precious papers back into the big boxes, because they may be useful again in other circumstances.

For example, I'm especially fond of a sentence from André Malraux's *Les Voix du silence.* "Doubtlessly one day, when confronted by arid expanses or land reconquered by the forest, people no longer will be able to guess the intelligence that man imposed upon the land by piling up the stones of Florence amid the swaying of the vast Tuscan olive groves." That can express a lot of different things. I can use it to show that man intervenes in nature and, through his transformations, creates something far finer. Another time I might use the same text to illustrate a historical point.

Here's another example, one involving Winston Churchill. You may not recall that the House of Commons was damaged beyond repair by German bombings during World War II. Well, after the war the House of Commons had to be rebuilt. The old building was badly equipped, too small, truly uncomfortable. Some people said, "Now let's build a big building, a push-button gem that's spacious and convenient." Churchill, who was prime minister at the time, opposed this in an impassioned speech. His argument went like this: "No, nothing should be changed in the House of Commons, because all of English history and all of our democracy developed within these walls." He also said, "In the French Chamber of Deputies and in the United States, you can move easily from one side of the hall to the other without ever being very sure about showing your political allegiance. In England, if you're a member of the Tory party, you sit on one side, and if you're a member of the Labour party, you sit on the other. A move from one side to the other is truly very difficult. I know something about that, for I've already done it twice in my life!" And Churchill added, "Of course, there's not enough room for everyone, but that's not the important issue." And he summarized his views with this sentence: "We shape our buildings and afterwards our buildings shape us." When discussing medicine, I can use this quotation, to stress that the surroundings that we create influence what we become. I can use it to compare the English people with the French. I can use it for all kinds of purposes. Indeed, I *have* used it at least four or five times in the books I've written. It's one of those sentences that made me stop. So I immediately memorized that sentence, and now it's a part of me.

I find it hard to define exactly what influences me in these evocative sentences, these sentences that say far more than the words that make them up. This is why I read every day. Every evening I arrange a pile of books and magazines beside my bed and look through them. I'm not implying that I read them, but I go through them and, as I said, if I spot something that impresses me, I jot it down.

ESCANDE: Is it important for a scientist to read a lot of things? Go through a lot of books? Do you have to read a great deal in order to come up with ideas, or does the idea come once you've forgotten everything?

DUBOS: Well, reading just doesn't work for some people. Take a man like Avery. He read very little and worked on only two or three problems in his scientific life. His reading was related solely to the problems he was working on, and he never talked about anything else. Yet he was extremely creative from a scientific point of view. I've known other great scientists who, like Avery, knew very little—even on scientific questions—outside their own field of interest.

ESCANDE: Monomaniacs, in a way?

DUBOS: If you wish. Indeed, perhaps that's the best way to be a scientist. And yet, if I was successful in science, it was by doing a bit of the opposite. My reputation was made as a result of two discoveries, gramicidin and adaptive enzymes, and as a result of my work on tuberculosis. But perhaps I'm to be criticized for not having continued to work on those three subjects, for having stopped too soon.

ESCANDE: I'd like to interject my opinion here. I don't think there would have been any advantage in your continuing to work on those questions. The essential problem in the world of research is letting people who know how to discover things make breakthroughs. And once the breakthrough has been made, it's letting those who know how to analyze a field in depth explore the new path that's been blazed. I think it's really useful to allow these intuitive creators, to use Monod's expression, to work in different fields successively, rather than make them spend their time increasing their knowledge about one single problem. I think you were right to spread out into several fields. But what do *you* say? Do you regret what you did? Deep down, do you regret it?

DUBOS: As a matter of fact, I've no regrets from a self-centered point of view. Obviously my life was much richer as a result, and my influence broader. Because I've really exerted an influence in the United States, and that always astonishes me. It's a fact that when I lecture at a university or elsewhere, the room is never big enough! Still, I rarely present specific facts. I don't really teach, in the sense that I don't contribute definitive information. Yet there's no doubt that my words can affect people's thoughts and attitudes.

ESCANDE: Along that line, what do you think about provoking people to think, as Socrates did while walking through the streets of Athens and saying to the first person he met, "Do you really think you're right? Think about it, suppose you're mistaken?" Is it important for you to travel all over the world in order to make people think? And in your opinion, how does a person make ideas progress? By using those ideas as a weapon? Or by telling people in a Socratic way that they may be mistaken, that perhaps they ought to do some thinking?

DUBOS: Well I for one am persuaded that the only way to exert an influence is not to use arguments as weapons, to make categorical assertions, but to employ the parable as much as possible. I'll describe a recent experience that happened to me in September 1978. I was supposed to close the ceremonies for the three hundredth anniversary of the Academy of Göteborg, Sweden. I was to be the final speaker. I was eager to show that, despite everything that people say today, human intervention in nature hasn't been exclusively destructive and in many places has even been extremely beneficial. Human beings have built extraordinary things. The places in the world that man loves most are the places that he himself has completely transformed. That was to be my message, my explanation. Well, I thought of that sentence of Malraux's about Florence. I think those words are extraordinary, although their content is blurred and imprecise. During those three days of lectures, this was the only time that people applauded with enthusiasm.

But I'll give you the finest example of all. In my opinion, Christianity took root not so much because of its truth but because Christ spoke in parables. In fact, he says just that in the book of Matthew: "Therefore speak I to them in parables." For only parables can express the full complexity of things. If you define your subject too precisely, you'll stifle its living richness, but if you tell a story, then your listeners will sense everything that the story may contain. I truly believe that this is the way to make ideas grow. In addition, I confess it's much more entertaining.

But let's get back to Malraux, for whom I have great admiration. I've observed that his texts often are terribly diffuse. *Les Voix du silence* is full of attempts to show that Greek or Hindu statues resemble a Florentine Virgin, if seen from a certain angle. Undoubtedly there was a lot of cheating going on there, but we're so marvelously enriched by it! I too believe in these invariant aspects of humanity that are revealed in seemingly different forms which really are only each culture's way of expressing something very profound and common to all men. Stated as I've just put it, this idea sounds frightfully banal, but it seems marvelously enriching when the same need or the same way of perceiving problems is expressed by a parable, a story, or a picture, which seems so different on the surface. That's the case with those Hindu or Greek faces that reappear in the features of a Florentine Virgin. With the help of certain parables, I'd like to transmit a certain way of thinking—for example, the concept that man's essential nature can be seen in his need to face the horizon, a need that's satisfied by the straight allées in Le Nôtre's French gardens or the vast lawns in the English gardens. For me, these are different ways of expressing the unity within human nature.

ESCANDE: Man, therefore, is never free. He continues to be guided by needs that haven't changed since the Stone Age. Dare we hope that one day man will attain his liberty?

DUBOS: Liberty? Liberty is something that can't exist per se, and here's why: no organism—no organism, be it an animal, a human being, or a society—no living organism exists solely as an entity. An animal or a human being also exists and above all exists to the degree that it forms a part of a whole. In like manner, the social group is a part of a larger whole. So existence necessarily implies constraints. Liberty, you see, means collectively accepting certain constraints. Which boils down to saying that you shouldn't look at something for itself alone; you should see it only in terms of the relationships that it forms with everything else.

This is the precept that has guided me throughout my life, from childhood to this very day. And this is the precept that I ask you all to reflect upon.

The Doctor and the Damned

St. Martin's Press
New York

Grateful acknowledgment is made to the Federation Nationale des Deportes et Internes Resistants et Patriotes, for permission to use Fernand Leger's Resistance Behind Barbed Wire; *and to E. P. Dutton for permission to quote from David Schoenbrun's* Soldiers of the Night.

Design by Andy Carpenter

Library of Congress Cataloging in Publication Data

Haas, Albert, 1911–
The doctor and the damned.

1. Haas, Albert, 1911– . 2. Physicians—France—Biography. 3. Holocaust, Jewish (1939–1945)—France—Personal narratives. 4. World War, 1939–1945—Underground movements—France. I. Title.
R507.H18A33 1984 940.53'15'03924 [B] 83-1607
ISBN 0-312-21478-2

First Edition
10 9 8 7 6 5 4 3 2 1

ACKNOWLEDGMENT

I must express the special thanks I owe to Sheila Sperber Haas, my daughter-in-law. Her skills as a writer, plus her sensitivity to the personal qualities and subtleties of my experiences, were instrumental in giving them life on the printed page.

To my wife Sonja, who survived;

To my youngest sister Vera and her three-year-old son, who both died in Auschwitz' gas chamber;

To the memory of my sister-in-law Paule, who died from the long-term effects of her sufferings in the camp;

To all my comrades, those who survived and those who perished;

To Howard and Gladys Rusk, whose friendship helped restore my faith in human nature;

To my children, François and Sheila, who are my future;

To Alexander, my grandson, who has given me such pure joy in my later years. May he never experience the extreme cruelty of which men are capable.

FOREWORD

I met the author of this book in Europe in 1949. He had recently conceived and established two precedent-setting rehabilitation centers, one in France for patients who had contracted tuberculosis and other respiratory diseases in the concentration camps, and one in Germany for variously handicapped, displaced former camp inmates. They were the first institutions ever to offer comprehensive rehabilitation services of this kind. After our meeting, which developed into a lasting friendship, I invited Dr. Haas to come to the United States on a professional visit. Not long after, he became an American citizen through a special Act of Congress. We have been lucky to have him. Dr. Haas set up this country's very first medical facility for comprehensive care and rehabilitation—medical, psychological, social, vocational—of the pulmonary handicapped, people suffering from TB, from asthma, from emphysema and chronic bronchitis, and people who have undergone extensive thoracic surgery. And in the years since then he has continued his valuable contributions to the total care of those who must live with chronic respiratory handicaps.

During the early years of our growing friendship, I learned of Dr. Haas' experiences in the ten years before our first meeting. They are finally on public record in *The Doctor and the Damned.* This is truly a modern saga, a heroic and moving narrative detailing the incredible experiences of a young doctor in the French Underground, and later prisoner in a number of Nazi concentration camps. The book is a vivid and extraordinary document by a physician who managed to retain his human dignity and even preserve the values of his profession in appalling conditions. This vital record of the best and worst in human nature is distressing and painful, yet also thrilling, challenging, and humbling. It is a book that everyone should read.

Perhaps the most frightening aspect of this war's events is that too many people seem not to have absorbed any lasting sense of horror. Despite public knowledge of the unspeakable atrocities committed by the German state and many of its citizens and friends during World War II, men's desires to act out their capacity for cruelty toward their fellow men continue unabated. Greece in its civil war, the Soviet Union, Cyprus, Biafra, Ethiopia, Ireland, Argentina, South Africa, Lebanon . . . this post-war list is, sadly, just a sampling. Although the scope of each further assault on human freedoms and lives has been circumscribed compared to Germany's global and systematic efforts at genocide, the intent grows from the same primitive, brutal, and subsuming desires for power and supremacy. If there is any future hope of changing this course, victims both past and present must continue to speak out. For this reason, *The Doctor and the Damned* is a book that everyone *must* read.

Howard A. Rusk, M.D.
Distinguished University Professor
New York University Medical Center;
Chairman
World Rehabilitation Fund

INTRODUCTION

This introduction to the wartime experiences that I survived will be brief, because the content really speaks for itself. My story focuses on that part of my life circumscribed primarily by the Second World War, first as a French army physician and active member of the French Underground, and then as a political prisoner and physician in a succession of German concentration camps. It opens with the antecedents that led—some directly, some indirectly—to my Underground involvement and eventual deportation to Germany, then Poland, and finally Austria.

Although Germany's concentration camps carried out a philosophy that was unthinkably beyond our most terrible imaginings, my existence there was mercifully tempered somewhat with humor, with friendship, and with some extraordinary incidents of human kindness. But an essential aspect of camp experience was powerfully magnified for me because I was responsible for the medical care of my fellow prisoners. No one could be a saint in the daily, and often primitive, fight for survival. But for me, those times when the basic values of my profession—which I still cherish deeply—wrenchingly contradicted my instinctive responses to the needs of survival, the insoluble moral dilemmas that I faced tore me deeply. Even today, I still question decisions I made during those years.

It is important for the reader to keep in mind that a great many of the extraordinary things that I did, or that happened to me, did not seem so to me at the time. It was an era when the extraordinary was often taken for granted, and only the ordinary and the unthinkable had the power to shock.

A heartbreakingly tiny fraction of deportees survived the concentration camps. All national governments whose survivors returned home attempted to ease the burden that was to be part of each survivor's lifetime legacy, but the government of France stands above all, both then and now, as the most caring, active, and generous in its recognition of our scars, and of the wounds that can never fully heal.

My experience, as spelled out here, is not complete. The impact of incarceration in those infernal camps is so profound and far-reaching that to be most fully understood it should be followed through the former inmate's adaptation to a "normal" life. But that is another story.

Albert Haas

RESISTANCE

I

I

I was dreaming that I was home. I was busy in the living room with Sonja and François, my wife and baby son, as if we had never been separated. We were preparing a party to celebrate our son's first birthday. The German occupation of France had not yet invaded our small apartment in Nice. The feelings of warmth and security were almost as palpable as the abundance of food set out on our table. It was a festive display unaffected by the shortages of wartime. Yet underneath it all drifted a vague sense of threat.

The sight of Sonja holding our infant son in her arms reassured me. I wondered out loud when her sister and mother would arrive for the birthday party.

There was a knock at the door. I could not move to answer it. My heart began to pound. Anxiety knotted my insides.

Voices called out. They were not speaking French. . . .

I began to wake up. I fought to hold on to my illusion of reality. The impact of my dream receded slowly, leaving me too disoriented to comprehend my dreary room or the words being shouted through the door.

The guttural language crystallized into meaning: "*Häftling Arzt, aufstehen!*" ("Prisoner-physician, wake up!") The SS *Sanitätsdienst* wants you! There has been a work accident. Bring your instruments—you will have to operate!"

I awoke from my dream into my living nightmare. It was late November, 1944. I was a doctor in the concentration camp Gusen II.

I collected my surgical equipment, professional tools carefully accumulated since my arrival at the camp. They had been gathered by theft, piece by piece, whenever an opportunity arose—a few battered hemostatic clamps, tweezers, scissors, a small saw, three butcher's knives, a barber's sharpening strop, gauze, needles and catgut, a woman's white silk gloves, rubbing alcohol, an oilstone, two syringes, and some precious ampules of the German anesthetic, Evipan.

I guarded these items with my life, including the two homemade reflectors I attached to a kerosene lamp to augment the feeble bulb hanging from the ceiling of my operating room. A fellow inmate risked his life to smuggle four of these makeshift reflectors out of the machine factory in exchange for ten cigarettes and a half portion of bread. He was caught with the third reflector and sentenced to death for sabotage of the German war effort. He was hanged in front of the entire camp. The inmate who performed the execution received five cigarettes from the SS for his work.

I hastened through the dark night to the *Häftling Krankenbau.* We called it the HKB, the inmates' infirmary. Grotesquely emaciated patients, awakened by the commotion, watched me from the shadows. My operating room was a twenty-five-foot square area delineated by four beds surrounding a narrow table, a niche marked out in an immense barrack filled with tiers of crude wooden bunks.

A body lay on the table. SS Sanitätsdienst Schultze told me what had happened. The victim had been struck by a heavy piece of metal accidentally dropped from a crane. His entire left side was smashed. I recognized the man, a Dutch prisoner in the camp. He seemed already lifeless aside from his laborious breathing. My two inmate "assistants," Bournasell and Pedro, helped me to examine his shattered body. My kerosene lamp revealed a missing left ear, a crushed left shoulder in a lake of coagulated blood, and a smashed left leg. Jagged pieces of bone protruded everywhere from the knee down, and the leg bent like a piece of rubber hosing as Pedro lifted it. There was also severe internal bleeding. The dying man's raspy breathing and barely perceptible pulse were his only vital signs.

After a brief conference with Pedro, SS Sanitätsdienst Schultze ordered me to proceed at once to amputate the dying man's smashed leg.

In the concentration camps the life of an individual inmate had no value to the German authorities. But they prized scrupulously maintained records, as I well knew, and that required a properly filled out death certificate. I was ordered to amputate a dying man's leg to satisfy Gusen II's bookkeeping needs. If I disobeyed, the Sanitätsdienst would not hesitate to kill me. I answered, "*Jawohl, Herr Hauptscharführer.*"

Before starting to operate I had a split-second decision to make. Should I use some priceless Evipan on an apparently unconscious and dying man, or save it to barter for life-sustaining favors? I decided to save it, and prayed that the man on the table would die before I began to cut. As a doctor in Gusen II, I had to make such terrible choices almost daily.

I looked around for a makeshift tourniquet and noticed the rope holding up Bournasell's pants. He lent it to me, holding his pants up with one hand and using his free hand to help me tie the tourniquet above the knee, beyond the point where the fractures stopped. I made the semicircular skin flap incision. Bournasell held the tourniquet. A sudden whistling sound and a gasp broke the quiet. I looked over at Bournasell, then down at the victim's unseeing eyes.

There was no appreciable bleeding as we released the tourniquet. I turned to Sanitätsdienst Schultze. "Permission to talk." He nodded. "This man is dead," I said. "There is no point in completing the amputation."

"Finish the operation! Do what you're told!"

I dared not hesitate. In the midst of this horror the SS Sanitätsdienst had created a circus act. He was the trainer. I was the animal who must obey him and sever the mangled limb from the dead body. Soon we even had a clown to amuse our spectators.

After I severed the leg the SS Sanitätsdienst ordered Bournasell to hold it up for his inspection. As Bournasell raised the smashed leg, his cordless pants fell to the floor. He stood naked from the waist down, his white rear end in the spotlight thrown by the kerosene lamp reflectors as he held up the bloody leg like a sacrificial offering.

A living skeleton in a neighboring bunk suddenly splintered the silence with a smothered chuckle. He was echoed by a sound not quite like human laughter as a second skeleton joined him. The ghostly laughter spread like wildfire, even to those who were unable to see what had happened.

The scene was Dantesque. Skeletal heads with huge expressionless eyes peeked at us over bedboards, a horde of skulls grinning from the infernal depths.

I stood motionless—the well-trained animal done with his performance, waiting for "*Abtreten,*" the order to leave.

The maimed corpse with its death stare seemed to say, "I have given you everything. I can give no more. Count me out."

I remembered my first autopsy in medical school. Our pathology professor formally introduced the cadaver to us. "Gentlemen, this lifeless body was very recently a thinking and feeling human being with ambitions, accomplishments, troubles, and failures. Even though our failure allowed his life to end, his dead body now helps us to learn how and why we failed. An effective healer must remember his failures and recognize his shortcomings."

In the vanished days of my former life, the dead were treated far more kindly than the living were now treated in the Nazi concentration camps.

"*Abtreten!*"

I looked up. The SS Sanitätsdienst was ordering me to leave. His outstretched hand held a cigarette, the animal's reward for a good performance.

As I left the HKB the sun was coming up behind the mountains surrounding Gusen II, chasing the shadows from the camp's deserted streets. I lit up my reward and inhaled deeply, wondering how much longer I could survive in this hell.

I was being held here as a political prisoner—an officer of the defeated French army and an active member of the Resistance. I was still alive as a prisoner-physician . . . only because my Nazi masters did not know that I was a Jew.

Would I be here now if I had not decided to leave Hungary—my father's country—to join my mother and sisters in France? Or if I had not decided to become a physician? Or if I had not married a daring woman whose sense of injustice was as keen as my own? Or would I have felt compelled to fight the Nazis just as vigilantly no matter where I had found myself and no matter what else I had been doing?

I decided to become a physician when I was still a boy. My mother's father, who had migrated to Hungary from France, was a doctor. There had been a doctor in every generation of his family, and I wanted to carry on the tradition.

Since my mother's family was French, she and her children automatically received French citizenship. My two sisters and I were born with a dual nationality. We grew up speaking Hungarian, French, and German, the latter a legacy from the Austro-Hungarian Empire. We spent our childhood in the village of Zala Lövö, near the present-day Yugoslavian border, where my father's family had lived since the sixteenth century. In Kormend, the town closest to us, storks built their nests in chimney tops just as they did in our fairy tales.

When my parents divorced, my mother returned to France with my sisters and settled in Versailles. My father remarried and took me with him to Budapest, where I completed my schooling. I steeped myself in the atmosphere of a highly cultured and convivial city.

I remained in Budapest to attend medical school, where I also joined the chess and boxing teams. I was good at both, and enjoyed competition. Chess I had loved for years. Boxing was a recent acquisition to protect myself against the rough anti-Semitic gangs that roamed the university neighborhood.

At the end of my third year I contracted tuberculosis, and spent the next eight months in Budakeszi, a TB sanatarium in the mountains close to Budapest.

In the sanatarium I managed to finish my medical studies, and I met a patient, Sidonia Adler, also from Budapest, one of the loveliest and shyest women I have ever known. She was about my age. Her hair was black and her eyes just a shade lighter, the color of black coffee. After I returned to Budapest I wrote to her regularly.

My experience in the sanatarium had awakened in me a strong and persistent fascination with chest diseases. I interned in a hospital that had a TB ward. Soon Sidonia was transferred from the sanatarium to the hospital where I was working. Her condition had worsened considerably.

I visited Sidonia every day. Her fragile beauty and gentle nature captivated me. I grew to love her. I knew that she loved me, and though I knew that she had only a few months to live, I had the idea that happiness could achieve what medical treatment had failed to do. I thought that I could give Sidonia the strength she needed to prolong her life. I proposed to her and she accepted. We were married at the conclusion of my internship. The wedding ceremony was performed in the hospital at the bride's bedside. Sidonia died six months later.

I became deeply depressed. Finally I had to run away from everything that reminded me of Sidonia. I left Hungary and joined my mother and sisters in Versailles, where I stayed until I had completed a residency in pulmonary diseases.

As a French citizen now residing in France, I had to take care of my military obligation. After my residency I entered the French army and was sent to Indochina. I was discharged after only a few months. I had picked up infectious hepatitis. I was hospitalized and then sent back to France for convalescence. I returned to Versailles to rest and think about my future.

I wanted to leave Europe for a totally different environment. In the fall of 1938 I enrolled in the Belgian Institute of Tropical Diseases in Antwerp and put my name on the waiting list for a job with a leper colony in the Belgian Congo. I imagined that I would find reward and fulfillment working for the African lepers.

One afternoon in late autumn, while strolling down Antwerp's Avenue Kaiser, I saw a young woman standing in front of a movie theater with some other girls. She eclipsed them all. I stared at the most beautiful girl I had ever seen, with vibrant dark hair against a red straw boater. Her carriage was elegant, her legs were lovely, and her derriere was wonderful. Her face reflected intelligence and a mischievous sense of fun. She was bursting with life! She disappeared inside the theater and I walked on, but her face haunted me for months.

In those days most students did their studying in their favorite café. My haven was the Café Imperator. One typical day I glanced up from my books and found myself looking directly into that face. She was with a group of friends at the table opposite me.

What should I do? In those days it wasn't at all proper to approach a stranger of the opposite sex. I stared at her and she smiled at me. I tried to gesture that she should meet me at the rear of the café by the telephones. She got up with her friends and left. Had I offended her, or had she just not understood what I was trying to say?

For the entire next week I lived in the Café Imperator whenever I wasn't in class or on rounds. I looked up from my books so constantly that I finished no work at all. On Saturday night she returned, again with friends. I smiled hello and resumed my sign language. This time it worked!

We met in the foyer outside the rest rooms and telephones. Her name was Sonja Nadel. I was thrilled to discover that she had been attracted to me when I had signaled to her the week before. She worked full time in a lawyer's office. I made a date with her for lunch on Monday. That was just the beginning.

We had lunch together every day, and we spent the entire weekend together. I learned that Sonja had returned from Rumania to her native city of Antwerp less than a year before we met in the café. Her four-year marriage to the son of Rumanian Catholics had just ended in divorce. The two had met as fellow students at the Institut Supérior de Commerce, Antwerp's school of business and commercial law. Romance blossomed despite the violent objections of Sonja's Zionist family. They married right after graduation. Sonja moved to Rumania with her new husband, and quickly added this language to her already considerable linguistic repertoire. She easily found secretarial work in the upper echelons of several different American multinational corporations during her stays first in Ploesti, next in Turnu-Severin. Then Sonja's pregnancy with their first child ended in a stillbirth. When the marriage fell apart not long afterward, there was nothing in Rumania to hold her there.

I wanted to be with Sonja constantly. It was astonishing to have found a woman whose philosophy of life was exactly the same as mine—work hard, play hard, live each day as it comes, and don't worry about tomorrow. We found so much to talk about, and even more to make us laugh.

Two weeks later I told Sonja that I wanted to marry her. It seemed inevitable to me, but although Sonja accepted my proposal she didn't see any need to rush things. Marriage could wait. It was early 1939, springtime, and we felt we had plenty of time.

Then, in September of that year, World War II began. I decided to stay in Antwerp and finish my second year at the Institute for Tropical Diseases before returning to France to join the army. My decision was as much romantic as it was practical. Sonja had become the most important part of my life.

Sonja's father and uncle were ardent Zionists. Her father, who had died shortly after World War I, had been a close associate of Theodor Herzl. Sonja's uncle formed a committee to aid the Jewish refugees now arriving in Belgium after escaping the Nazi invasion of their own countries. There were quite a few German and Austrian physicians among these refugees. I helped them as much as I could, frequently taking them with me to Sonja's mother's house for dinner and always on the lookout for odd medical jobs for them.

Despite the changes that were going on around us, the war didn't disturb the personal rhythm of our relationship. I continued to pick Sonja up at her law office twice a day, for lunch and then in the evening. I would whistle a few bars from "The Donkey's March" under her office window to let her know I had arrived, and she would come down to meet me. In the evenings we went to the movies or a good concert. On the weekends we escaped to the countryside to walk through the quiet woods, row across the lake, or play tennis. But the shadow of Hitler's war so heightened our sense of the ephemeral nature of our future that we never wasted time making plans.

2

On April 29, 1940, the postman delivered my orders to report immediately to the French army. Sonja came with me on the train from Antwerp to Lille, the first French city across the Belgian border with an army induction center. I was signed up as a Service de Santé ("Medical Corps") physician with a lieutenant's rank and placed in a battalion of Engagés Volontaires pour la Durée de la Guerre ("Volunteers for the Duration of the War"). These battalions had been formed for naturalized French citizens and for foreigners living in France, including foreign legionnaires. Although the legion was based on French soil, Frenchmen were not permitted to join.

Sonja returned to Antwerp. I spent two days with my mother and sister in Versailles before reporting to the military center in Sathonay, near Lyon. Because Sathonay also housed a foreign legion post, the area was crawling with soldiers of all kinds and nationalities.

My orderly, Bertrand Gabriel, was an escapee from the French justice system who had used a false Swiss citizenship claim to gain refuge in the foreign legion. When war began, he had joined the Engagés Volontaires pour la Durée de la Guerre. Bertrand, formerly a successful pickpocket, was one of society's misfits. But he was superbly gifted at survival once the normal conditions governing daily life broke down. These special talents—and his evident pleasure in using them—would soon endear him to us as, time and again, he was to make our lives immeasurably easier and more comfortable under exceedingly difficult conditions.

My mornings at Sathonay were occupied with patients in the infirmary and my afternoons given over to military training. Evenings were spent with my newfound friends and chess partners, Michel Lubchansky and François Hauser, both psychiatrists, whom I would meet again years later in an extraordinary setting.

I was barely settled into my routine when, on May 10, the blitzkreig began with the German invasion of Belgium and the bombing of France's major airfields. I was sound asleep that night when the booming reverberations of explosions awakened me. Barefoot and in my underpants, I ran out to the assembly ground. Our antiaircraft guns were unmanned and the area was deserted except for a lone figure stumbling through the pale dawn. I ran over to help the wounded soldier and discovered a legionnaire too drunk to walk.

I heard the unmistakable sound of a German plane approaching. I stood alone and helpless, shaking with fear. I panicked and grabbed the legionnaire by an arm, pulling him along as I ran to the nearest antiaircraft gun to hide under its steel shield. I lost my balance, tripped, and fell on the gun's trigger. The drunken legionnaire landed on top of me. Our weight set off the antiaircraft gun. The sudden deafening noise was terrifying. The gun barrel swung wildly as it fired into the sky, the powerful vibrations sending shock waves through my body.

When my ears finally stopped ringing I became aware of death rattles from the legionnaire. My legs were wet. I couldn't move them. We had both been shot. He was dying; I was paralyzed!

What extraordinary relief to discover that the legionnaire was only snoring in a drunken stupor. What amazement to discover that I had shot down a German plane! What mortification to discover that my legs were wet because I had soiled my pants in my terror.

And that is how a hero was born!

After the raid, everyone gathered in the assembly ground. The camp commander praised my heroism to the skies. He wanted to know if I had some special request he could grant me to express his appreciation of my heroic deed. I was still white as a sheet and could not stop trembling. Trying to mask my embarrassment, I asked for a clean pair of underpants.

The legionnaire showed up in the infirmary the next morning to thank me for saving his life. He introduced himself as Monsieur Gangle, a fellow Hungarian. I thanked him in return for the protection his body had given me. Monsieur Gangle felt my gratitude was unnecessary but admitted that he would greatly appreciate a temporary sick leave. Under the circumstances it seemed too small a favor to refuse, so I gave him three days if he promised to stop drinking and stop by for a daily checkup. He kept half of his promise. He came in for his checkups.

Despite Gangle's heavy periodic drinking, he possessed some intriguing, though undefinable, quality for me. I invited him to stop by the infirmary and visit whenever he had the chance. He began to drop by from time to time. I was astonished to learn that Gangle was serving his fifth tour of duty in the legion. Normally, legionnaires were discharged with French citizenship after one five-year hitch. But Gangle had left his home and family behind the day he had suddenly ended his successful smuggling career in the border area between Austria and Hungary. He had run away for good after shoving a customs officer into the river. The argument could never be smoothed over—Gangle's opponent was dead before he had ever reached the water!

But Gangle was truly happy in the legion—most of the time. He had long ago come to regard this fraternity of fugitives as his family, and wanted to get away only when he was drunk. Then he would always come to me for a sick leave. I never had the heart to refuse him, or to send him to *la gnouf* (military prison) as regulations required. I enjoyed Gangle's effervescent conviviality. And as I grew aware of his loyalty and impressive resourcefulness, we became friends.

A few days after the May 10 bombardment, the Germans overran Belgium. French military leaders were weak and indecisive. Essential war materials were unavailable or defective. The casualty list was spiraling. French intelligence was honeycombed with infiltrators, and the pro-Nazi Fifth Column was demoralizing the civilian population with rumors. The army had already begun to retreat in the face of German attacks.

I asked my commanding officer for a transfer to the front, but not from any patriotic sense of duty. I was worried about Sonja. I had had no word from her. As far as I knew she was still in embattled Belgium, and serving on the front would be a legitimate way for me to get to the Belgian border.

On May 18 King Leopold of Belgium surrendered and the Germans broke through French and English defense lines. I ran to my commander. What was delaying my transfer? He stared at me.

"Are you crazy?" he said. "We have lost the war already, so why do you want to go and be a hero? If you want to be killed by the enemy, at least wait for the Germans to come here to you!"

"My fiancée is in Belgium, and I can't sit around here on my ass!"

With characteristic French sensitivity to such matters, my commander replied at once. "Why didn't you say so? That is another kettle of fish entirely!"

My transfer arrived the next afternoon!

Bombing attacks on our airfield and the high number of injuries among our civilian neighbors had kept me up most of the night, giving me a scary taste of battlefield medicine and time to realize that I could never find Sonja in the chaos of war. By the time I was handed my transfer orders I had a fine case of cold feet. I didn't want to go to the front, but I was confronted by an uncomfortable dilemma. As the "hero" who had shot down a German plane a week earlier, how could I admit that I was really a coward?

My commander was delighted that I was going to the front. Every medical unit had just been ordered to send ambulances and personnel, and I was his ready-made volunteer. I accepted my transfer orders. I was stuck with it.

We started north in the first week of June, unaware that the Germans were about to attack Paris. I was traveling with three ambulances and drivers, two medics, blankets, tents, and a well-stocked first aid kit. Bertrand Gabriel, my orderly, and M. Gangle, my hard-drinking legionnaire friend, were two of my drivers. I had been given the authority to requisition gasoline and building space, though only for the treatment of military injuries.

We drove through the vineyard regions of Burgundy and Champagne, passing through a fertile springtime countryside lush and green with budding life. It was unexpectedly peaceful.

The roads on the way to Châlons-sur-Marne were almost deserted except for an occasional military column traveling our way, some of the soldiers without helmets and wearing civilian shoes.

At Châlons-sur-Marne the Gardes Mobiles ("National Guard") suggested we go to Chateau Thierry because the heavy military activity there had resulted in numerous casualties. A doctor would be very useful. But they warned us to be careful of the trigger-happy Stukas, the terrifying German dive bombers.

Their warning frightened me. I decided to head toward Rethel instead. As we drove on I had an eerie sense of history repeating itself in reverse. This entire area of the Marne valley had been the site of the two great battles that had saved France from defeat before the Germans in World War I. The victorious French commander during the second battle had been Maréchal Pétain, who was now about to betray his country to the Germans.

About twelve miles on our way we ran headlong into the massive and chaotic civilian and military retreat. Abandoned vehicles littered the roadside ditches. The roadway was a sea of people and family pets fleeing on foot, in automobiles, in horse-drawn wagons, in trucks, and on bicycles. It was an impasse, so we had to take the route to Chateau Thierry after all.

But we quickly encountered an identical situation. Disorganized military columns, deaf to their officers' entreaties, were shoving civilians out of their way as they fled south. It was total chaos. We had to join the retreat.

My orders were to look after the military wounded. Soon we had collected so many wounded soldiers fleeing the battle zones that we had to jettison tents and blankets to free our ambulance roofs for those we could not fit inside. We began inching our way back south.

Stukas approached, filling the sky with the noise of a violent thunderstorm, suddenly diving, spewing death over the unprotected people clogging the road. Children were trampled to death as everyone rushed to the roadside ditches in crazy panic. The Stukas seemed to rain bullets forever, but in minutes it was over. Bodies lay everywhere. The wounded were screaming in pain, pleading for help.

One of my ambulances was on fire. I ran over and saw the driver slumped over the steering wheel, but the searing heat forced me back. I finally got close enough to grab the door handle, but the door wouldn't budge. My skin started to burn. M. Gangle pulled me away and threw me to the ground just before the explosion.

I was the only physician around. My medics, two drivers, and I busied ourselves with the wounded. When I had time to look around again, the road was virtually deserted. Except for a few people who were too dazed by shock to comprehend what had happened, almost everyone who could walk had abandoned their possessions and fled into the surrounding countryside. Those who remained were mostly the dead and dying.

Small children sat on the ground by their lifeless parents. Parents embracing bloody children in their arms approached our ambulances, begging for help. What words can describe my pain as a man and a doctor? I had to tell them that I did not have the equipment to help their child, or that help was useless because their child was dead.

The inert bodies of a man and woman were faithfully guarded by their pet dog who seemed to beseech me with heartbreaking whimpers and sad eyes to help his masters. What could I do to help the dead?

I walked over to caress the dog for his loyalty and give him the piece of sandwich in my pocket. I wanted to see—hope against hope—if either of the people might still be alive. As I came near the dog bared his teeth and snarled, protecting his masters against what he took to be a menacing stranger. I put the food on the ground and returned to my ambulances.

An army truck appeared. The soldiers slowly collected the corpses littering the road. The dog threatened them as they approached his master and mistress. The soldiers ignored him. The frantic dog bit one of them and another soldier shot him. I ran over to the sergeant in charge to explain that the dog had died because of his loyalty to the people who had cared for him. Would he do me the favor of placing the dog's body on the truck so that he could remain with his master and mistress?

The army truck drove off with the three corpses—dog, master, and mistress—lying side by side.

Night fell and waned into dawn. I was still working nonstop in a frenzied effort to care for the endless numbers of wounded soldiers who were piled on every available surface of our two remaining ambulances. Our food rations had long since disappeared. We were so lucky that Bertrand Gabriel was prodigiously talented at foraging the countryside for provisions.

As we began moving south again I gave up my seat to the wounded and walked by the ambulances, dazed by fatigue, numbed by despair. During a stop to tend to the wounded, a medic caught me injecting morphine into a dead man.

The vicious rhythm of the retreat continued. Whenever we heard planes approach we raced to the roadside ditches. Afterward we added more wounded and dying to our ambulances. Although we were now regularly abandoning the corpses on the road, we still had so many wounded to carry that I continued on foot as the ambulances crawled along the thronged roadway.

Just north of Lyon, we stopped in a meadow by the banks of the Saône River. This was my first chance to rest since the retreat had begun, so many days before, but I found no relief. I was too overcome by exhaustion and despair. Then time blurred. I retained little sense of what I was doing or how long it took us to reach our camp at Sathonay. After we finally arrived, our comrades there told us that Paris had fallen and the French government had been given over to Maréchal Pétain. The Sathonay infirmary was in the midst of evacuation. My patients were added to their load. Finally I could sleep.

I awoke to find a letter from Sonja that had arrived the day after I left for the front. She had written on May 10th to tell me of the German attack and that she, her mother, her sister, Paule, and a few close friends would be taking the train for France that same day. But it was now weeks since Sonja had written that letter, weeks of the Stukas occupying our skies. I knew firsthand how the Germans were bombing everything that moved. I had terrible visions of Sonja's train lying mangled somewhere in the Belgian or French countryside.

In the morning my men and I hit the road again. Six days later we arrived at our final destination, Aix-en-Provence. I fell asleep the moment after I was shown to my bed and woke up forty-eight hours later. I could barely walk. I counted twenty separate blisters on my right foot. The blisters on my left foot were beyond counting.

I later learned that I had been nominated for the *Croix de guerre* for my service to the wounded during the retreat. I was astonished. It seemed to me it was the circumstances in which I had found myself that were extraordinary, and not what I had done.

While I had slept, Maréchal Pétain had signed an armistice on the evening of June 22 that sliced France into two sections. The German-occupied zone included the entire Atlantic coast and everything inland except the poorest part of the country in the south and southeast. The French armed forces at home and in North Africa were to be demobilized, and the entire French fleet was to be disarmed and laid up. France was a conquered country.

Returning to the Institute of Tropical Diseases in Belgium was out of the question for me.

And I had no idea where Sonja was.

3

Retreating soldiers were still streaming south, regrouping in temporary garrisons to await their demobilization. The Medical Corps appointed me medical chief for the Twelfth and Twenty-first Volunteer Regiments. They were now quartered in the villages of Trets, Fuveau, and Penier, outside of Aix-en-Provence.

I had to set up an infirmary in Trets and try to solve the terrible sanitary problems that had developed during the retreat. Soldiers on their way south had looked for hiding places that provided both safety and food. That meant barns and pigsties. Large numbers of men were returning with lice and contagious skin diseases picked up from living with the farm animals.

We had no local treatment facilities, and I had to come up with some way to convince the overwhelmed disinfection center in Aix-en-Provence to accept my soldiers. We couldn't keep them quarantined forever.

So I made up a phony list. I put down every soldier who needed disinfection as a suspected carrier of a serious and highly contagious disease. I gathered them all in Trets and marched them to the municipal hospital in Aix-en-Provence. Our column was so massive that the tail was still standing in Trets when I arrived at the hospital steps. Our frontal assault was successful. All of them were admitted.

I soon gained the cooperation of the local mayors and populace in return for the medical care I gave them before their local physicians returned from the war. I was particularly appreciative of the generosity shown by the butcher, the baker, the grocery store owners, and, last but certainly not least, the head of the local wine cooperative.

Toward the end of July I got a letter from Sonja. The postmark on the envelope was French. Since mid-May she and the companions with whom she had fled Belgium had been imprisoned in a detention camp just outside of Mende, only 150 miles away from me! The train on which they had escaped was crowded with frightened refugees from other German-occupied countries who had initially sought asylum in Belgium. Sonja, who spoke a number of European languages, had tried to help some of the dislocated passengers. The edgy Frenchmen who worked on the train convinced themselves that she and her traveling companions were German spies.

In a frightening twist of fate, Sonja had been forcibly removed from the train when it arrived in France and put in a military camp for spies, Communists, and drug dealers. Her only hope of getting out was to marry a French citizen. I wrote at once to the war ministry for authorization to marry her.

During the nearly two months that I waited for the answer, the enlisted soldiers were discharged. A semipermanent camp was set up for those who had no homes to return to, primarily Jewish and political refugees who had come to France when the war broke out to fight the Germans. This demobilized group, which included a large number of Spanish Republicans, was administered by the foreign legion from their main post in Marseille. The Germans had permitted them to retain operational headquarters there.

When my authorization arrived I wrote to Sonja that I would come to Mende as soon as was humanly possible. Then I spent several increasingly discouraging days in Marseille attempting to find a way to reach Mende on the hopelessly disorganized railroad. Finally I resorted to a technique that had worked for me once before. I publicized *amour* as my urgent reason for travel. And again, all French barriers instantly melted! I was unofficially put on the very next train, a string of freight cars transporting farm animals across the country. I arrived in Mende after two days and nights in their earthy and aromatic company.

It was September 27, 1940. I exchanged a brief but emotional greeting with Sonja. The camp administrator gave her permission to leave the camp for our wedding ceremony, which would take place the next afternoon, and said she could stay out until the following morning. But I was crushed to learn that she would have to remain in the camp after the wedding and could not return with me to Trets. In our naiveté we had assumed that our marriage would automatically guarantee her freedom. Now we learned that Pétain's government had made no provisions for releasing a suspected spy no matter what the change in circumstances. We would have to think of another way to get Sonja out.

We wanted privacy and decided to walk the few minutes into town and go to my hotel room. Arm in arm, we strolled past the dilapidated camp buildings and through the unattended gate in the barbed-wire fence. A guard whistled at us. We thought he was being friendly, but as we reached the hotel a gendarme came tearing after us and placed Sonja under arrest. The charge was attempted escape from the detention camp!

We had to return to the camp under guard. The camp administrator was in a rage. I explained that since Sonja had been officially permitted to leave the camp for our marriage ceremony and our wedding night, we didn't realize that she still needed his permission to leave the camp for a little while that afternoon. The camp administrator remained furious. Why hadn't we come back when his guard whistled at us? Now it was my turn to be angry. "What stopped your guard from putting his warning into a few simple words? Does he think I'm somebody's pet dog who comes running whenever he's whistled at?"

The camp administrator, an unpleasantly self-important man, was beside himself. First I had ignored his authority, then I had insulted one of his men. Although he had to drop the ridiculous escape charges against Sonja, he found a way to get even. She had to be back in camp four hours after our wedding ceremony.

Sonja and I were married at 12:30 P.M. the next day in a hurried five-minute ceremony performed by the justice of the peace in the *mairie* of Mende. Our papers were missing the documentation of Sonja's Rumanian divorce from her first husband, but none of the officials cared. It was wartime. The ceremony was witnessed by those who had been arrested with Sonja (her mother, her sister, Paule, her friend Margo, and Margo's mother), and by several Greek girls Sonja had become friendly with in the camp.

In the golden Indian summer afternoon we walked from the *mairie* back to my hotel. Our wedding feast, which was to have been an evening celebration, had been moved to 1:00 P.M. because of Sonja's deadline. Despite the severe wartime food shortages, nothing was spared for us. The table was beautifully set. We began with pâté de foie gras and champagne, then sat down to a succulent veal roast dressed with fresh vegetables from the *patron*'s garden. Dinner ended with real coffee and a moist and tasty wedding cake baked by the *patron* himself.

As Sonja's deadline approached we left the group of celebrants and walked back to the camp. The guard waiting at the gate took her arm and closed the gate behind her. I stood and watched her walk away. I expected her to stop and turn to me to share one last look, but she just continued walking away from me. I fastened my eyes on her receding back. I couldn't leave without a last farewell. I still waited, watching her through the barbed wire. Just before she disappeared into her barrack she turned and waved to me across the distance.

I picked up my bags at the hotel, said my good-byes, and began my journey back to Trets. I passed much of my traveling time daydreaming about Sonja. I wandered back through our early days in Antwerp and drew fantasies of our life together once she was finally out of the detention camp. I savored in memory the pleasures of the eggs I had been given for breakfast that morning and the superb dinner that had capped off our wedding. They had all become rarely tasted luxuries since the war had begun.

Back at Trets my camp commander, Major Schwab, called me into his office. During my brief absence the government had decided that the homeless army volunteers would be organized into semipermanent labor battalions to take care of things like roadwork and basic construction. Now I could either return to my family in Versailles or remain in the demobilized army as physician for the four labor camps setting up in our area of southern France. One would be in La Ciota on the Mediterranean coast, with a sister camp in the outskirts of the port city of Marseille. Another was slated for the seaside marshland of La Camargue with a sister camp located between the sun-baked cities of Arles and Tarascon.

If I stayed I would still be subject to military command. But I would also receive my modest lieutenant's salary and would not have to live inside the camp or wear a military uniform. My housing would be paid for by the army, and I would be able to buy my food through the military cooperative. That was a guarantee of reasonable prices and a reliable supply of scarce items. The demobilized army would give me a lot of freedom.

It appeared to be the best alternative, but I wanted to wait for Sonja's arrival so she could participate in the decision. It seemed like no problem, because I had until the end of October to decide. But by mid-October Sonja was still in the detention camp. She would never have gotten out if not for the intervention of a childhood friend of hers who had become the aide-de-camp of the king of Belgium. He liberated the entire group on the pretext that they were being returned to Belgium to stand trial. He took Sonja to Marseille to meet me, while her family and friends immediately left for Perpignan farther south. When his telegram arrived, I was struck by the irony of my wife spending her honeymoon voyage with another man.

Once Sonja and I were properly reunited, we sat down to discuss our immediate plans. For France the war was over, but my army job was still the best spot for us for the time being. It would protect us from the disorganization and food shortages wracking the country. It would also, we assumed, protect us from the Germans' plan to deport France's Jews.

We were too distressed about what was happening to France and too committed to the anti-Nazi cause to remain passive observers. We knew that the shelter of my army position could not last forever, but it would give us the time we needed to find a way to join General de Gaulle and the Free French in England.

For us the war was far from over. Our work in the Resistance was about to begin.

4

On November 10 Sonja and I were transferred to Tarascon. I had already supervised the moving of the infirmary from Trets to the Tarascon-Arles area where our camp workers had begun improving the road between the two cities. Sonja found a temporary room for us at the Hôtel de la Place. I bought a bicycle for transportation and began my daily rounds and the task of equipping the infirmary to care for inpatients who were not sick enough to need hospitalization.

Before I had even had time to establish an affiliation at the local hospital, the camp commander's messenger woke me one night to come and see a worker who was vomiting blood. My examination disclosed a gastric ulcer about to perforate. I asked for the name of the best surgeon in Arles. The name Dr. Jean Gorodisch was given to me, along with the warning that this extremely acid-tongued man detested being disturbed in the middle of the night.

I gave orders for my patient to be transferred to the Tarascon hospital immediately. I drove off in the small camp truck and pulled up at Dr. Gorodisch's door at 3:00 A.M.

I rang his bell several times rapidly to communicate urgency, and waited. The door was opened by a man in his forties. His eyes, a full head above mine, radiated contempt. He looked down at me through the eyeglasses perched on his nose and barked, "Who the hell are you to wake me up in the middle of the night?"

I stammered that a patient of mine desperately needed his surgical skills.

Dr. Gorodisch continued to bawl me out. "A gentleman introduces himself before starting a discussion! And don't you think my wife also values my company at night?"

Fortunately for my own composure, I noticed a twinkle in his eye. I apologized for my unintended rudeness and introduced myself before describing my patient's condition.

Dr. Gorodisch cut me off again. "I'm not interested in your dissertation. I can make my own diagnosis, and you're just wasting precious time!"

He gave me a brandy to steady my nerves while he got dressed and was ready to leave in no time.

At the hospital in Tarascon he agreed that immediate surgery was essential. We called in the nurse who had typed the worker's blood. She paled at the sight of Dr. Gorodisch, terrified at having to tell him that the hospital was out of the patient's blood type. When Dr. Gorodisch discovered that none of us had the blood type, he announced, "Well, I do. Dr. Haas, take two hundred cc's of my blood so that we can begin surgery! Meanwhile, the rest of you can find another donor."

While we were draping the patient, an orderly volunteered the additional blood.

I assisted Dr. Gorodisch. I was awed by his skill and his comprehensive grasp of the situation. And his approach to finishing the operation was unique. Instead of ordering a lowly assistant to close the incision and sweeping out of the operating room like a prima donna leaving the stage, he stitched it up himself, then left for his home with a mumbled good-bye.

And I went home to grab a nap before returning to work. The telephone woke me just as I fell asleep. It was Gorodisch's secretary. The doctor expected me in his office at the Hôpital Mixe (Municipal Hospital) in Arles at eleven-thirty, as soon as he finished his morning schedule of surgery.

I almost told her to tell her inconsiderate boss to go to hell, but before the words left my mouth I remembered that he had also been up until five-thirty or so, and he must have been back at the hospital by seven or eight o'clock. I calmed down somewhat and accepted his invitation.

I arrived at his office still feeling somewhat miffed and anticipating a brusque reception. But Dr. Gorodisch greeted me with a charming smile and a warm handshake, complimenting me on my diagnosis and handling of last night's emergency. Then he introduced me to the hospital superintendent and secured permission for me to admit my patients there.

We left the hospital together and went to his private clinic (named Clinique Nicholas, after his infant son) for lunch. After discussing France's political situation for a while, Gorodisch suggested that Sonja and I move to Arles to take advantage of its superior medical facilities and social opportunities. He recommended that we stay at the Hôtel du Sud, a small pension close to the Arles-Tarascon road operated by a retired army officer and his wife.

Sonja and I found a comfortable room there and moved in several days later. The Hôtel du Sud was the place where we finally made contact with the Resistance.

One day, after we had settled in at Arles, Sonja accompanied me to the Hôpital Mixe in Arles where I now visited patients every morning. We ran into Gorodisch, who invited us to lunch at his clinic. At the table, Gorodisch eyed Sonja and me pointedly. "You obviously stayed in the army because you are Jewish."

"And what's wrong with being Jewish?"

"Plenty, under the German occupation! Don't be so sensitive. Some of my best friends are Jewish—including me!"

Gorodisch suggested that we form a small clandestine group that would include some of our new acquaintances at Arles. Each member would seek to establish a connection with the Resistance network that was rumored to be forming, a connection that we knew would lead us to the Free French in England. Gorodisch decided to put me on his staff with a small salary. His clinic would provide us with a secure and private daily meeting place that would not attract the notice of the many Pétain spies in Arles.

I questioned Gorodisch about the political leanings of our hotel *patron* and his wife. He felt they were harmless, but he was very uneasy about one of the residents, a colonel who continued to wear his French military uniform although he no longer appeared to be working for the army and certainly didn't have any other kind of job. Every morning he went out walking with his wife and a pet bulldog. He spent most of his afternoons fishing. Gorodisch suspected that the colonel was really spying for Pétain or the Germans. He fished in a secluded river outside the city where no one ever caught fish. Suspicious-looking strangers often appeared on the lonely riverbank for a few minutes of whispered conversation.

I was amazed at how intimately Gorodisch knew the colonel's activities. He laughed. "When you count half the city—including the police and the government secretaries—as your patients, you hear a lot of gossip. And boy oh boy, can they gossip!"

Sonja and I returned to our hotel, excited about our vague plans to sniff out the colonel's real identity. This was the kind of adventure that Sonja thrived on.

The hotel *patron* and I had early on discovered a shared interest in stamp collecting. That evening I invited him to our room to talk stamps over coffee and brandy, planning to guide our conversation toward the colonel. My discreet attempts were unsuccessful. Sonja sat quietly, a smile on her lips. Suddenly she lost patience with my unproductive subtleties and asked pointblank, "I noticed an army colonel and an elegantly dressed woman leaving the hotel with their dog this morning. Do they rent a room here?"

"Ah," said the *patron*, "you mean Colonel and Mme. Jeantrel and their English bulldog, Monsieur Bouboule."

He was impressed by the colonel's dog. If you called the dog simply "Bouboule," he growled and refused to move. When you addressed him formally as "Monsieur Bouboule," he came running.

I asked what the colonel himself was like. The *patron* was puzzled by him, since he should have been in Avignon where all the other career military engineers had been assigned since the German occupation. The colonel was a quiet and austere man who arose at six o'clock every morning to walk his dog. Breakfast was always sent up at seven-thirty on the dot. After breakfast he went out—to do what, no one ever knew. His wife remained in their room most of the day. Once in a while he received a phone call in the middle of the night. Then he would leave the hotel in civilian clothes and disappear for a few days.

The *patron* added that the colonel went fishing almost every afternoon.

"What a coincidence. I love to fish," I said, which was not true but might enable me to spend time with him. "Unfortunately, I don't have a fishing rod here."

Sonja burst in helpfully, "Perhaps you could borrow one from the colonel?"

I looked inquisitively at the *patron*, who offered to arrange an afternoon tea to introduce us. If we would provide the supplies, he would invite colonel and Mme. Jeantrel and ask his wife to bake some pastries. As usual I would make the coffee. Sonja grinned at me. Her eyes seemed to say, "You see? You have to dive right in if you want things to happen!"

So the next morning I ordered coffee, flour, and sugar at the army food cooperative, and located some fresh butter at a neighboring farm. When Sonja and I entered the *patron*'s apartment for our rendezvous, Monsieur Bouboule walked over to sniff us inquisitively before settling down by the colonel's feet. From this vantage point he followed my every movement with his eyes. I was introduced to Mme. Jeantrel, dressed for tea at the Ritz, then to Colonel Jeantrel, who shook my hand energetically.

The ladies started talking about clothes. The colonel asked me about fishing. I admitted that I really wasn't much of a fisherman. It was more a way for me to find quiet time alone, meditating while the worm swam around at the end of the line. I said I sorely missed these peaceful sessions, but without a rod I had no pretext. The colonel offered to lend me one and invited me to join him at the river where he fished. I accepted. I wanted to bring up France's political situation and start to feel him out right away, but I couldn't think of an unobtrusive way of doing it. I decided to postpone it for one of our fishing dates.

The colonel also turned out to be a chess enthusiast. We decided to play a game in the next few days. Sonja and Mme. Jeantrel made a window-shopping date for the next afternoon. Monsieur Bouboule, appreciative of the tidbits of brioche and croissant I gave him, became my friend.

I went fishing with the colonel as often as my schedule permitted but learned only that he was a very kind, though distant and introspective, person who enjoyed literature and classical music. We discussed both at great length.

As the long autumn drew to a close, my Arles friends and I were still feverishly and fruitlessly looking for a connection to the Free French. When winter brought snow and the mistral winds, the colonel and I switched from fishing to chess. One of my few successful attempts to talk politics with him occurred when I asked why he still wore his military uniform.

"I'm waiting to find out whether or not I can still remain in the army. My military office is nearby in Avignon, but I prefer to live here in Arles because of the annual music festival."

"If that's how you feel about the French army, why don't you join de Gaulle in England?"

"It's too late for that," the colonel replied.

He seemed suddenly uncomfortable and angry with himself, as if his answer had slipped out accidentally. At our next chess game I tried to draw him out again with a discussion of de Gaulle and Pétain, but he did an about-face. He spoke of Pétain as a great patriot and de Gaulle as a traitor.

I decided to invite the colonel to visit the Tarascon-Arles workers' camp. Maybe there I could get him to open up again. He seemed pleased with the invitation and said he would come as soon as he returned from an imminent official visit to Marseille. We had coffee together before he left. In the midst of an innocuous conversation he caught me totally by surprise.

"Since you're Jewish, why don't you try to get to England?" he asked me.

I admitted that Sonja and I were frightened of the Germans and hoped to get to England eventually. He thanked me for my honesty but warned me not to be so open in the future. He seemed genuinely sincere, yet the interchange left me extremely uneasy. His abrupt move to Avignon three days after his return from Marseille only heightened my anxiety.

About a week later, as the end of 1940 approached, I had something else to worry about. The area labor inspector suddenly arrived to inspect my infirmary because of anonymous complaints about conditions there. I was angry and frightened. Here was proof that Colonel Jeantrel had reported that I was Jewish. This inspection would be just the start of the harassment in store for me. The labor inspector refused to state the complaints lodged against me or who had made them. Instead he began poking methodically into everything, obviously looking for a fault. I began to simmer but kept it inside. He found everything in order until he discovered one single ampule of morphine that we couldn't account for. Only one person other than myself had a key to the medicine cabinet. The inspector immediately demanded to know if the man was Jewish.

I exploded. "What business is it of yours what this man's religion is? He fought in the army for France, now he's working for our country, and that's what's important!"

The labor inspector answered contemptuously, "You don't know what a Jew will do to gain a few pennies."

I angrily insisted that I would stand responsible for the missing ampule.

"Don't worry," he responded, "you can be sure you'll hear from us about it!"

When the inspection finally ended at noon, I prepared to leave for Arles to meet Sonja for lunch. The inspector insisted that he had unfinished business to discuss with me. He refused my invitation to talk during lunch with my wife, so we went without her.

At lunch the inspector launched into a stream of invective against the Jewish people, ending with his hearty approval for Pétain's appointment of the anti-Semite Darquier to head the Bureau de Question Juive. With that remark he leveled his gaze at me.

"And are you Jewish, Haas?"

I felt myself flush as I stammered, "Yes."

"I know," he said. "It's obvious that you are Jewish and trying like hell to hide it!"

I jumped to my feet in a rage, called him a damned Fascist, and stalked out of the restaurant with my face on fire and my blood pounding. I rode my bicycle home.

I was still furious when I told Sonja what had happened. I was sure that Colonel Jeantrel had denounced me as both Jewish and anti-Pétain. I felt I should go into hiding until I realized that I would be guilty of desertion from the army and so supplying my enemies with a solid reason to court-martial me. For three tense days I waited for the other shoe to drop. On the fourth day I was ordered to appear before Colonel Louis Deronne in Marseille.

Sonja wanted to come with me, but I thought she should stay home. If I didn't return in the evening she would know that my worst fears had materialized. Then she would still have time to protect herself by secretly moving in with Gorodisch in Arles, or going to my mother, who had moved to Villon.

I was early for my appointment in Marseille. My stomach was in knots and my hands were wet by the time I was ushered into Colonel Deronne's office. The labor inspector from Arles was there. He staggered me by standing up and stretching out his hand in an unmistakable gesture of friendship, then apologizing for the other day's unpleasantness. He explained that he was actually a member of the Resistance! Although he and his comrades had been reliably informed of Sonja's and my desire to work with the Free French, they always had to be on guard against infiltrators.

I was astounded to learn that these people had detailed dossiers on all of us, including Sonja and myself! The inspector smiled.

"We are not amateurs. Colonel Jeantrel belongs to the Deuxième Bureau, and he is one of our crack recruiters. He weeds out the genuine from the fake, then we do a final check to be as sure as we can possibly be."

Colonel Deronne arrived, and the inspector introduced me as "the doctor who was recommended from Arles."

I found myself racked with doubts as to *their* genuineness, unable to banish the fear that this was all a phony setup to trick me into betraying myself and my friends. I decided to play dumb until I felt more capable of evaluating my situation.

In response to the inspector's introduction I said, "I don't know what in the world you're talking about." Colonel Deronne realized my suspicions and tried to reassure me.

"I ask for your word of honor," he began, "that whatever we say in this room remains between the three of us."

He pulled out the files on Sonja and me. Trying to read upside down, I saw detailed notes on our background, code numbers, and the word *oiseau* ("bird") written on each. The effect was the opposite of comforting. The expression *C'est un drole d'oiseau,* the French equivalent of "He's a queer bird," convinced me that they had marked me as a suspicious character and would drag me from this room straight to the military prison.

Colonel Deronne went on. "We are under General de Gaulle's orders to organize a Resistance network in France. We're looking for the right person to head up a local group forming in the Alpes-Maritimes (Nice and Cannes). We feel that you are the person we've been looking for."

I was told that each small Resistance group would operate independently and generally be kept ignorant of the others in the area. The network was going to be coordinated from London by the Bureau Central de Renseignements et d'Action ("Central Office for Information and Action") under the direction of Colonel Passy. "Passy," of course, was a cover name. Colonel Jeantrel was one of Passy's adjutants and Passy was de Gaulle's right hand. General de Gaulle, as commander in chief of the Free French, had assumed direction of the Forces Francaises Interieures, their underground military branch inside France.

"If you accept our offer we'll have you transferred to the Côte d'Azur as physician to the labor camp forming in Mandelieu about three miles northwest of Cannes. And we also recruit women, in case your wife is interested."

I still pretended not to understand what they wanted of me. Finally Deronne said, "You don't believe us yet. When we contact you the next time we'll give you sufficient proof of our sincerity." The interview was over.

I returned home to Arles feeling as if I were emerging from an uneasy dream. Sonja thought it was wonderful, a real-life spy story. I sought her advice as to what we should do. With her characteristic fatalism, she felt we should wait for their promised proof to materialize. So we waited. The New Year came and went. Contact was not made again until mid-February.

I received a call from Lieutenant Konrad at the labor camp in La Camargue, a politically sympathetic man with whom Sonja and I were quite friendly. There was an emergency case at the camp, and would I come right away? Since emergencies usually meant a patient to bring in to the hospital, I drove out there in our ambulance truck. Lieutenant Konrad was waiting for me outside the camp. He laughed as I climbed down from the truck.

"There is no patient. It was just a pretext to get you out here! What's your decision?"

"What decision are you talking about?" I said cautiously. I was still playing my stalling game.

Lieutenant Konrad took me into his office and opened a well-camouflaged cabinet in the wall. It contained radio transmission equipment.

"What's your decision about Colonel Deronne's offer? I have to send him your answer right now. If you say yes, we'll get your transfer to the Côte d'Azur in the works. Then once you've gotten a preliminary organization started there, you and Sonja will be sent to England for special training."

The dreamlike cloud descended again. Yet despite this intense sense of the surreal, a germ of trust was beginning to replace my suspicions. But I wanted one more definitive piece of proof that this wasn't a well-designed trap. I was soon given it.

The lieutenant took me into a small back room, opened a cabinet door that had blended in with the surrounding wallpaper, and revealed a collection of machine guns, plastic explosives, and foreign currency.

"This, my friend, is our regional supply center."

I gave him my answer—"Yes!"

"Glad to have you with us." He shook my hand enthusiastically.

My answer was radioed to Colonel Deronne and I was committed, though I still had to wait for my transfer to Mandelieu to come through.

I asked Lieutenant Konrad if I could confide to my special circle of friends what had happened to me. He advised me to be cautious.

"Gorodisch is clear, but the others in your Arles group still have to be thoroughly investigated. We prefer to handle all of it from our end. And one more thing—no more outbursts of temper when you hear anti-Semitic remarks!"

I would have to be careful.

My more immediate problem was that I had been called out ostensibly on an emergency sick call. My visit had been entered in the official records. To avoid arousing suspicions I needed a sick patient to bring back with me. Lieutenant Konrad generously loaned me his adjutant, suggesting that I teach him to simulate appendicitis during the ride into Arles. I called Gorodisch to enlist his cooperation in hospitalizing the fake appendicitis and discharging him after a miraculous recovery.

When I got back home I shared with Sonja all that had happened, stressing our need for caution. I ended with the pièce de résistance, that we were to be sent to England for special training. Her eyes lit up. "This is an occasion to go out and celebrate!" And celebrate we did. We went to our favorite restaurant for dinner. We toasted our new future so many times that we could barely walk through the door when we left to go home. We held each other up as our unsteady feet negotiated the suddenly treacherous cobblestones. We giggled all the way back.

Now we had to wait again. The time passed so slowly. We got a new commander at the Tarascon camp, an unpleasant and overweight Corsican with tufts of dark hair on his fleshy hands. He was a man who indulged his two overriding passions—the Catholic religion and the local brothel—in that order every Sunday without fail.

We took an instant and profound dislike to each other. Lieutenant Konrad sent me a warning note, cautioning me to watch my step because the new commander was a dangerous Pétainist. But despite my best efforts our relationship deteriorated steadily, climaxing with an incident that I was convinced would wreck my transfer to Mandelieu.

One evening my nurse's aid, Lazarre, called me about a laborer who had just come into the infirmary complaining of severe chest pains and dizziness. I instructed Lazarre to hold the man in the infirmary overnight and keep an eye on him. Lazarre called me back a few hours later when he found the man unconscious and breathing with difficulty. I ordered Lazarre to rush the patient to the hospital in the ambulance truck, and promised that I would be waiting for them at the hospital gate. I threw on my clothes and bicycled to the hospital, arranged for the admission, and stationed myself at the gate to wait for the ambulance. It never arrived.

Sonja suddenly appeared half an hour later, out of breath from running. Lazarre had called just after I left. The truck and the driver were gone. There was no other way to get to the hospital. The night guard, overhearing us, suggested that I call the police for the loan of a paddy wagon to pick up my patient from the infirmary.

I arrived at the infirmary to find the ambulance truck parked in its usual place. Lazarre ran out, highly agitated, to tell me that the patient had died. And while he lay dying, our ambulance truck and driver had been parked outside the local brothel waiting for our new commander and his friends to finish their pleasures.

The next morning the commander ordered me to begin sleeping in the camp to prevent any more such incidents. I refused, and threatened to report him for causing the death of a man under his supervision. He backed down and curtly apologized, making no attempt to keep the intense anger and dislike out of his voice. I feared that he would get even with me by blocking my transfer to Mandelieu. But I wasn't using my common sense. He didn't want anyone around who could interfere with his access to the brothel. So my transfer suddenly arrived in no time at all! In March we left Arles for the Côte d'Azur to begin our new lives.

Our last few days in Arles were spent with our friends. Gorodisch and I had grown close, and he was the one whom I was saddest to leave.

5

By the middle of March 1941 the occupying Germans had already begun their planned deportation of Jews from France. All of the towns on the Côte d'Azur were subject to frequent neighborhood dragnets. Everyone had to show their identity cards. Men also had to have their genitals examined. By the time Sonja and I arrived in Mandelieu, we were prepared. Lieutenant Konrad had given us doctored identity cards that omitted our Jewish classification. He also gave me a medical certificate stating that my circumcision had been done for medical reasons when I was a child.

To all outward appearances I was in Mandelieu as physician for a newly organized labor camp containing the Twenty-seventh Group of Travailleurs Étrangers—mostly made up of Spanish Republicans—under the command of Captain Montagne. Some of these Spanish refugees were loaned out to help on surrounding farms. The majority of them worked for a pro-Pétain slave driver who owned a charcoal company in the Esterel forest. He wanted as many unpaid workers as he could get to help him meet the greatly stepped up demands for charcoal brought about by the severe wartime gasoline shortage. Automobiles were now equipped with two ungainly tanks on either side of the hood that used a gasifying process to convert charcoal into a usable fuel.

I had to visit all of the Spanish workers twice a week and supervise the health care of the men in several smaller groups in the area. I established hospital privileges in the Hôpital Mixe in Cannes, where I became friendly with Dr. Buso from neighboring La Napoule. He had been the town's only doctor since the war had begun. He was always tired from his conscientious efforts in trying to handle all of the work. Dr. Buso asked me one day how I felt about moving to La Napoule and sharing his practice. As long as I discharged my military medical responsibilities, I was clearly free to do whatever else I wished. Dr. Buso's offer seemed like a good idea to Sonja and me. Many of Dr. Buso's patients were farmers and their families. I liked the idea of a country practice. The people were often interesting, and they tended to pay their doctor bills with chickens, butter, and eggs. Even though we had access to the army cooperative in Fréjus, rationing was in effect and quantities were often severely limited. And not least in our decision was our growing affection for Dr. Buso. We rented a small apartment in La Napoule that overlooked the Buso family's backyard.

Our one unanticipated problem in establishing ourselves safely on the Côte d'Azur was La Napoule's priest, a rabid anti-Semite. He sympathized with Germany's plan for ridding France of her Jews, and informing on his neighbors was to him the ultimate act of patriotism. Whenever this priest met me in the street he talked eagerly about the most recent "Jew traitors" he had denounced to the Germans.

Sonja and I realized immediately that our regular Sunday absence from his church would soon raise his suspicions. I let Lieutenant Konrad know we urgently needed to document ourselves as Unitarians. Since the only churches in our area were Catholic, our lack of desire to attend Sunday mass would be perfectly acceptable.

Once Sonja and I had familiarized ourselves with La Napoule, I sent word to Lieutenant Konrad, who was still in La Camargue, telling him that we were ready. He came to La Napoule to prepare us for making contact with our local comrades in the Resistance.

In two weeks we had to start listening to the BBC broadcasts regularly. Our orders would be broadcast the night before we were supposed to meet our unknown comrades. We would be identified by our code names. I was *oiseau* ("bird"). Sonja was *canard enchaîné* ("chained duck").

We were to listen nightly until we heard the phrase *l'oiseaux est pret à voler avec le canard enchaîné* ("the bird is ready to fly with the chained duck"). This was the signal that we would be contacted at 1:00 P.M. the next afternoon at a preselected café in Cannes. Sonja was to be holding a copy of *Le Canard enchaîné*, a popular newspaper known for its trenchant political satire. A stranger would approach and ask, "Where did you find the *Canard enchaîné?*" We were to answer, "We picked it up at the Arles railway station before our train left." The final response would be, "It is difficult to get it in this part of France."

We had to choose a suitable café where our contacts could find us. Sonja and Lieutenant Konrad met me at the Hôpital Mixe in Cannes and we walked around the neighborhood until we found a large place that was well frequented. While Sonja and I took a table, Lieutenant Konrad inspected the building. Escape provisions were adequate. Exits led in three different directions. This was part of our instructions. Whenever we would have to meet in a risky situation, we had to have more than one direction in which to scatter and run so at least one person had a chance of escaping.

I wondered if our first contact would be with one or two people. We were told to expect two, most likely a man and a woman. That appeared most normal and would attract the least attention.

Before returning to La Napoule we chatted over coffee for a while. Lieutenant Konrad expressed his private hope that his next mission would send him to Normandy, where he grew up and where his family still lived. That evening he took us to dinner at a little bistro, and we returned to our apartment to listen to the Free French broadcast over the BBC.

We turned on our radio and found the frequency. At first we heard nothing but static. Then the opening notes of Beethoven's Fifth Symphony announced the broadcast. "*Ici la France libre. Les français parlent aux français. D'abord les nouvelles, quelques messages personnels.*" ("Here is Free France. Frenchmen are speaking to Frenchmen. Before the news, here are several personal messages.") After the message *les pecheurs rentrent à la maison* ("the fishermen are going home"), Konrad suddenly sat up in his chair. It was his coded order to leave from Cap d'Antibes the next morning on a submarine to England.

While we waited for our contact to be scheduled, my job was to look for potential recruits for our group. I did not have to stray far from home. Dr. Buso and his family were violently anti-German and anti-Pétain. That included the Buso's two daughters—Huguette, about sixteen and still in school, and Mimi, about twenty-four, an intense and elegant Mediterranean beauty who managed a women's custom dress shop in Cannes. Dr. Buso hated their two enemies equally. Mme. Buso hated Pétain even more than the Germans because of the food shortages. And Mimi hated the Germans with particular fury because they had killed her fiancé.

Sonja and I ate dinner with the Busos several times a week. They provided the food. I brought cigars for Dr. Buso and cigarettes for Mimi from my ration coupons since I wasn't much of a smoker myself. After dinner we relaxed over coffee and Mme. Buso's freshly baked cookies. I remember the aroma of Balto cigarettes and Diplomat cigars mingling with the fresh smells of the garden and the sea. And every time our conversation touched upon the Germans, Mimi's fury exploded.

Mimi was intelligent, resourceful, and adventurous, and very, very angry. I was certain she would make an excellent recruit, but I was unsure about the best way of approaching her. Sonja volunteered to speak to Mimi directly, but first I decided to find out a little more about her background. I learned that she was actually the daughter of Dr. Buso's first wife, who had died, but that no jealousy or favoritism divided the family. I felt that pointed to a stable background. And since her fiancé's death, Mimi was without serious entanglements. To me, that meant she would be fully committed to the Resistance.

But Sonja and I realized that we still had to be cautious. We arranged for Georges Fischer—a dear friend from Arles who was already working in a Resistance group in Chambéry, near the Swiss border—to meet Mimi. He soon got the formal okay from our superiors, and Mimi was thrilled when I told her that she had been accepted as an official member of our group.

We now had our first recruit. I gave her the job of keeping her eyes open for additional new members, and she went on to become a crackerjack recruiter. She brought us people from the mayor's office and from the *gendarmerie* ("municipal police") and even placed two of our Resistance agents in the militia, the French version of Germany's Brown Shirts and Italy's Black Shirts.

Two weeks after Lieutenant Konrad's visit, Sonja and I began listening to the BBC broadcasts. Each night we listened excitedly, but time passed and there was still no code message for us. Our anticipation and disappointment grew nightly. We became tense and restless. After a seemingly endless series of frustrating broadcasts, the introductory bars of Beethoven's Fifth Symphony were finally followed by: "*L'oiseau est pret à voler avec le canard enchaîné.*" Tomorrow was our big day!

Sonja met me at the hospital the next afternoon to walk over to the café. I was in despair. I had scoured the city in vain for a copy of the *Canard enchaîné.* The government had just suppressed its publication, and there wasn't a copy to be had anywhere! Now we had no way of meeting our contacts. Our superiors in London would be terribly disappointed in us, and they might even change their minds about giving us any real responsibility.

Sonja was undaunted. "They must know that the newspaper isn't available. If we fail this time, they'll simply set up another plan."

"Well," I said, "we're already here. Let's at least have a cup of coffee, and maybe a good idea will come to us while we sit."

We found an empty table on the terrace and ordered coffee from a passing waiter. Sonja's eyes suddenly lit up: "What do the French call a cognac-soaked lump of sugar?" I began to laugh. "It's called a '*canard*'!" Sonja was really pleased with herself. "The people coming to meet us will know we couldn't find the *Canard enchaîné.* It's logical that they'll look for some kind of related signal."

As I thought about the clearest way to put Sonja's idea into action I noticed a small, very ordinary-looking blond woman passing the café, idly studying the people seated on the terrace. Could such an anonymous-looking little woman be one of our contacts?

A very tall man walking past the café for the third time also caught my gaze. He was very thin and wore baggy clothes that obviously had once belonged to a fat man. Had he recently lost a tremendous amount of weight or had he bought his clothes in a secondhand shop? I didn't care for his face. The long, thin nose and little moustache sitting on top of such a scarecrow body didn't give him a trustworthy look. Was this man one of our contacts?

The waiter came with our coffee. I had a plan ready. I asked him for some extra lumps of sugar, whispering that my pregnant wife had quite a sweet tooth and that he would receive a handsome tip for the favor. He brought me ten more lumps. I made a theatrical display of feeding Sonja each lump, dunking it ceremoniously in my coffee and raising the spoon with a flourish before dropping each makeshift "*canard*" in her mouth. When the waiter passed by again I loudly called out for him to bring me a copy of the *Canard enchaîné.*

The small blond woman walked quickly across the terrace and disappeared inside the café, where the telephones were located. The tall man followed on her heels. I now suspected that they were our contacts. If I were in their spot, unable to follow my initial instructions but finding a likely looking couple using the password, I would want to confer together privately or call a superior before doing anything.

A few minutes later the tall man came into view behind the café's window and took a seat directly behind the terrace. But his gaze seemed to go beyond us, focusing on the Isle Ste. Marguerite off the coast. We had only one lump of sugar left. Sonja looked at it with distaste but swallowed it. Nothing happened.

In a last attempt I started to drum on the tabletop with my fingers, tapping out the distinctive da-da-da-DUM of the Fifth Symphony's opening bars. Before I could finish the small blond woman dashed over and hissed in my ear, "Stop it!" In a gentler tone she asked Sonja if she liked "*le canard.*" Sonja replied, "Yes, I like the *Canard enchaîné,* but I couldn't find a single copy when I was at the Arles railroad station." At that the blond pulled up a chair and joined us. She had run over so abruptly because she was frightened that my tapping would give us all away. She admonished me for my stupidity and signaled to the tall man inside to join us.

We all shook hands and got down to business. The blond introduced herself as Madame Billard. She used the cover name of Mme. Raffia, taken from the raffia handicrafts shop that she ran in Nice. The scarecrow in baggy clothes was Monsieur Ramoine, chief architect of Nice. His cover name, Captain LeGrand, picturesquely reflected his height.

The Resistance group we were joining was the Groupe Surcouf, named for a submarine commander from Toulon who had already become a martyr in the struggle against the Germans. Our Groupe was part of the larger Réseau Marco Polo. Mme. Billard would be our main contact and *boîte aux lettres* ("mailbox"), the conduit through whom all messages to and from us would pass, sometimes via a go-between. Ramoine took charge of our *réseau's* security, always alert for possible danger and responsible for whatever protective action needed to be taken. I told them about Mimi Buso, our first recruit. They already knew about her.

I asked when we could expect to go to England for our training.

Ramoine answered, "Just as soon as I can arrange for your submarine passage. I'll meet you in Nice soon to explain the procedure for the submarine pickup."

Mme. Billard followed up.

"When you need to meet me, send me a letter addressed to my cover name at the *poste restante* (the French post office box). Use the central post office in Nice. Let's decide on several meeting spots. We'll assign a number to each. In your letter, which should be as innocuous as possible, write 'over' at the bottom of the page just preceding the page number indicating the meeting place you want. If you want to meet at 'one,' don't write 'over' on any page. We'll make this café number one, the Hôpital Mixe here in Cannes number two. Number three will be the Cintra Vogad café on the Place Masséna in Nice, and number four will be the bar or terrace of the Hôtel de Paris in Monte Carlo.

They left. Mme. Billard was back a moment later. "I forgot to tell you one thing. Get yourself a month's furlough so you don't arouse suspicion by suddenly dropping out of sight for the four weeks you will be in England."

Next morning I spoke to Commander Montagne about my furlough. I encouraged his sympathy for my request by pointing out that, though I was entitled to one, I hadn't yet taken a furlough from my work, not even for my honeymoon. Now that everything seemed to be settling down, Sonja and I were hoping to get away for some time together. A furlough was fine with Commander Montagne as long as I found someone to cover the medical service during my absence. I had already lined up Dr. Buso.

Soon after our café meeting, M. Ramoine took me to a house in the Vieux Port in Nice and introduced me to a young dockworker named Jean. He would take us to Cap d'Antibes and hide us there to wait for the submarine that would take us to England. Jean did not look any more like my notion of a Resistance fighter than M. Ramoine or Mme. Billard did. He appeared no different from any of the others working around the harbor, a man in his mid-thirties wearing a beret and gray waterproof garments. Jean made a rendezvous to meet Sonja and me in three days at 8:00 A.M. sharp.

When Sonja and I arrived at the Vieux Port to meet Jean three days later, we felt that we were beginning our formal initiation as Resistance fighters. We were incredibly excited. Jean took us to a beachfront house on Cap d'Antibes. The owner hid us in his cellar. We waited.

Night fell. No one came. At last the owner of the house appeared. The rendezvous with the submarine had to be postponed twenty-four hours. The Germans had been alerted to the plan to smuggle some Resistance workers to a submarine waiting off the coast. German guards now saturated the entire area. We had to stay in hiding.

The next day at noon our host visited us in his cellar. He brought a welcome meal and good news. We would be leaving at midnight. Sonja and I played cards to pass the time. Our host came down at around eight o'clock in the evening to confirm our midnight departure. He suggested that we try to get some sleep. To our amazement we actually managed to doze.

We were abruptly shaken awake. Hurried instructions were whispered to put on the fishermen's clothes our host had brought with him. We quickly got into our waterproof outfits and funny-looking little hats. We followed him out to the beach to meet the fisherman who would row us out to the submarine. We were posing as his eldest and youngest sons. The fisherman warned us not to speak if anyone stopped us, since neither my Hungarian accent nor Sonja's female voice could possibly belong to a French fisherman's son. We pushed off from the beach. The shoreline disappeared, blending immediately into the moonless black surrounding us.

We headed out to sea. I listened to the rhythmic sounds of waves lapping on the receding shore, the oars slicing in and out of the inky water, the rise and fall of my own breathing. Without warning, the fisherman pulled in his oars and pointed. In the dark I could make out the still darker silhouette of a submarine and a light flashing "Stop" in Morse code.

The fisherman reassured us, but Sonja and I felt extremely uneasy. I was afraid that we were walking into a trap. We waited. Something separated from the submarine and moved toward us. Was it Germans coming to arrest us?

It turned out to be a small inflatable boat that would bring us out to the submarine. The inflatable was manned by a Frenchman. Could he be a collaborator who would take us straight to the Germans?

To my immense relief, the submarine commander was also a

Frenchman. He greeted us warmly and showed us to our quarters below. Then our first underwater voyage began.

Sonja, with her highly adventurous nature, adjusted easily. I felt I had been buried alive. Breathing became so difficult I was convinced that the oxygen content in the submerged vessel had fallen well below normal. I was assured that it was identical to the air above. I realized that I was claustrophobic. Gradually, during the several days it took to reach the English coast, I adapted to my underwater environment.

The submarine surfaced off the coast of England and traveled in darkness for a few more miles. Then the same inflatable boat took us ashore.

We had absolutely no idea where we were. Again we waited. A car met us, and then we learned that we were about 120 miles from London. We were driven to a private home in London's Soho district where we spent the night.

The next morning at breakfast, we had some time to study our surroundings. Our London hosts were a very pleasant couple in their fifties. The furniture and even the odor in their apartment reflected the same homey and settled air they gave off. Opening their home to French spies didn't seem to fit with their character, but the couple appeared very comfortable with it all. Since they didn't speak French or German, Sonja's fluency in English was very welcome. I enjoyed their conversation even though I was not able to participate in it directly, but I must admit that their cooking was terrible. Happily, I didn't yet know that not a single one of our meals in England would rise above this abysmal level.

After breakfast our hosts gave us the Free French military uniforms that had been delivered for us. The uniforms were for identification while we were in England. Even though the uniform clearly marked me as a member of the Free French army, I didn't much care for it. I had always disliked uniforms. At home I rarely put on the army uniform that my lieutenant's commission entitled me to wear.

Sonja's reaction was just the opposite. She loved the idea of dressing up in an army uniform. She even went to a London tailor to have the fit of her uniform improved. Her one disappointment was the French hat, which lacked the dashing elegance of the English military cap. So she bought one for herself at one of London's military clothing stores.

Shortly thereafter, a young French lieutenant picked us up and took us to an official building in Whitehall. He left us in an empty room. It was blindingly lit. The door to the room next to us stood open. That room was shrouded in total darkness. Our floodlit room was bare except for a small table and two empty chairs confronting that black and mysterious open doorway. We sat down and waited . . . for what?

A voice suddenly rang out from the impenetrable darkness. "*Bonjour, mes amis.* I am General Paul Guivante de Saint-Gast, chief of the Réseau Marco Polo to which your local unit Surcouf belongs. Welcome to England, and welcome to my private office!"

Then the voice went on to explain very matter-of-factly that we were going to be thoroughly trained in the art of self-defense and a broad range of skills relating to sabotage and intelligence work. If we decided at any point that this work involved more danger than we could tolerate, we were free to transfer to a less risky branch of the Resistance. General Saint-Gast hoped we would stay with this branch because our languages and my medical skills would be invaluable in the fight against the Germans.

The voice from the dark soon concluded, "I am sorry to have to meet you under such anonymous and impersonal conditions. But this way, if you are ever arrested you will not be able to divulge any useful information about my identity. *Adieu, mes amis!* And good luck."

Silence returned. We were covered with goosebumps.

The very next day we started our four weeks of intensive schooling at the Resistance training center in London. We began with jujitsu and the arts of sabotage, including the use of plastic explosives. Our espionage teachers taught us to decipher architectural plans and to read upside down. They trained us in the arts of observation, they taught us how to steal, and they drilled us in responding cooly to unexpected dangers. We were regularly warned about one danger in particular, the British-directed French Resistance units that were also being formed. We were told that their real purpose was to feed false information to the Germans, and that their members were unwitting pawns in the process. We had to stay clear of them or we would endanger our own lives.

For our final lessons we were driven out to an airfield in Sussex to learn parachute jumping. Our very last task was to memorize the codes that would enable us to decipher our orders from the cryptic messages of the BBC evening broadcasts.

Sonja found the four weeks very exciting. She tended to be a much better student than I was, particularly in parachuting. I was hesitant and clumsy, a headache for my instructors. My jujitsu was at least passable, but it left me tied up in knots despite my outward calm. My student boxing experience wasn't much help. But there was one subject in which I outshone Sonja. I left her in the dust when it came to reading upside down.

During this period we met an English couple who shared a common language with me. They were Jewish, so with their Yiddish and my German we had little difficulty communicating. It almost got me into trouble, though. An English bobby overheard us talking about some items in a shop window. He called his superiors and reported a group of German spies. I didn't know anything about it until I was accosted and hauled into British intelligence by an agent who told me he had been tailing me for days. There were some red faces when my identity was cleared up!

Our training came to an end. We turned in our military uniforms. We were sent back to France by air and parachuted down near Grenoble. A member of our local group was waiting to drive us back to Nice. We felt so much more grown-up than the naive young couple who had left for England only four weeks earlier.

It was the summer of 1941. We were ready to go into action.

6

Over the next year and a half we were given a variety of missions to carry out. Our instructions usually arrived locally or via the BBC coded broadcasts. We were called back to England a few more times to meet with Saint-Gast for special briefings. My reactions to the Resistance work surprised me. Danger didn't frighten me nearly as much as I had thought it would. My work was unexpectedly exciting. And it became an invaluable education in learning to analyze people and develop the ability to think one step ahead of them.

I became affiliated with the Hôpital St. Roche in Nice. I had got myself appointed to the medical staff that examined the young French men and women from our area who had been chosen for forced labor in Germany. The French physician in charge cooperated with me in declaring most of these candidates physically unfit for work in the labor camps. Those who had to go received instruction in sabotage.

Sonja and I distributed false identification papers and ration cards to new members of our Resistance unit. We worked with Ramoine to warn comrades who were in danger of being exposed or arrested. We shadowed people who were suspected of collaborating with the Germans. We were always on the lookout for any scrap of information about German troop and naval activity.

Mimi Buso and I put together a successful charade to transport wounded Allied pilots to freedom. Whenever the need arose, I used a German ambulance from the hospital. Mimi dressed in the distinctive blue-and-white striped German nurse's uniform, and we would drive the pilots to Chambéry, near the Swiss border. From there, my friend Georges Fischer and his Resistance comrades would smuggle them across the border into Switzerland.

On two different missions I was parachuted over Austria with equipment to set up clandestine radio transmitters with Austrian Underground fighters. The first time I worked with a bishop in Vienna.

The second time I parachuted down in Seefeld, outside of Innsbruck, to work with a highly respected hematology professor at the medical school in Innsbruck. When I first met the man he was already in his early fifties. In my experience the professor seemed old to be running such risks by working in the Resistance. I was impressed by his courage and soon came to respect his judgment.

My group and I planted explosives in cafés frequented by the German soldiers. We hoped to demoralize the Germans and at the same time inspire our fellow Frenchmen with the proof that an underground Resistance movement was actively fighting for their freedom. I was in charge of these efforts. It was an irony for me as a physician, but it was also an extension of my deep commitment to fighting and destroying our enemy.

One Resistance mission sent me to Belgium to form a liaison with a newly formed réseau and help set up their operations. I was parachuted over German lines. My landing was smooth but I picked a very bad spot. I landed right at the feet of two German soldiers. They came running and aimed their rifles at my chest. I shouted in German, pretending to be a German officer in disguise. They tentatively lowered their guns, but they began to doubt my story. It was their lives or mine. I raised my machine gun and killed them. My legs began trembling so badly that I had to sit down for a while.

Sonja and I rarely worked together on missions, and we never spoke about our separate jobs until they were over. If either of us was caught, we would not be able to pass along information that could endanger the other. But we did work as partners during our series of parachute jumps to drop American and English weapons for the Alpine Resistance center at Vercors, near Grenoble.

Our last jump of that mission marked the end of Sonja's career as a parachutist. She was already several months pregnant. She had managed to keep it a secret even from me until the final jump was over!

While all of this was going on I continued my medical work for the Spanish Republicans in the French labor camps. I was still sharing Dr. Buso's caseload as well. It was essential for me to maintain my cover as an ordinary physician doing the best I could to carry out my responsibilities to the sick.

Sonja's mother and sister were now living near us in Nice. Thanks to our Resistance contacts they had been able to move up from Perpignan as soon as we got false identification papers for them. Not long after they arrived, Paule, Sonja's younger sister, joined another of the local Resistance units. She had recently become engaged to one of the men she worked with. We never let the closeness of family ties mitigate our caution. Neither of us ever discussed our Resistance work with Paule, and we never knew what she was involved in.

Life for most Frenchmen was becoming more and more difficult. Just about everything was now strictly rationed in order to cope with the ever worsening shortages plaguing the country. The French franc's former dominance in the marketplace had been usurped by more precious commodities—cigarettes, coffee, and razor blades. People with gold teeth suddenly considered themselves very fortunate, because gold had become the one stable medium of exchange for essential food and medication. Dentists did a brisk business in extractions. A lot of people stopped laughing in public to hide the embarrassing gaps in their teeth.

The Germans were methodically stealing everything they could get their hands on. "*Les boches,*" the disrespectful nickname from World War I, had been replaced by "*les doryphores,*" the potato bug that quickly devastates entire crops if it is not kept under control. The patients I had taken over from Dr. Buso now paid me almost exclusively on a food-for-service basis. I was lucky to be working as a local physician in an area of farmers and fishermen.

Our son was born on October 9, 1942. Sonja and I had already moved to a larger apartment in La Napoule, but our close relationship with the Buso family was unaffected. Dr. Buso was Sonja's obstetrician, which made François' birth very much a family affair.

We gave a little party to celebrate this momentous event in our lives. The abundant table was provided by the fruits of my local medical practice. The milk on the table was my honorarium for draining M. Pluvieux' axillary abscess. The butter and eggs in the freshly baked cake had been given to us for the delivery of the Dalmasso baby. The flour and chocolate, as well as the coffee, were the result of my treatment of M. Delmas' herpes zoster. The chicken and potatoes were my payment for setting M. Weber's broken leg.

On November 8, less than a month after our party, General Eisenhower led the American and English forces in a successful assault on North Africa. On November 11 the Germans extended Nazi rule over the whole of France. There was no more unoccupied zone.

I was recalled to England for the last time. Sonja stayed home with our infant son. Our submarine route had been exposed and could no longer be used, so I had to go by plane despite continuous German surveillance of the skies over France.

As I sat in the now familiar floodlit room, Saint-Gast's disembodied voice spoke to me about an extremely important mission that was part of the overall preparations for an eventual Allied landing in France. To further this mission my task would be to infiltrate the Todt organization, which had now established a central office in Nice. Todt was the German paramilitary group that had been designing and building defense structures along the entire French coastline against a possible Allied invasion.

Saint-Gast's disembodied voice briefed me and explained why I had been chosen.

"We feel that you have the best chance of all our men to get inside Todt," he said. "You speak perfect German. You are a physician, which tends to inspire confidence. There will be no clue to your Jewish background. Hungary is allied with the Nazis, so being Hungarian-born will be in your favor. And you will have a lot more flexibility than a Frenchman would in inventing a convincing cover story. Your wife should also try to infiltrate Todt. With her German and gutsiness, she can do us a lot of good. Once you get inside, keep an eye out for a good contact in the *Sonderdienst.* It's the army's propaganda branch. They get all the details whenever the Germans develop a new weapon or design a new fortification. Good luck."

Back in Nice I worked out a plan. Many of the German military frequented a bistro on the Avenue de la Victoire. I went in one day and struck up a conversation with a junior officer. I told him that I was a Hungarian physician trained in the treatment of tropical diseases, and had emigrated here in search of work in the French colonies. Now the French were persecuting me for my pro-German sympathies. I had trouble earning a living, and I had a wife and small child to support.

I found a very sympathetic ear. My new "friend" soon introduced me to those in the higher echelons of the Todt organization and the Sonderdienst. I played my role well and the social rapport I developed with them as a pro-German Hungarian led naturally to the introduction and inclusion of Sonja and her sister, Paule. Sonja and I soon became accepted members of their social circle.

I became especially friendly with one man in particular. He was a kind and decent person, not anti-Semitic. He despised the Gestapo and everything it stood for. He was a survivor of the Stalingrad debacle that Hitler's troops had suffered on the Russian front. He had come away from that experience with the realization that Germany was ultimately going to lose the war. Now he was far from the front lines, using his writing skills with the Sonderdienst. He couldn't have been a more perfect contact for me.

One day he offered to help me out in any way he could. I told him that I needed a regular job and that I wanted most of all to help the German war effort by working for the Todt organization. A friend of his who had fought with him in Stalingrad now headed the main Todt office in Paris. He arranged for me to travel to Paris with him to see this man. From there we were given an introduction to Todt's medical director, who hired me on the spot to care for the health of the German soldiers and French civilians working in the Nice branch. My identification papers (which had been provided by Lieutenant Konrad) had already been approved by both the French and German authorities, so there was nothing to keep me from starting my new job right away.

I thanked everyone profusely.

By the time I began my job as physician in the Todt organization, I had already established a very amicable relationship with Hauptbauleiter Baumgartner, one of the military architects supervising construction and camouflage of the Atlantic Wall. When I ran into him at Todt we stopped to chat. I knew that Hauptbauleiter Baumgartner liked Sonja. I mentioned how interesting she found his work. Baumgartner had a brilliant idea. He called Sonja and asked her if she would like to work in his office. "With pleasure!" she replied. With that, Sonja entered the inner sanctum.

I had been working at my new job for only a week or so when

Baumgartner told me that Peter Schreiner of the SA had asked for certification that Sonja and I weren't Jewish. The SA *(Sturm Abteilungen)* were Hitler's fanatic Brown Shirts. Sonja went to see Schreiner. Our identification papers were insufficient. Certification had to come directly from the Gestapo.

Hauptsturmführer Recsek headed the Gestapo in our area. Sonja made an appointment to see him at Nazi headquarters in the Hôtel Hermitage in Nice. Their meeting went smoothly. Then Sonja arranged for me to meet him to fill out the papers necessary for my certification.

I walked into the office of this major official of the Nazi secret police expecting a very stiff interview with a cold, businesslike military officer. Hauptsturmführer Recsek was wearing elegantly tailored civilian clothes. He jumped up from his desk to greet me with a warm and open geniality. I was totally unprepared for the immediate and mutual sense of rapport I experienced. He behaved like a warmhearted, intelligent English aristocrat, the kind of man one could always count on.

I saw Recsek more and more frequently after that initial meeting. He was always extremely polite and well-spoken, a quick-witted and well-educated man who had taught history at Breslau University before the war. Recsek always treated me as a colleague, even calling me "Professor."

He was very much the Prussian nobleman. I never saw him in military uniform, but he kept himself in top physical condition and was always immaculately groomed. Recsek's Prussian discipline was softened by a sense of humor and by the tenderness that always seemed to illuminate his eyes. But his goodwill did not extend to Jews. He was very vocal about his infallible ability to "smell a Jew." Yet he never questioned that I was anything other than an Aryan physician from Hungary, an ally of Nazi Germany.

Hauptsturmführer Recsek actually became one of my close friends. We played tennis regularly, and he was a frequent visitor at the apartment we soon moved to in Nice. We spent many after-dinner hours together discussing European history and philosophy. Sonja and I were careful never to question Recsek about his work. We could not afford to risk arousing his suspicions. But sometimes, after a few drinks, he would take us into his confidence and tell us about whatever spy he had recently arrested.

Often when Recsek would stop by after work to chat with us and relax over a drink, François would be running around the apartment without a stitch of clothes on. We had taken the precaution when he was born of not having him circumcised. Recsek called François "the most beautiful Aryan child" he had ever seen and seemed to love him as a son.

Recsek was the first Nazi to become an intimate part of my life. You could say I knew him well, but that wasn't true. To know a man well is to see all sides of his character. Recsek showed me his good nature, but there was more, as I was to discover. He was a Gestapo Hauptsturmführer. No one achieved such rank in the Nazi regime without having a profound streak of cruelty and a ready willingness to use it. Recsek's frequent allusions to his hatred of the Jews should have forewarned me of what lay beneath his benevolent surface.

Sonja's fluency in a variety of languages made her very useful to the Todt organization. She became personal secretary to Hauptbauleiter Pracht, the man directly responsible for construction of the Atlantic Wall. Thanks to Sonja, central intelligence in London had copies of all the plans before the first stone was even laid.

It didn't take Sonja long to gain access to the safe where Todt's top-secret blueprints were kept. Being married to a physician, she knew that a man with prostate problems has to urinate very soon after he drinks. Hauptbauleiter Baumgartner, who kept the combination to the safe taped inside a secret desk drawer, had a prostate condition. Sonja manipulated him into inviting her for a drink in his office. His trip to the bathroom gave her just enough time to photograph the combination with the Minox hidden in her pocketbook.

Infiltration of Todt created a variety of opportunities for Resistance work. My knowledge of the daily entry password gave our unit's saboteurs nightly access to Todt's construction areas. In my medical capacity I was able to replace a number of the pro-German workers by diagnosing their state of health as requiring long-term convalescence. Members of our own unit took their jobs. Our agents assigned to building fortifications secretly mixed seawater with the cement so that the defense wall would look solid but crumble in a short time. The more Resistance members we were able to get into Todt, the more numerous our sources for gaining and cross-checking information became. Todt had assigned me a car and driver. I soon sent the driver home for a lengthy convalescence and Jackie, one of my comrades, got the job.

I was also helping Ramoine to track down and identify stool pigeons who had infiltrated our unit, and I continued my task of organizing the placement of plastic explosives in cafés frequented by German soldiers. Sometimes, by strange irony, I was called in by Todt to help tend to German soldiers who had been wounded in these incidents.

Although Sonja, François, and I still held on to our apartment in La Napoule, we had recently moved to Nice, occupying rooms in the Hôtel Napoléon, where most of the Todt personnel lived. We got Paule a job with the German radio station in Monte Carlo. We continued to consolidate our friendships with members of the Sonderdienst and the Todt organization. We were living daily in the midst of our enemies.

Sonja's work almost involved us in disaster. She had stolen some fortification plans from the Todt safe overnight to be copied and sent to London. I gave them to a comrade to deliver to my *boîte aux lettres,* and he was to give them to Ramoine to copy. Our comrade must have lost the plans almost as soon as I handed them over. The next morning a pro-German Frenchman called Sonja's office to inform the Todt authorities that he had found German architectural plans for what appeared to be a major military installation. He intimated that a spy must be working in their office.

By sheer luck it was Sonja who answered the phone when he called. She told the informer that he deserved a reward for his service. She took his name and address and told him not to leave his house. Then she called Ramoine. Ramoine retrieved the architectural plans that were meant for him in the first place. The "reward" he delivered silenced the recipient forever.

After that one unsettling incident, our work went smoothly. I fondly recall a business trip Sonja made with Hauptbauleiter Baumgartner and Reichsminister Albert Speer. They had asked her to accompany them on their inspection tour of the new camouflaged coastal defenses at the French and Italian border cities, Menton and Ventimiglia. Sonja returned with a detailed description of buildings that looked like seaside villas from outside. Inside were bunkers furnished with long-range cannons. Reichsminister Speer shook Sonja's hand at a formal Todt dinner held in Nice after the inspection tour. He publicly complimented her as the most intelligent and industrious of all the secretaries in the Todt organization.

Late in the spring of 1943 a man we had known for a while came to Sonja with a disturbing story. The man was a white Russian who dealt in the black market. He heard the story from SA Peter Schreiner, who was one of his regular customers. Peter Schreiner was the Nazi who had requested our Gestapo clearance after I began working at Todt. He had recently informed Gestapo headquarters in Toulon that Sonja and I were too interested in military installations. A Gestapo agent visited Hauptsturmführer Recsek, who had laughed out loud at these allegations. He knew his tennis partner as well as one could know another man. Recsek guaranteed that Albert and Sonja Haas were loyal to the German cause. Sonja and I were thrilled at Recsek's response. We thought it was proof that we were safe from suspicion.

One evening three or four months later, we were sent word to meet Ramoine in the laundry near our hotel on our way home from work. The laundry was owned by a couple in our unit and was one of our information drops. Ramoine had gotten wind of a menacing rumor that we had been denounced to the Germans. As soon as Sonja and I returned home a sympathetic young German soldier stopped by to tell us that Paule had been arrested that afternoon at the radio station. He told us that Paule turned out to have been working in the Resistance. Someone from her group had been caught and confessed the names of his comrades. They were all arrested that same afternoon. He was glad the unit had been broken, but he felt very badly for Paule and for us. He shook our hands warmly before leaving.

This was it. God knows what the Gestapo interrogators might be learning about us from Paule at this very moment. I told Sonja that she and Francois would have to go into hiding at once. She didn't want to leave me. She did not think we were in such immediate danger. I told her office at Todt that she was ill. Cautiously I went about my business. I had to feign horror at my sister-in-law's Resistance connections and manacle my overwhelming desire to move heaven and earth to get her out of jail. The only thing I could do without arousing suspicion was send Paule a change of clothes and a parcel of food every day.

No one bothered me and nothing out of the ordinary happened. My terror began to dissipate. After several of these very ordinary days I told Sonja to come out of hiding.

One evening we went to the Cintra Vogad on the Place Masséna to transmit an architectural plan to my *boîte aux lettres.* No one appeared. As we left the café a stranger brushed past me in the street and slipped a note into my hand. The coded message read: "Be careful! You are being followed!"

We went straight to my office at Todt to remove any potentially compromising papers from my desk. Jackie, my driver, was with us. I told him to inform Ramoine that Sonja and I were in serious trouble.

When Jackie picked me up the next morning he told me that Ramoine would see me in the afternoon at the Hôpital St. Roche. He would come ostensibly for fluoroscopy. The morning slowly passed. At the hospital I put on my white coat and dark goggles and went to the fluoroscopy room. Ramoine was brought in shortly afterward. He had contacted our superiors in London, and they wanted us to carry out just one more mission. We were to find out the type of fortifications and gun emplacements being planned for the airfield just north of Nice. Then they would smuggle us to the Spanish border and we could escape to England via Spain and Portugal.

I agreed to do it if Ramoine promised to arrange for our son's protection in the event that Sonja and I were arrested. I wanted Mme. Billard to take François to my family in Villon. He would be safe there.

I briefed Sonja on our last mission, including the precautionary arrangements I had made with Ramoine to ensure François' safety. The very next day she "borrowed" the plans of the airfield fortifications. My assistant, André, took them to Ramoine for copying and forwarding. André returned them to Sonja early the next morning. She safely replaced them in the Hauptbauleiter's office before the workday began.

At the end of the day I went to Todt headquarters in the Hôtel Westminster on the Promenade des Anglais to pick up Sonja. I had an uneasy premonition. Two strangers who eyed me as I passed through the outer office looked like Gestapo agents. They weren't, but their presence unsettled me. Sonja brushed my fears aside. "Come on," she said, "let's go home."

François and Herta, his governess, were waiting for us at our apartment in the Hôtel Napoléon. All seemed as it should. But Jackie called me before dinner to tell me that as he pulled away from the curb after dropping us off, he noticed a black Citroen following us. It was probably a Gestapo car. I did not sleep well that night. I expected the Gestapo to break into our rooms at any moment.

The next morning I was a bundle of nerves. Sonja pointed out that when the Gestapo arrested people they either waited for them at home at the end of the workday or they broke in during the night. Our undisturbed night seemed proof of our safety.

Our breakfast was sent up. Herta brought François in to spend some time with us before we left for work. I could not shake my preoccupation with our danger. Sonja complained that I was ignoring François, who was being unusually endearing that morning, and chided me for still fearing that the Germans were about to arrest us. Just at that moment the phone rang. I hoped it would stop so I wouldn't have to answer it. My heart was in my mouth as I picked up the receiver.

It was the Wehrmacht hospital. Would I come in immediately to assist the French surgeon who would be operating on German soldiers injured by a Resistance bomb? I knew that members of my own unit had set it off in a café the night before. Evidently I was still unsuspected.

Sonja went to her office in the Hôtel Westminster, François was left with his governess, and I went to the hospital. During surgery two SS guards, guns drawn, pushed their way into the operating room. The surgeon turned white and asked me if I knew what was going on. The SS men pointed their guns at me. Not a word was spoken. Finally I broke the silence.

"Put your guns down. Our scalpels aren't loaded! Why are you interrupting your comrade's operation? What do you want here?"

What they wanted was to arrest me. A Wehrmacht physician came in to replace me in the operating room. The SS guards escorted me out. A black Citroen took us to the Todt headquarters, where Sonja was already under SS arrest in her office. From there we were both taken to Gestapo headquarters at the Hôtel Hermitage.

I turned to look at Sonja sitting next to me in the car. She was wearing her high-collared dark green coat flecked with light green and her dark glasses. She was so nearsighted that she was almost blind without those glasses. I joked that with her turned up collar and dark glasses she really looked like a spy.

Sonja whispered to me, "Pracht, my boss, tried in his own way to let me know what was happening. I didn't realize. All morning he kept insisting that I take the day off. I kept telling him 'No, my husband is planning to meet me here for lunch as soon as he gets out of surgery.' He finally gave up trying to get me out of the office."

We were taken to Hauptsturmführer Recsek's office at gunpoint. I was sent in first. There were no warm greetings, no benevolent smiles. My tennis partner was obviously enraged and had already condemned me. I knew the penalty for espionage—execution by firing squad. The story Sonja and I had decided we would tell if we were caught seemed worthless at this moment, but it also seemed my only hope. I confessed to Recsek that I was a Jew. I confessed that I had hidden myself right under my enemy's nose becuase it was the last place they would look for a Jew.

Recsek began to shout. "Don't lie to me! We have proof that you are a French spy working for de Gaulle!"

"I'm not a spy," I protested. "I'm Jewish. I can prove it to you. I'm circumcised."

"That's not proof of anything!" To my astonishment, Recsek contemptuously unzipped his fly to show me his own circumcised penis. He zipped his fly back up. The case was closed.

I should have realized that my confession would be unbelievable to a man like Recsek. Capturing a French spy would be a big boost to his career in the Gestapo. And he simply could not swallow the fact that he had befriended a "non-Aryan." How many times had I heard Recsek boast to his colleagues of his unfailing ability to smell "the stink of a Jew"? How could Recsek possibly acknowledge, even to himself, that he had befriended and been fooled by a Jew? It would never occur to him that I could be both a spy and a Jew.

He ordered that Sonja be brought in. He asked us if we knew Paule Nadel. "Of course," Sonja replied. "She's my sister." I tried to mask my fear and joked that I never mixed into family affairs. Recsek took my humor for sarcasm and angrily slapped my face. Sonja and I repeated my story that we were simply a Jewish couple trying to camouflage ourselves. Our story fell on the ears of a deaf man. Recsek had caught a pair of dangerous French spies, and that was enough for him.

Sonja was taken away. I was removed to the Villa Trianon, the Gestapo's infamous interrogation headquarters.

Francois' governess had taken him to the park near our hotel that morning. Where was he now—on his way to my family in Villon or in the hands of the Gestapo?

I was convinced that I was finished. I was going to be shot because I was a French spy.

My Gestapo interrogation began with a beating that broke my two front teeth and left my face practically unrecognizable. I refused to talk. They held my head under water in a bathtub. I felt myself fading. They hauled me out and hung me up by my thumbs, which both dislocated. I was terrified and on the verge of hysteria.

Recsek came in while I was hanging. He eyed my disfigured face calmly. He quietly announced that he knew all about me and my miserable past, including my work in the local Resistance group. He demanded the names of my comrades. I couldn't endure any more torture. I decided, almost with a sense of relief, to tell him. He slapped my face before I could utter a word.

The slap shocked me. Then rage dominated my terror. I thought I could appeal to Recsek's powerful identification with the military aristocracy. I told him that if he really knew all about my past he must know that I was a French officer. He should expect me to uphold the officer's code of honor and loyalty. As a fellow officer he should appreciate the fact that—if I really were a spy—I would never confess and betray my country. I talked calmly and coherently despite my frightening pain and desperation. I had never suspected that I could function this way. "Even so," I added, "my only crime is in being Jewish and trying to stay alive by pretending to be someone I'm not."

My appeal seemed to work for the moment. Recsek ordered his Gestapo henchmen to cut me down. I plunged into unconsciousness.

I woke up in a cell. Almost immediately the interrogation and torture began again. They split my toenails. I told my persecutors that I had nothing to say. If they planned to kill me, stop wasting time and get it over with. They continued to beat me. I continued my silence.

Even today I can't adequately explain why I was able to resist their torture. Was it really Recsek's slap in the face? Was I afraid the Gestapo would shoot me as soon as I gave up my secrets? Was it the knowledge that betraying my comrades would start a domino effect, destroying the Resistance movement in the entire south of France and costing the lives of hundreds of good and courageous people? Would I have held out, I wonder, if I had known what awaited me?

In early October I was transferred from the Gestapo prison near Nice up to Fresnes, a prison southeast of Paris. To my astonishment, the group of Resistance prisoners in my transport included several men I had known but never suspected of working in the Underground. At Fresnes we were all separated and sent to solitary confinement cells in the prison's north wing.

7

Long-term solitary confinement crumbles your defenses. When you are unable to exchange even an occasional word face to face with another human being, the pain of total isolation grows unbearable. You become your own torturer, focusing on the very things guaranteed to undermine your stability. I had no idea what had happened to Sonja and François. My fears and fantasies about them became my constant companions, almost driving me out of my mind.

I was obsessed with wondering how Sonja had stood up under torture. I knew that the Gestapo were specialists in mental torture and I feared they would exploit her severe myopia in the method they devised for her. Had she told them anything? Would she have given the Germans my family's address in Villon in the hope that François, if he had safely arrived there, would be brought to her?

During my second day in solitary confinement at Fresnes, indistinct sounds of voices puzzled me. I discovered that they came through the pipes in the interconnected heating and plumbing systems. These pipes provided prisoners with an effective communication system. The toilet in each of our cells served as both loudspeaker and microphone. We eventually dubbed it the LIB, Latrine Information Bureau! News was broadcast every morning at about 10:00 A.M., introduced in the manner of the BBC broadcasts by tapping out the opening notes of Beethoven's Fifth Symphony. We learned who had been taken to Gestapo headquarters in Paris for more interrogation, who had been sent to military court and with what results, who was in the latest transport to Germany, and who had just been taken to Mont Valérien to face the firing squad.

Our communication system also covered military progress and

always included morale-boosting rumors of German retreats and Allied advances. This ritual dose of optimism helped us to keep our spirits up. The daily contact, however indirect, probably helped a lot of us hold on to our sanity.

I hadn't yet come up for trial before the Nazi-dominated French tribunal, but I had already been served with the judicial summary of my crimes. I was accused of sabotage, terrorism, assassination, and high treason. This last charge, the primary one, meant that I had committed espionage in order to deliver France into the hands of her enemies—the Allies.

Since my case was to be tried under French law, I was entitled to legal representation. The attorney appointed to defend me made his extreme distaste for defending a "traitor" very clear. It was the only topic he discussed at our twenty-minute meeting.

He assured me that my case was open-and-shut. I deserved my fate. I was an Aryan who had acted like a despicable Jew and sold out my country for profit. He knew that my accomplices and I wanted the Allies to win the war so that we could grab a postwar fortune from the scavenged holdings of patriots such as himself.

"You will find your postwar reward at the end of a rope," I told him. "Nazi Germany is never going to win this war."

He recorded all of my comments diligently. I would hear them repeated at my trial, being used by the prosecution to prove that I was a hardened criminal.

My days in solitary confinement passed very slowly. Apart from that one meeting with my defense attorney and the daily LIB broadcasts, I had no verbal contact with other people. I eagerly awaited my daily half-hour walk in the prison yard, always trying to find a pretext for delaying the return to my cell. Occasionally I was able to pass a word with my guard or one of the men in my exercise group, but the opportunity was rare and inevitably frustrating. Sometimes the Quakers, whom we called the angels of Fresnes, distributed food and provided a welcome break in the monotony.

To keep myself occupied I tried to catch some of the fleas that infested my cell, with the idea of putting together a flea circus. I never found out whether I had any talent as a trainer because they always escaped before I had a chance to start. At least they afforded me some amusement, which is more than I can say for the roaches, rats, and bedbugs that also infested my cell.

The rats weren't too bad once I managed to overcome my repulsion. Most of the time they just scrambled noisily around in dark corners, holding tactical meetings about invading my box of crackers from the Quakers or occasionally nibbling at my toes. The roaches, although eternally silent, never ceased to disgust me. My skin crawled every time I took a cracker from my box and found a roach crouching underneath it.

I faced my major enemy, the bedbugs, every night. I crushed them, pretending they were my human torturers, but after exhausting the entire SS staff, the Nazi government, and the French collaborators, there were still enough bedbugs to cause me many sleepless nights.

"Traitors" were not allowed mail privileges. I had no way of contacting my family to let them know I was still alive and no way of finding out what had happened to Sonja and François. If my family and my Resistance comrades knew that I was in France, maybe someone would manage to come to my trial. The sight of a familiar face would mean so much. I wanted to send a letter to Mme. Billard under her cover name at the *poste restante* in Nice, but to accomplish this I needed a go-between to get my letters out of the prison.

Gustave, the French guard who brought my meals and took me on my daily walk, had recognized my Hungarian accent. One day he told me that his wife was also Hungarian. In their several prewar visits to Hungary he had liked the people so much that he and his wife actually planned to retire there after the Germans won the war. My having his wife's accent brought pleasure to his walks in the prison yard with me. He began encouraging a relationship of sorts between us.

One day Gustave brought me a piece of home-baked Hungarian strudel. His eyes twinkled as he surreptitiously handed it to me in my cell. As our friendship grew I thought he might understand if I approached him with the worries that plagued me about my family, but I remained hesitant about trusting him until the day he told me that my lawyer had requested a meeting with me, indicating that my trial was near.

"I will be sentenced to death at my trial," I told him, and I asked him to accept the gift of my personal belongings. Then I asked if he would be kind enough to mail a letter to my family and to bring me any answer that I might receive.

I was unprepared for Gustave's reply. "I've been planning to suggest that to you," he said. "You can give me a letter to mail, or tell me your family's address and I'll carry it to them personally."

I didn't want to reveal my family's address. I said it would be easiest for him to mail the letter to the *poste restante.* Gustave gave me his own home address so that any reply could be sent there. The next time he came to my cell he brought me writing materials. The morning afterward I gave him my letter. Gustave mailed it—or so I hoped.

My trial took place in the first week of November 1943. The judges in high hats and flowing robes sat majestically in their lofty seats, intent on proving their loyalty to the Germans by treating me as a traitor to France. The state's attorney recited my sins, concluding that I was a criminal who unquestionably deserved the death sentence.

My lawyer took only fifteen minutes. He said that I had acted in good but misguided faith as to France's real interests. He cited the fact that I was a doctor and pleaded my concern for human life. He read my *Croix de guerre* citation describing the many lives I had saved at the risk of losing my own. Then he threw me on the mercy of the court.

I looked up to gauge the effect of his speech on the judges. Two were engrossed in conversation. The third seemed to be waking up from a nap. After two and a half minutes of deliberation the presiding judge delivered my death sentence.

He set forth the details of my forthcoming execution by firing squad at Mont Valérien and emphasized the court's humanity in permitting me an honorable execution instead of the common criminal's death by guillotine. I had no recourse to appeal. I was entitled only to see a priest for the cleansing of my soul.

I heard the gavel fall. My lawyer said, "It's finished. I can't help you now."

I was numb. I tried to stand up. My muscles did not want to support me. This couldn't be happening to me! I should have stayed alive as a coward rather than die as a hero! My family might never know what had happened to me. I would certainly never see any of them again. My guard Gustave's gloomy face filled my vision. I suddenly felt urgent pressure in my bowels, which were about to respond to my fear as they had when I fell on the antiaircraft gun in Sathonay. Gustave got me to a toilet just in time.

Sitting on the toilet I reached a decision. I would take my fate into my own hands and hang myself right then and there. I looked up for a spot from which I could suspend my belt, which had been returned to me to hold my pants up for the trial. It occurred to me that I had no idea which would truly involve the greater suffering —strangulation by my own hands or being shot by someone else.

I thought of the agonies of patients I had seen slowly suffocating from lung diseases. Hanging myself would definitely be the worst way to die. Anyhow, Gustave would probably hear my strangulated gasps and pull me down before I died, making it no more than a useless gesture. My death would wait for the firing squad. I hoped that the six bullets fired simultaneously would leave me no time to realize that I was dying.

Then I thought, still sitting on the toilet, that if I died by firing squad I would become a national hero. François would be proud to be my son. How proud would he be of a father who had hung himself in the prison toilet? I pulled up my pants and left. Death deserved a more elegant setting.

Gustave tried awkwardly to comfort me by talking about the lousy state the world was in and how difficult life was with the future so grimly uncertain. At that moment I didn't give a damn what shape the world was in so long as I could stay alive in it. He felt the Allied military advances should give me hope. I didn't have the energy to point out that my execution would be over long before the Allies could possibly reach me. The maximum time between sentencing and execution was ten days, the time needed for Pétain to rubber-stamp my sentence. They would not have offered me a priest if I were going to be around for a while.

Back in my cell I faced another decision. Should I die as a Jew or as a Christian? I realized that I was stuck with my Aryan role. If I asked the prison officials for a rabbi, the SS Soldiers would arrest and deport him, or else he would be a collaborator who would try to pry out the whereabouts of my family in order to save his own.

A priest would be just as dangerous. Stool pigeons among Christian men of the cloth were often provincial priests from the lower rungs of the clergy. Their concern was not the protection of their own families but hunting down the Jews.

(Later I learned of many notable exceptions. Monsignor Raymond, for example, the bishop of Nice, hid Jewish families throughout the war. The Reverend Jacques Brunel, a barefoot Carmelite, was sent to Gusen II because he refused to give up the Jewish children hiding in his monastery near Fontainebleau.)

My next problem was notifying my family of the court's decision. I had to write to Mme. Billard and utilize Gustave's services again, even though I had no proof yet that my guard had really mailed my first letter. This time I asked Gustave to put carbon paper in with my writing materials. I intended to send a copy of my letter to Ramoine under his code name.

I had only a few more days to live. I sat with a blank mind in front of a blank piece of paper. I was at a loss to write the very last letter of my life. What could I write that wouldn't sound tragic or frightened? I decided not to address the letter to my mother, who was suffering from cancer and had more than enough sadness to contend with.

Should I write a long letter to my son, apologizing for deserting him and asking him not to feel bitter about my death? I wanted to tell him of my pain in realizing I would never see him grow up. But was there any point in writing of such things? By the time François would be old enough to read my letter, he would no longer remember me. I would not even have a grave for him to visit.

My anguished thoughts turned to Sonja, so beautiful and courageous when I last saw her, comforting me despite her own danger. I was still so worried about her dark eyeglasses. I was terrified the Nazis would smash them, and she would be almost blind without them. As the blank piece of paper continued to stare up at me, I decided to write my last letter to her.

My hand trembled uncontrollably. It was impossible to announce to the person I loved most in this world that I had reached my irrevocable end. Trying to confront the reality of my own death terrified me. And it hurt me very deeply to realize that if Sonja survived and eventually read my letter, I would not be there to comfort her in her grief.

It grew dark. I still had not written a word, not even "Sonja *cherie.*" I could not blot out the image of myself standing in front of the wall at Mont Valérien, staring into the barrels of six German machine guns.

Perhaps if I could place myself in the center of the group of condemned men I would have a few extra seconds of life before the bullets reached me. Or maybe they would run out of bullets. Or maybe the Maquis, the left-wing guerilla arm of the Underground, would attack the jail and liberate us.

I had watched other men leave their cells for the firing squad still clinging to their magical hopes. They were really being transferred to another jail to ease overcrowding here, or they were being secretly exchanged for a captured German officer. Some had believed it so strongly that they exchanged addresses and favorite cafés with their prison buddies so they could meet after the war. Now I was comforting myself with similar fantasies of the miracle that would save me.

Finally I lay down on my cot and fell asleep. When I was awakened by the light from the ever present electric bulb in my cell, it was still dark and silent outside.

I was confronted by my piece of blank paper again. One phrase blotted out everything else, turning around and around in my head like a phrase on a broken record: "This is the end . . . this is the end . . . this is the end . . . this is the end . . . this is the end . . . this is the end. . . . I tried to stop it by walking around my cell, but my trembling legs wouldn't support me.

I lay down again. I poked myself to simulate a bullet's impact. It was a ridiculous exercise that did nothing to calm me. I pinched myself as hard as I could to see if I could tolerate the pain. I was too numb from fright to feel anything or even to remember the pain I had endured during my torture. I came to the realization that my terror and anguish came as much from not knowing what it was like to die as from knowing that my death was so near.

I looked out of my little window at the lightening sky. Footsteps sounded outside my door. Perhaps I had fallen asleep without realizing it and continued to struggle with my terror in my dreams. I looked at the sheet of paper still unblemished by a single word.

My guard entered with breakfast—black coffee made from roasted chickpeas and a piece of heavy black bread that tasted like glue. Gustave looked around carefully to make sure no one was watching. Then he closed my cell door and approached with his finger to his lips in a gesture of silence. He pulled a letter from his pocket—a letter from Mme. Billard! It was still sealed, proof of his discretion and respect for my privacy. His features expressed compassion and pleasure as he watched my trembling hands open the letter.

Mme. Billard spoke of my comrades' pride in me. They had heard the details of my torture and wondered how I had endured it without betraying them.

François was safe with my mother. Sonja had been slapped around a little but not tortured. She and her sister, Paule, had both stuck to the story that they were Jews trying to hide. They were finally believed because, less than a week later, their mother had been informed on and and arrested as a Jew. The three of them had first been sent to the Hôtel Excelsior in Nice, a regional holding center for Jews. Then they were sent up to the holding camp at Drancy, near Paris, where they were at this moment awaiting deportation to Auschwitz.

I read the name Auschwitz, but it made no special impression on me. I assumed, like Mme. Billard, that it was a Jewish labor camp. Sonja would be okay there. She was strong. Hard work would not bother her. The great thing was that she was alive and safe.

Mme. Billard's letter gave me such an immeasurable shot of optimism that I not only pulled myself together but practically convinced myself that I would be liberated in just a few short days. My son was safe. My wife had escaped torture and was alive and well with her family in a camp right near Paris—no less than a miracle. And if a miracle had happened once, perhaps there was another one waiting for me.

The morning LIB broadcast interrupted my thoughts. I listened to talk of a rumor that the Germans had just stopped executing Resistance prisoners. Instead they were going to send us to forced labor camps in Germany. None of the scheduled executions for the past week had taken place. My miracle!

An optimistic letter was easy to conceive but still painful to write because of the physical aftermath of my interrogation in the Villa Trianon. My thumbs were very weak and sore, and my shoulders were still swollen and extremely painful. The multitude of facial bruises and swellings was gradually disappearing, but my broken front teeth produced a whistling sound whenever I breathed through my mouth or talked. At least I had been able to lessen the pain from my split toenails by cutting my shoes into open-toed sandals. The physical pain, combined with my intense emotions, made the act of writing extremely difficult. I took great care to make my words legible.

As I was writing, an unscheduled LIB broadcast reported that twelve prisoners had just been taken away for execution in reprisal for the assassination of a group of German soldiers in Paris. Five of these men belonged to the group in which I was taken to the courtyard for my daily walk, men I had been able to speak a word with from time to time. The executions were fact, not rumor. I stopped writing my letter.

Later that day, after word spread that the executions had taken place, a monk among us, Père Bonaventure, asked us to pray for the slain men. The Hebrew words of a childhood bedtime prayer came into my head: "*Shema Yisroël adonai elohenu adonai echod.*"

I was not a religious man, but I was weeping as I uttered the prayer over and over and over again. I did not know whether I was crying for the twelve dead men or for myself.

Gradually I took hold of myself. I finished my letter to Sonja. I tried to write honestly and directly of what was going to happen to me, what I felt for her, and my boundless gratitude for the lifetime of riches she had already given me. I hoped for the strength to bear my execution with courage, and I concluded: "I am so happy you are alive! I am sure that your strength and intelligence will help you find a way out and hold on to your faith in the Allies. The war will soon be over, and you will return to France and be reunited with our son."

It wasn't exactly what I had wanted to say, but at least I had said something. I closed it with tears I couldn't hold back. I would never see François and Sonja again, never hold them in my arms.

I sealed the letter, planning to give it to my guard in the morning. I lay down with it placed safely under my pillow, trying to recall as much as I could of the short time Sonja and I had spent together. I tried to recapture our lunches and long weekends in Antwerp, our hurried marriage ceremony, our submarine trips to London, awaiting the birth of our child, then looking at our son for the first time.

What a scene it had been when Sonja told me she was pregnant! We were in our last parachuting of weapons to the major Resistance center at Vercors. As our jump approached she promised to tell me something very important once we were safely in hiding for the night. After an uneventful jump and weapons pickup, the

Maquis chief took us to the mayor of nearby Les Abrets, who hid us in a primitive shack in his cornfield.

We were lying side by side on a narrow cot when Sonja asked if I recalled the promise she had made. Her voice was warm and soft. Her eyes glistened in the light from our small lamp. She laughed and whispered to me, "You know, we just made history!" I was puzzled. We hadn't done anything new on this mission.

Then Sonja said, "We didn't only parachute weapons today. We parachuted our future baby!"

I was thunderstruck. She had risked her pregnancy to make her last parachute jump with me. What a chance she had taken—but that was Sonja!

"Why didn't you tell me?" I asked. She replied, "You are strong when it comes to yourself, but you fall apart when anything out of the ordinary happens to anyone close to you. You could have blown your mission over Innsbruck, and our jump today."

I took off her dark glasses and kissed her eyelids. My wife was now three months pregnant, and I was going to be a father!

The sun was shining through my cell window. Footsteps rang in the hall. Gustave was bringing breakfast. I took the letter from under my pillow and hid it between my jacket and shirt so that I could slip it to him without being noticed by the inmates who helped in distributing the morning meal.

My cell door opened. I looked up and froze with an unfinished smile on my face. This was not Gustave's day off . . . why was his relief guard here? I was never happy to see this unpleasant old Pétainist even on his appointed day. His unexpected presence today frightened me. Had Gustave been caught helping another prisoner and given me away? This old man had been wounded at Verdun during World War I and decorated personally by Pétain who, to him, still remained the infallibly wise military hero. He had enthusiastically supported Pétain's submission to the Germans. He was convinced that Germany, out of respect for Pétain, sincerely wanted France to become a great power again. Ergo, *we* were France's real enemies.

The old man came in whistling cheerfully. I asked him what he was so happy about. He told me that a great Resistance network had been wiped out. Vercors, an Underground nerve center, was no more. It had been a training camp, a hiding place, a major administrative headquarters for the Maquis, and the operational center for their industrial sabotage activities. Sonja and I had gotten to know the area and the people during our weapons parachuting missions there.

Triumphantly, the old man told me that members of the fascist French militia had infiltrated the Maquis. Their information enabled the Germans to carry out a land-and-air assault. The Maquis had been taken by surprise. Gustave, my friendly guard, had been sent with the detachment to bring the few survivors to Fresnes. He was expected to return with them in a few days.

I took small comfort in the thought that a lot of German lives must also have been taken. I wondered if any of my friends had survived, and what had happened to the Americans who had trained the Maquis to use the weapons we dropped.

The next day, Sunday, took an eternity to pass. Monday morning the footsteps outside my cell preceded Gustave's familiar, but very sad, face. He had just returned from Vercors. His face told the story.

I gave him my letter to Sonja. He told me to tear it up. I was leaving Fresnes right after breakfast! The blood drained from my face. He handed me a list of my belongings for my signature certifying their return to me. I was trembling in fear. He tried to reassure me. "You are not going to be executed! Some of the older prisoners have to be transferred to another jail to make room for all the new ones coming in. That's all!"

I tried to compose myself. I thanked Gustave for inventing a lie to calm me down, but I knew very well that my end had come. My one hope was to get it over with quickly. Gustave kept repeating that I had no cause for worry. I didn't believe him. How easy it was for him to say! I asked him to hold my letter to Sonja until after my death and to do me one last favor. He was to include a note describing my execution but omitting any mention of cowardly behavior. I gave him my belongings, including my wristwatch and several hundred francs. He shook my hand slowly. (When I returned to Paris after my liberation from the concentration camp and went to the office of the Free French army to take care of my papers, I was given a package that had been brought in with my name on it shortly after the German surrender. The description of the man who left it matched that of my former guard, Gustave, although the note in the package was signed with another name.

In the package I found my watch and 297 francs. The note explained that Dr. Albert Haas had given these items to the writer before his transfer from Fresnes. I tried everything I could think of to locate him, even hiring a private investigator, but he had either signed the note with a pseudonym or had already moved to Hungary with his wife.)

As I dressed I realized that my solitary bug-infested prison cell had really been a shelter for me. Here, at least for a short while, I had been safe. Convinced as I was that I was embarking on my final trip in life, I poignantly regretted leaving these familiar and comforting surroundings. I bid an emotional farewell to the bare walls, the open toilet, the fleas, the rats, the roaches, and the bedbugs. Before leaving, Gustave embraced me, tears in his eyes.

Then I was handed over to the Germans.

8

In the central yard three waiting trucks were surrounded by SS guards. Two were closed and ready to go. My arms were handcuffed behind me and I was pushed up into the third truck. I found a place to sit on one of the benches. The men already seated in the semidarkness looked familiar. I must have seen them in my cell block or during my daily walk. Other prisoners still lined up outside were shoved in with us, making thirty in all. Two armed SS guards joined us, and the doors were closed.

I heard the trucks and motorcycles start up, then we began to move. In the murky shadows it was impossible to identify my companions. Everyone was staring sadly at the floor. A prisoner who seemed ill asked for a glass of water and received a whack with a gun barrel that bloodied his forehead. His handcuffs prevented him from wiping his blood out of his eyes.

The SS guards posted by the doors prevented verbal communication and my handcuffs prevented hand signals. I spent my time listening to the noises filtering in from the outside, trying to figure out the direction in which we were heading. I could hear voices and sounds from cars driving alongside us. Our frequent stops had to be for traffic lights, so we were going through a string of localities. Then I heard airplane noises and realized that we must be near Orly.

Our stops became more and more frequent. I felt instinctively that we were passing through Paris. The word "Italie" emerged from the guards' momentary conversation. We must have entered Paris by the Porte d'Italie. Then the converging sounds of an increasing number of automobiles were followed by the hollow sounds of wheels passing over a bridge, and I realized that we must be driving over the Seine. But the route to Mont Valérien and the firing squad did not go through Paris. I felt a tentative ray of optimism, the first I had experienced since I found out that I was leaving Fresnes.

About an hour or so into our journey I smelled something reminiscent of smoke from a locomotive. Not long afterward I heard the sounds of a train. We stopped for a few minutes. I heard orders being given in French and German and the sounds of a heavy metal gate opening. The truck started moving again, slowly. About fifteen minutes later a German voice shouted "*Halt,*" and we came to a dead standstill. Again I heard a shrill mixture of French and German, trains moving slowly and coupling, whistling, then the puffing of a locomotive. The SS guards had exchanged very few words during the trip. Since the van stopped they had remained silent and motionless except to put out their cigarettes. They appeared to be under strict orders to keep us ignorant of where we were going.

I continued to listen for informative sounds. A scraping kind of noise brought to mind the image of the heavy door of a freight train sliding open. A voice yelled, "*Aufmachen*" ("Open up"), and keys rattled in the lock of the truck's rear door. Blinding sunlight burst into the dark interior and we could hardly see. The SS guards used their gun butts to help us climb out. A number of men who couldn't keep their balance with their hands cuffed behind their backs fell to the ground, some twisting an ankle or breaking a leg. I jumped forward to help a man of about forty who wore a monk's habit. As I turned my back to him and knelt to reach out my hands an agonizing pain shot through my shoulder and head. An SS guard was beating me. He didn't stop until I stood up. Fortunately, I had been able to help the clergyman to his feet. His eyes expressed gratitude. An SS guard jammed the sole of his boot into the small of my back and pushed me forward. The three other prisoners who had stepped forward with me were also bloodied in their attempts to help our comrades.

I looked around me and recognized my surroundings. We had been taken to the Gare du Nord, the Paris railroad station that served all trains traveling north. A station clock showed the time to be 3:00 P.M. I was stunned that a fairly short trip had taken six or seven hours.

There were many more than our three vans from Fresnes, perhaps fifteen or twenty in all, indicating that prisoners had been brought up from several prisons. Many vans stood open and empty, having discharged their human cargo. About one hundred armed SS guards were ranged the length of a freight train standing in the station.

My group was led to an open freight car. Rough pushes and punches accompanied the commands of "*Schnell, schnell!*" ("Hurry, hurry!"). Those who had difficulty climbing in were assisted with angry kicks and the whack of a gun butt. When our car held around forty or fifty men, the command came to close the sliding door. We found ourselves back in darkness, our sole source of light a useless air vent in the ceiling.

Someone outside shouted that the car wouldn't be opened for a while. Anyone needing to urinate or to have a bowel movement would have to use the garbage can that had been put in the car as a toilet. We hadn't eaten since our early breakfast, but we were too anxious to care about food. It was difficult to stand or sit comfortably. We had been packed in like sardines with our hands still handcuffed.

The coupling of freight cars jolted us violently. We all listened intently. The locomotive switched into forward. Its whistle split the air. The train began to move. In the next instant the silent bodies packed into the freight car came to life with an overwhelming urge to talk. Everything held back in solitary confinement for so long exploded in a counterpoint of voices describing wives, children, parents, arrest, and torture. The unassailable barriers that normally underscore our separateness dissolved. The realization that this train was bearing each of us to the same uncertain future brought us so closely together that we seemed to share the same heartbeat. I have never, before or since, experienced such an instant, complete, and all-embracing sense of community.

One by one my fellow passengers resumed their silence. They seemed drained by the intensity of their first chance to talk. The train rolled on, the blackness in which we sat relieved by the occasional light that dimly illuminated the interior of our car.

As our eyes grew accustomed to the dark I noticed the clergyman I had helped sitting near me. He returned my sign of recognition with his sad eyes and laboriously worked his way to my side to express his gratitude more fully. Then he asked, "Aren't you the '*toubib*' ("sawbones") from Solitary R16?" The clergyman was Père

Bonaventure, the Dominican monk who had been my neighbor. He told me his story as we sat in the darkness of the sealed freight car.

He had joined the Resistance, hiding weapons and wounded fighters from the Maquis in the monastery. When the head of his local network was caught he assumed leadership. An arrested comrade revealed under torture that his chief was a monk. The Germans immediately ordered the abbot to give him up, threatening to incinerate the monastery and everyone in it if he weren't in their possession within twelve hours.

The abbot spoke with Père Bonaventure, revealing that he, too, was in the Underground. The monastery housed several Underground workers and sheltered many refugees and wounded. The monk was able to convince his superior that the survival of the Resistance movement itself was of far greater importance than his personal survival. So Père Bonaventure gave himself up to the Germans, who then randomly selected five monks to aid them in searching the monastery for refugees. Fortunately, they discovered nothing.

Père Bonaventure had withstood the most barbaric torture, including having acid injected into his urethra. The Germans finally executed the five innocent monks who, in the moments before their death, spoke in Latin to their colleagues of their contentment in dying for a just and moral cause. Père Bonaventure was thrown into a temporary cell and beaten daily as he continued to maintain his innocence. When the marks of his torture finally disappeared he was transferred to Fresnes. A mock trial resulted in a death sentence that was automatically sanctified by the Vichy government. The rest of his story, of course, I knew.

After I shared my story with him, I talked about my terror of the moment just before being shot. My fears of flying were the same. I was not afraid of crashing but of those minutes just before as I fell helplessly through the sky. Père Bonaventure's advice was not to worry about the terror of those few minutes. After my death I certainly wouldn't remember them!

Then he gave me some information that temporarily lifted my obsession with death. Before we left Fresnes, Père Bonaventure had called in the prison chaplain to hear his confession before he was executed. The chaplain agreed to do it although he felt it was unnecessary. Reliable sources had told him that no more of the condemned Resistance fighters were to be executed. Hostages and known Communists were the only ones whose sentences would still be carried out. The rest of us were being taken to Compiègne-Royallieu, a French military base midway between Paris and the Belgian border that was now the German holding center for men awaiting deportation to the concentration camps. Now it was obvious why we had been taken to the Gare du Nord.

I became absorbed in my own thoughts, unaware of time passing until the rhythm of the wheels began to change as the train slowed and switched tracks. Perhaps we were approaching a station.

The train continued to slow to a standstill. Outside men shouted in the now familiar mixture of German and French, then our door slid open. The station sign confirmed our arrival in Compiègne. The hands on the station clock stood at 11:30.

The armed SS guards ordered us to jump down. I stood on the platform under a cold, fine rain, the weather making the residue of my torture unusually painful. My hands were numb from the handcuffs. I had terrible hunger pains and a pounding headache. Fifteen or so SS guards held machine guns on four or five hundred prisoners lined up in the rain by the railroad tracks. As we were all ordered to line up by fours, armed German soldiers and SS guards spread themselves along the length of our column. An officer heading the column shouted the order to march.

The dreary night was dark and cold. The icy rain showed no sign of abating. The few dim lights of Compiègne barely outlined the wet, deserted street in front of us. The entire populace was behind locked doors, the wooden shutters in all the windows so tightly closed that not a crack of light escaped. The buildings were like so many giants with closed eyes, as if they were ashamed to acknowledge our suffering. A phrase from one of the poems of Paul Verlaine passed through my mind: "*Il pleure dans mon coeur comme il pleut sur la ville.*" ("Tears fall in my heart as rain falls on the town.")

I marched in a daze, an automaton waiting to be awakened from a bad dream. My makeshift sandals and blood-caked socks were soaked through from the rain. My toes throbbed. Our overcoats and jackets became sodden, the dampness penetrating to the bone. Each time I stumbled Père Bonaventure, walking beside me, helped me keep my balance. I felt myself to be beyond endurance.

Hopelessly overwhelmed by the futility of my efforts to survive, I announced my sudden resolution to sit down and refuse to budge. Whatever would happen, would happen. Père Bonaventure insisted that the SS guards would murder me there in the street as an "escaping" prisoner. God's hand had spared me from the firing squad, therefore I had been saved for a purpose. I didn't have the right to squander my life for the pleasure of these assassins. He exhorted me to think of my family. He supported me with his words, his voice, his touch. I kept on walking.

Strange lights appeared on the horizon, looking much the way I have always imagined the aurora borealis to be. They increased in intensity as we marched. About twenty minutes more brought us to Royallieu, a military base located on a broad plateau, surrounded by extensive grounds. The lights I had seen were high-powered searchlights encircling the complex of buildings, their restlessly moving columns of light scrutinizing the night sky like giant arms trying to catch an invisible fly.

At the gate we were ordered to halt. An SS guard walked up to the entrance. Several silent minutes later the gate swung open. We marched into the armory, directed toward one of a number of low white buildings standing in a neat line. We halted while an SS guard unlocked the door. We entered a long, dark corridor. The guard turned on a bright flashlight and led us to another locked door. We waited while it was opened, and then were herded into an unlit, unheated room. The door was locked behind us.

One of the searchlights from outside lit up the large room periodically. Each sweep disclosed three straw pallets piled in a corner. The returning darkness engulfed us so quickly that we succeeded only in stumbling into one another as we all tried to reach them. Someone stepped on my throbbing toes, and I screamed.

Père Bonaventure called out for everyone to stand still. He suggested that three men go to the corner during the next sweep of the searchlight and, when it returned again, carry one pallet to the opposite side of the room. Three more would take the second straw pallet to the right-hand wall on the following two tours of the searchlight, then another three would take the last pallet to the left-hand wall. Once this was accomplished, we realized there were no blankets. We tried to get the attention of whoever might be in the hallway. The guard who came to see what our shouting and door-pounding was all about informed us that we had to do without blankets until the central supply stores opened.

With Père Bonaventure's guidance we spread the straw out to give everyone a modicum of protection from the damp, cold floor. It was two or three in the morning by then. We still hadn't been given anything to eat or drink. I was numb, too drained and exhausted to care about my hunger or the pain in my big toes or my rain-soaked clothes. I dropped to the straw-covered floor and immediately fell asleep.

My sleep was torn by a vivid nightmare of torture and struggle. When I woke up, my head was pounding. Someone was sponging my forehead with water. I sensed people gathered around me. My eyelids were too heavy to open. I felt myself burning up and then freezing. When I finally opened my eyes a disembodied face was bending over me through a haze, speaking with an urgency I couldn't associate with myself.

"He has a very high fever. . . . We have to get a doctor!"

The next time I was aware of my surroundings I felt pleasantly warm and dry. I was lying in a bed with clean sheets, covered with a military blanket. Panic quickly replaced my sense of security when I discovered that I couldn't move my arms or legs. Anxiously I lifted my head. My arms and legs were tied down. I realized that I had been very ill, but I could not remember becoming sick.

The man in the bed next to me shouted, "Nurse, he's moving! He opened his eyes!" A voice answered, "I'll call the doctor." This was followed shortly by the appearance of a white-coated, pleasant-looking man in his fifties.

"I am prisoner-physician Dr. Reiner, *mon cher confrère.* I must admit that I never expected you to recover from your pneumonia."

I was totally bewildered. I was in Royallieu, a German holding camp for political prisoners. The day after my arrival I had been brought from the quarantine barrack into the infirmary delirious with pneumonia. I had now been in the infirmary for ten days and had finally reached and passed my crisis the previous night.

My nose began to itch while Dr. Reiner was talking. I tried to scratch it on my pillow since I couldn't use my hands. Dr. Reiner cut the gauze that was restraining me. He apologized for having had to tie me down and told me that I had jumped out of my bed and tried to run away to my family at every opportunity. It had taken two male nurses to subdue me.

I scratched my nose, then touched the contours of my face. Dr. Reiner brought me a mirror. I saw a gaunt stranger—a bearded middle-aged man whose suffering eyes were underlined with deep shadows. The last time I had seen my face was my last morning in Nice as a free man.

Dr. Reiner told me that I almost hadn't gotten to the infirmary in time. There was no provision for emergency admissions. He had the authority to treat only outpatients right away. Everyone else who required medical treatment had to wait until their infirmary admission was approved in person by the German military doctor when he made one of his occasional visits to the camp.

Fortunately for me, a German soldier had shot himself in the foot that afternoon while he was cleaning his gun. Dr. Reiner had called the military doctor and used the opportunity to get me admitted to the infirmary. By then I was no longer conscious. Père Bonaventure had carried me there on his back.

Each day Père Bonaventure had sneaked out of the quarantine barrack to visit me. When I showed no improvement after several days, he had given me the last rites—just in case. Although he didn't know my religion, the good monk had decided that it certainly wouldn't hurt me, and at least he could recommend that God treat me well when I arrived.

When the news spread that I was out of danger, the joy of my fellow prisoners touched me beyond words. Since there was full freedom of movement in the camp except for newly arrived prisoners confined to the quarantine barracks, I had a constant stream of visitors. Friends and strangers from all over—Frenchmen, Americans, Englishmen, South Americans, Jews, Communists—came to see me with gifts of precious food sent by relatives or Quakers.

Dr. Reiner, formerly a lung specialist at the Vienna Medical School, was a kind and conscientious man. He brought me real coffee and milk every morning and used whatever pretexts he could find to keep me in the infirmary.

As soon as I was able to get around, I used my relative freedom to familiarize myself with the Royallieu camp and search out an escape possibility. I kept out of the way of Sergeant Kuntz, the dried-up weasel who actually ran the camp. Royallieu's commander preferred the comforts of town and willingly handed his camp over to Kuntz' slow-witted, sadistic ministrations. I learned that it was this man, formerly a spy disguised as a Paris café waiter, who had refused Dr. Reiner permission to call the German doctor about my admission to the infirmary.

Royallieu was actually three detention camps separated by barbed wire and watchtowers. Camp One was for Jewish prisoners, Camp Two for French political prisoners, and Camp Three for non-French nationals. I was in Camp Two, the most heavily guarded of the three, encircled by a chain of watchtowers with armed guards and sentinels patrolling the areas in between. Floodlights played over us continuously at night. Escape looked impossible.

Visiting the different barracks, I got to know people from every walk of life—professionals, artists, shopkeepers, laborers, members of government—and made many wonderful friends. Our initial enthusiastic attempts at planning an escape waned as we realized the futility. We turned our efforts to organizing evening entertainment programs to help us pass the time. We experienced a profound and genuine sense of fraternity and equality despite our lack of liberty . . . or perhaps because of it.

On the Sunday morning after my crisis, Père Bonaventure brought me the news that our group was to be deported to Germany the next day. We embraced, exchanging addresses and promising to meet again when this madness finally ended. We were cheerful and optimistic as we speculated on what Germany held in store for us.

When I saw Dr. Reiner later that morning I sensed that he was upset. He had been ordered to discharge me. I was to leave on the Monday morning transport. He gave me a small bottle of vitamin pills from his pocket and told me he was sure the strength that had pulled me through my bout with pneumonia would sustain me through whatever might happen. At that moment a German soldier arrived with a list. The four names he called off included my own.

The four of us returned to our barrack and were taken to a large bare room. About thirty people sat on the straw-covered floor. Most of them had blankets, brought by family or friends or obtained through barter. Suitcases were piled in a corner.

The room chief greeted us. "You are the last ones. . . . Now our car is complete!" I sat down in a corner. Someone brought me a blanket and a letter, both from Père Bonaventure. The letter explained that the was giving me one of the two blankets he had gotten from the outside. I could worry about returning it when the war was over.

Shortly before 11:00 A.M. a German soldier came in and selected four men to fetch our food—a tasteless soup, a French-like bread, and ersatz coffee. The soup and coffee were stone cold. I didn't have a bowl of my own. I had to wait for someone else to finish his dinner and loan me his.

After my meal I sat down to read the newspaper I had with me. A corpulent blond man in his forties sat down next to me and introduced himself before I could get past the headlines. I couldn't believe my eyes! In front of me was Dr. François Ottinger, one of the most brilliant surgeons and teachers in Paris and the chief of surgery at a famous teaching hospital, a professor of the old school who was known for his biting wit. I had attended his clinical rounds and a few of his operations during my own residency, marveling at the agility of his sausagelike fingers.

Dr. Ottinger told me that he had been denounced as a Jew who would not wear the yellow star. He had been arrested and beaten. He was not Jewish, having descended from one of the great Huguenot families. He had expected to be released as soon as he provided documentation of his Protestant descent. Instead, because of anti-German remarks he made when he was being beaten, he was declared an enemy of the Third Reich and sentenced to be "reeducated" in Germany.

Another physician joined us. Dr. Aarons was a Jewish ophthalmologist who, when the dragnet for Jews began, had joined the Maquis and moved to a small village where he lived as a leather tanner. This sprightly little white-haired man was imprisoned when the Germans arrested the male inhabitants of his community and burned the village to the ground.

The others in the room gradually came to our corner and introduced themselves. As we talked on, we discovered many mutual friends in the outside world. At five o'clock our conversations were interrupted by a German soldier who selected four men to get dinner, a repetition of the earlier meal with the addition of something resembling sausage. After restoring the soup kettle to the kitchen they returned with the news that the guard around our barrack had been heavily reinforced and all doors were locked tightly.

We speculated about the worst our future could possibly hold. None of our fantasies came remotely close to the reality awaiting us.

I awoke the next morning to a noisy commotion. A guard was shouting, "*Aufstehen und packen*" ("Get up and pack your things"). Dr. Reiner brought me a farewell package of analgesics, a few sleeping pills, a bottle of disinfectant, gauze and cotton, a roll of surgical tape, and Band Aids. We weren't allowed to carry sharp objects, so I asked how I would use the bandaging materials without scissors. "With your teeth," was the reply. Before he left me we embraced. (After the war I learned that Dr. Reiner had been in one of the last prisoner transports when the Germans retreated from France. I tried to find him when I worked in Vienna's French zone, but he had never returned. The Documentation Center in Arolson, Switzerland, informed me that he had never been registered with them. Therefore he had never officially entered a concentration or labor camp. He was probably killed during the transport or before he was registered at whatever camp he was sent to. He was a witness the Germans did not want to have around.)

I rolled the first aid supplies in Père Bonaventure's blanket. In the meantime our food had arrived, two loaves of bread and half a sausage each to last us for our entire trip, and ersatz black coffee for our bon voyage breakfast. In my blanket were the jam, chocolate, and dried biscuits that remained from my get-well presents. I unrolled my blanket and added the bread and sausage, removing one dried biscuit to munch with my coffee. I tore off a long strip of gauze and tied my blanket into a neat package. I was ready to leave.

With the order to "*abtreten*" we marched outside and lined up under a predawn sky that was still dark and strewn with stars. It was going to be one of those crisp, invigorating early winter days. We were counted and lined up four across.

"Forward, march!"

The floodlights were still on as we passed through the gate under the eyes of an SS guard and four Gestapo officers standing by a black Citroen. A group of soldiers with fixed bayonets, our escorts, lined up along either side of our column. We retraced our route to the station at Compiègne, passing through the little town that was still deserted and only dimly lit.

By the time we arrived at the station the sun had climbed into the sky, bathing us in a soft, golden light. My group was halted next to a cattle car with a large "D" on the side. I assumed that it stood for *Deutschland.* Had I known that, in fact, it stood for *Dachau,* it would have meant nothing to me.

We were herded in and given a garbage can to use as a toilet and a kettle of water for drinking. Then the door slid closed with a resounding metallic boom, a sound that still strikes fear into my heart today.

The last thing I saw of Pétain's *belle* France was the sunlit blue sky. The little window vents at each corner of the ceiling were nailed shut, leaving us in total darkness. We heard our door being locked, then "*Abtreten,*" then silence broken only by the whistle of an occasional passing train. Dr. Ottinger suggested that we each hold hands with the two people nearest us and exchange names so that we could feel the closeness of another human being.

Someone in our car began to sing "La Marseillaise." We all joined in, then hundreds of voices from the other cars swelled our chorus and drowned out all other sounds. We were hundreds of men momentarily touching in the inky darkness. The sudden silence and the feeling of loneliness that emerged when the song ended were unendurable. One by one the men around me began to sob. Tears ran down my cheeks. Even after the sobbing finally calmed down, no one uttered a word.

We heard a locomotive approach and felt the impact of coupling. Our train began to move. We were on our way.

The sounds of the wheels on the tracks gave the impression of substantial speed. We made brief periodic stops which, according to Dr. Aarons, were probably to allow military trains to pass, the only ones that had priority over the transport of human cargo. Our stopovers were reduced to a minimum to prevent opportunities for escape. Dr. Aarons described an unsuccessful attempt of his Maquis group to liberate a transport train while it was waiting for a military train to pass. The failure had cost the lives of some of his comrades as well as a number of prisoners. His story doused my last flicker of hope for liberation.

A young man among us, a real *titi Parisien,* a street hustler who claimed to have been arrested for distributing anti-German leaflets, announced that he had smuggled a knife on board. If someone helped him he could remove the board covering one of the window vents. Standing on another man's shoulders, he succeeded in prying off part of one shutter. It was dark outside.

Someone suggested that we each give our name and describe our Resistance role, but Dr. Ottinger hurriedly warned us of the Gestapo tactic of planting a stool pigeon in a transport train to learn what they could about Underground activities and workers. His warning soon turned out to be accurate.

When we stopped at Bar-le-Duc to empty our toilet and get fresh water, an SS guard approached our car and requested the man with the knife to come forward. The young man was silent. We were ordered out of the car, but a search of the prisoners yielded nothing. After a soldier found the knife in a corner of our freight car the SS guards, without a single word having been exchanged, went directly to the young man and took him away. We never saw him again, and we never found the stool pigeon who had denounced him.

The train resumed its journey. The next time we stopped we heard trains passing and German and French being spoken, and artificial light filtered through the partially opened vent. One man climbed onto the back of another and managed to enlarge it by kicking away the rest of the shutter. He peered out and saw that we were in Metz, at the border between France and Germany.

When the train started up again it picked up speed. The rhythm of the wheels on the rails finally put me to sleep, and the next break in the rhythm woke me up. We were pulling into the German city of Saarbrücken, where we emptied our toilet pail and renewed our water supply. Then we were sealed in our car, and the train moved on into the heart of Germany.

We were moving more slowly and stopping more often. Our tiny opening gave us a glimpse of our daytime surroundings and enabled us to differentiate between day and night. Each morning we stopped and the door opened to reveal the countryside. We jumped out to empty our toilet, but we weren't always given fresh water. After a seemingly interminable four or five days we arrived in a large railroad station. We were in Munich. An hour or so after uncoupling from our engine, air-raid sirens began wailing. When the raid was over they shrieked an "all clear." When the air around us was quiet again we were coupled to another locomotive and resumed our journey.

We halted again only fifteen or twenty minutes later. Someone peeked out and saw that we were still in Munich, but in a small railroad station. We sat for an hour or two before coupling to yet another locomotive. Its shrill whistle announced our departure. Before long we heard the brakes, and the train came to a rest. Our door slid open roughly and the air around us exploded with angry German voices shouting, "Get out! Leave everything in the car!"

Framed in the doorway were two SS guards and a group that was both frightening and comical, a pathetic knot of severely emaciated men wearing dirty striped uniforms and little matching caps. They were led by a robust man in a clean, well-tailored version of the same uniform. He had a black armband on which I could read the word KAPO. In his right hand he wielded a rubber hose (a *Schlag*) with which he "helped" prisoners hasten their descent from the train.

I saw a large sign on a pole—"KONZENTRATIONSLAGER DACHAU."

The date was November 1943.

ESCAPE FROM DACHAU

II

9

I had just slipped two pieces of chocolate into my pocket when one of the emaciated men who had climbed into the freight car came over to me. "Do you have a watch and any money on you? If you do I'll hold them for you and return them to you later. In a few minutes you're going to be stripped of everything you have, including the suit on your back."

I answered that I had nothing of value with me. Angrily, he kicked me in the rear. My attempt to answer in kind was cut off by a sudden sharp pain in my head. I turned just in time to catch sight of the *Kapo* and his rubber hose as he hit me across my face and shouted, "*Schnell, schnell!*" He kicked and pushed me to the door, and as I fell off the train he yelled after me, "You French pig!" I looked up and noticed a number and a green triangle on his jacket pocket, and a face above them that was devoid of expression.

Once off the train we were lined up and "encouraged"—with fists and curses—to march to a huge hangarlike building filled with rows of benches. Three men in striped uniforms holding barbering tools stood by each bench. We were ordered to undress, leave our clothes neatly folded, and find a spot to sit.

I was still wearing the clothes I had put on the day I was arrested. They were filthy and smelled terrible now, but they were intimate companions from my past life. I clung to the solace they gave me. Losing them now, in this insane and alien place, I felt desolate and abandoned.

As soon as I sat down on a bench a "barber" chopped off my hair, shaved my armpits, my pubic hair, and a strip down the center of my scalp. The second man looked into my mouth for gold teeth. A third examined my anus for hidden valuables. With a kick in the rear, he sent me into a shower room *("Brausebad")* along with the others who had finished this unpleasant routine.

Using the tiny piece of soap we had been given, we washed under water that alternated between ice cold and steaming hot. Still dripping, we were told to dress in our new clothes: a navy blue shirt with thin white stripes, broadly striped pants, jacket and cap, and clumsy wooden shoes. Our "new" clothes were not clean.

We left the bathhouse in our damp uniforms, lined up outside in the cold, and were herded into another bench-filled hangar. Gestapo officials were calling out names and conducting brief interviews. I sat down. I wished for the feel of my own clothes on my skin.

After an hour or so I heard my own named called. I went up to the interviewing table to be asked if Haas was my real name, who should be notified in the event of my death, and what my profession was. A number was assigned to me and inked across the skin on my chest. Then I was sent to join a group of prisoners that included Dr. Ottinger and Dr. Aarons. I also recognized a railroad worker, a journalist from the Communist daily *L'Humanité,* and a young Communist party functionary. I had met them at Compiègne.

We had been standing there for quite some time when a man in a clean, well-fitting uniform came over to count us. I noticed a blue armband with *Lagerschreiber* ("camp secretary") written in white. With the puzzling words "*Hier sind ihre Hasen*" ("Here are your little rabbits"), he handed us over to another nicely uniformed man whose armband bore the title *Blockältester.* His facial structure and shaven skull brought a gorilla sharply to my mind.

The gorilla led us to a crude white-painted building that was much smaller than the hangars we had been in. It was about seventy-five by thirty feet and was poorly heated by an old-fashioned potbelly stove. The barrack had two doors, the one we were entering through and one in the northern wall just opposite that was closed and locked. About thirty men were already installed in some of the roughly built, multitiered bunks lining the two long sides of the barrack. We were lined up to choose our bunks. The gorilla went into a private room at one end of the barrack.

I chose the bunk above Dr. Ottinger, and Dr. Aarons took an upper bunk very near me, right above the railroad worker. Our beds were no more than bare straw-stuffed sacks—a large one for the mattress, a smaller one for the pillow—but we had two blankets apiece. One of mine was marked with a large reddish-brown stain.

Hunger pains began to plague me, and I wished my chocolate wasn't sitting in the pocket of a jacket that no longer belonged to me.

The Blockältester, accompanied by two adolescent boys, re-emerged from his private quarters. "Welcome to the Hotel Altona," he told us, ironically referring to a famous Berlin luxury hotel. He said that we were in temporary quarantine. We would be out of here, dead or alive, quite soon. We had to obey him blindly. Any orders from the two *Stubendiensts,* his assistants, were to be treated as his own. Every man had to be in bed by 8:30 P.M. Anyone caught out of bed after the lights were out without a justifiable reason would be immediately and severely punished.

I stared at the Blockältester's big shaved head as he talked to us. His ears were tiny and his chin receded. I had never seen anyone whose head was actually as round and smooth as a bowling ball. Behind his back the other prisoners called him *Kugelkopf* ("bowl head"). He was so ugly!

He asked for six men to carry our dinner back from the kitchen. I volunteered in the hope that my command of German would enable me to gather some additional information about our circumstances. A futile hope. The two Stubendiensts accompanied us and commanded total silence.

We carried soup from the kitchen in a wooden barrel and bread in a large sack. Two tubs held ersatz margarine made out of coal and some artificial honey of indeterminate origin. Black coffee was ladled from a small barrel. The coffee was so terrible that my previous definition of an undrinkable cup of coffee was instantly meaningless. I invented a silly riddle for my friends. "This coffee has an advantage, a disadvantage, and a mystery. What are they?" I was gratified by a few chuckles as I explained: "The advantage, it has no chicory; the disadvantage, it was made without coffee beans; the mystery, why is it called coffee?"

We ate our dinner, then went to bed with heavy hearts.

I awoke from a dreamless sleep to the barking of the Stubendiensts —"*Aufstehen, aufstehen!*" This word, which means "get up," rapidly became the cruelest word in the German vocabulary. It cut relentlessly into my consciousness every morning, pulling me out of sleep like a knife with a salted blade. After we dressed, the Stubendiensts distributed small bars of soap with the letters "RJF" pressed into them. I didn't learn until after the war that they stood for *Reines Jüdisches Fett*—"Pure Jewish Fat." The recollection that I used to cleanse my body with the by-products of the millions of Jewish people who perished in the gas chambers sickens me even today. We always tried to steal the RJF soaps because they made better suds than the other artificial soap alternately distributed to us.

We walked to the washroom, stripped to the waist, and washed at the long line of faucets that stretched along a white enamel sink. The water was ice cold, and we had no towels. By the time we finished buttoning our shirts over our wet bodies, our fingers were numb.

As I put on my jacket I noticed a rusty stain on the back marked with several small holes. I asked one of the Stubendiensts why we had been given such dirty clothes. He grinned and told me the stain was only rabbit's blood and that, anyhow, the clothes here were never cleaned, just disinfected.

When we returned to the barrack the Blockältester lined us up and again ordered us to strip to the waist. One of the Stubendiensts carried in a tray with two bottles of black ink. The Blockältester used a pen to darken the numbers on our chests while admonishing us to memorize our new "names." From now on we would be addressed only by our numbers. Not responding immediately was tantamount to insubordination, which was tantamount to rebellion and therefore punishable by fifty lashes on the buttocks. These handwritten numbers were only temporary. If and when we were transferred from quarantine to the inner camp, we would be officially registered and issued permanent numbers to sew on our jackets.

We were divided into thirty-man work groups, all part of the *Aufraumungen Kommando* responsible for garbage collection and sanitation. Our Ukrainian Kapo and under-Kapo were armed with rubber hoses, the marshal's baton that gave them unquestioned jurisdiction over us during working hours. Speaking was *verboten.* If the rule was broken the whip fell until the Kapo was too exhausted to continue.

Happily, my group included Dr. Aarons and Dr. Ottinger. We were given large wooden containers to collect the garbage from the northern third of the camp, including the kitchen, the straw discarded from the prisoners' beds, and the soiled straw from the camp commandant's horse stables.

The wooden containers were built in the shape of a huge closed box with a door at one side. Each container was manned by six prisoners and hauled around on a cart. When our container was full we pushed it to the garbage trench outside the camp to dump it. Two SS guards accompanied us to the ditch. They were Croatian *Volksdeutsche* (people of German heritage who were born and raised outside of Germany) who spoke only to the Kapos. Then we returned to fill our container again.

As I helped hoist our first full load onto the cart, a fleshy blue-eyed man in my group asked me in a guarded whisper if I was French. He spoke to me in German with a strong Slavic accent. I concluded that he was Russian. His manner seemed arrogant and strangely out of place in these conditions. I nodded an almost imperceptible yes to his question. In a barely audible whisper, he asked if I were from Paris. I nodded again. He started to whisper that he was a Polish officer, but the Kapo approached us and he fell silent.

This Polish officer came to me again as we prepared to push our cart out to the garbage trench. As we bent down to push, the back of the cart hid our faces from the Kapo. The Pole resumed his whispered introduction. He was strangely effeminate in his movements and gestures, which were in marked contrast to his heavyset features and prominent chin. He had the prisoners' shaven skull, but the new stubble was obviously blond. He told me he had escaped to France in order to get to England and join the Polish army in exile there, but the fascist French militia had caught him in Strasbourg and handed him over to the Gestapo. They found an address on him for a friend in Sussex, England, an address that turned out to be a military training camp. He was sent to Germany for interrogation. He had been sent to Dachau two weeks before with about thirty-five other prisoners. Only he and four others were still alive.

We arrived at the dumping grounds outside the camp. I was very surprised to see the SS guards remain off to the side. While we worked they passed the time gossiping, smoking cigarettes and chatting with the Kapo. They didn't glance at us once. For a few minutes they even turned their backs on us while they shared a bottle. For a moment I had a vivid fantasy of escaping. But I had already heard warnings from other prisoners that the SS soldiers enjoyed tempting prisoners into trying to escape for the sport of shooting them in the back. The SS soldier would receive a citation for shooting an escaping prisoner, and the prisoner would end up as ashes in the crematorium. I pushed my escape fantasies away.

As we lined up to get our supper that evening, the heavyset Polish officer asked if we could eat together. I said I would join him at his bunk. His bunk was closer to ground level than mine, so I thought we would be less likely to attract attention. In single file we shuffled past a kettle from which a Stubendienst ladled out half-quart portions of soup, hot water with a few lonely scraps of potato peel floating near the surface. The other Stubendienst handed each of us fourteen unappetizing ounces of black bread.

I carried my supper over to the Polish prisoner's bunk and sat down. He formally introduced himself as Captain Peter Kalka of the Polish cavalry and gave me some basic information, the benefit of his two weeks of experience in Dachau.

I learned that the camps were governed by two separate hierarchies, the SS and the so-called Prominents *(Prominenten).* Kalka warned me about both of them.

The SS ran the camp. The SS camp commandant and his staff of noncommissioned officers were the camp administrators, the deathly procedure-makers and bookkeepers. The SS privates manned the watchtowers and the camp gates. They supervised the morning and evening prisoner roll calls and inspections. They were all impressively efficient in carrying out their various duties, which were directed to maintaining us in conditions that produced a maximum of degradation, required minimal expense for our upkeep, and ensured—given these considerations—the maximum labor output from each easily replaceable prisoner. Those to be replaced, and those of no use to start with, were exterminated en masse. It saved time.

The Prominents directly ruled our lives. They were the SS-appointed "trusties." They formed an amoral aristocracy of criminal convicts and some chosen political prisoners selected by the SS. In keeping with the Nazis' goals for running the camps, the greater the Prominent's innate cruelty the greater was his legitimized authority.

Each Prominent in this hierarchy was a tyrant to those below and a slave to those above, except for the camp *Lagerältester*, the supreme Prominent, who obeyed nothing but his own desires. But they all had unquestioned license to kill whomever they pleased, however they pleased. And as the intermediaries between the SS and their fellow inmates, they more than willingly carried out individual executions dictated by their SS superiors.

These Prominents exercised power over us everywhere we turned. The Lagerältester ("camp elder") was the supreme master. Immediately under him were the Lagerschreiber, who registered, assigned, and transferred prisoners, the *Arbeitschreiber* ("work secretary"), who organized the work *kommandos*, and the *Lagerkapo*, who made sure that all his orders and assignments were carried out, from work lists to morning and evening roll calls down to the master's whims.

Each barrack, or *Block*, was a kind of provincial fiefdom ruled by a Blockältester ("block supervisor," or "elder") appointed by the Lagerältester. The Blockältester chose his own assistants, a Blockschreiber and a Blockkapo and one or two Stubendiensts. The Stubendiensts were always good-looking adolescents, often Ukrainian, chosen ostensibly to do housekeeping chores in the barrack. Their real purpose, however, was to satisfy the sexual passions of their masters. They prostituted themselves to the Blockältesters and Kapos who gave the best favors and were extremely skilled at manipulating their lovers. Whatever capacity for human tenderness the criminal Prominents possessed was glimpsed only in these distorted relationships. Although a Stubendienst did not have much formal authority, having the upper hand in his sexual relationship with the master gave him unlimited power over the defenseless inmates living in his barrack.

In November 1943, when I arrived at Dachau, an undeclared war was being ruthlessly waged in the camp. Kalka explained it to me. The criminals and the political prisoners were fighting one another to get control of these Prominent positions of power. They would even kill an opponent to open up a spot for one of their own.

In the camp you could identify the criminals and the politicals at a glance. Every prisoner wore a triangular patch of colored fabric on his jacket pocket underneath his registration number. The color identified the nature of his offense. Everyone who had been emptied from the convict jails into the camps wore a green triangle. Those who had been arrested for their political opposition to Hitler's regime wore a red triangle. Asocials—those who refused to work to support themselves—wore a black triangle. Gypsies received a maroon triangle, conscientious objectors a purple one, and homosexuals a rose triangle.

Finally came the camp untouchables—the Jews, who were easily identified by the six-pointed yellow star formed with two yellow triangles. Jews arrested for political reasons as well were given a star formed by a yellow triangle on top of a red one. Prisoners arrested in countries other than Germany wore an identifying letter on their colored triangle: "F" (arrested in France), "B" (Belgium), "P" (Poland), and so on.

I did not yet have my colored identifying triangle or permanent number because I was still in the quarantine barrack. Once I went inside the main camp and was officially registered as a Dachau inmate I would get an "F" on a red triangle, identifying me as a French "political."

"You'll have to be careful," Kalka warned me. "The Prominents don't like Frenchmen. But thank God you are not a Jew. They are separated from us in the camp, and however bad it may be in our part of the camp, it's that much worse where the Jews are. Not only do they live much worse than we do, but the Jews are the most frequent targets for the Prominents' sadistic pleasures, and they're most often selected by the SS for the gas chambers."

And I had begged Recsek to believe that I was a Jew! Now I realized that my survival depended on his disbelief.

I asked Kalka when he thought we would be transferred inside Dachau's main camp. "I don't know how long it takes," he replied. "But first you have to live long enough. Politicals who are taken to the *Bunkers* first—the interrogation dungeons—for questioning are usually never seen again. Of course, there's no guarantee you'll survive even if you aren't sent to the Bunkers. Prisoners who become too sick or exhausted to work are automatically liquidated. It's German efficiency. They bring in more healthy ones and send the wrecks to the gas chambers. They take their hair, their gold fillings, and whatever else has value, then send the remains to the crematorium to make ashes for fertilizer.

"There's even a Kommando of prisoners working in the gas chambers. The SS call them the *Himmel* ("heaven") *Kommando*. They have to undress their comrades, lead them into the death chamber, and turn on the gas. Then they remove the corpses and collect the hair and whatever else the Germans need.

"And then again, you may die just because a Prominent is in a bad mood or wants to indulge his pleasure in killing."

"But don't people fight back?" I asked. I could not believe that they would simply allow themselves to be slaughtered.

"It's useless. A human life is meaningless here unless it's in a Nazi uniform. We are all less than human to them. If somehow you escape an attempt to kill you, there are more than enough of your fellow inmates who will do the job to curry favor with a Prominent or an SS guard."

I was having difficulty absorbing all that Kalka was telling me. But it was more believable than it had been the first time I had heard it. Two men from another Resistance unit in Nice had been arrested and sent to one of the camps in Germany. They had managed to escape and get back to the Côte d'Azur. Everyone, including me, had scoffed at their stories of the Nazi concentration camps. The people who produced Beethoven and Goethe could not be capable of such systematic, massive cruelty. It all seemed like a fantastic horror story calculated to gain our sympathies. Now, however, I realized that the former greatness of German culture apparently had little to do with its present-day ethos. How was I going to survive?

Kalka asked me what I had done before I was arrested. I spoke to him about Hungary and France, about my training and work as a physician. I said nothing about my Resistance work, just that the Germans had been misinformed that I was a spy.

Then Kalka told me the story behind his own imprisonment. He described himself as the sole survivor of the Katyn massacre near Smolensk. The Germans had claimed that this atrocity was perpetrated by the Russians, but Kalka told a different story.

When Germany attacked Poland, the Polish army, with its World War I weapons, was easily defeated. After the country was overrun, the intelligentsia and military officers were taken off to quarantine camps. Kalka and his fellow prisoners worked in a field shop repairing damaged German tanks and carriers.

In 1941, shortly before Germany's surprise attack on Russia, the Gestapo requested Polish volunteers for an anti-Bolshevik legion. Kalka, along with the majority of the Poles, refused to fight by the side of their German conquerors. The day after Germany attacked Russia, several large trucks guarded by Wehrmacht soldiers arrived at Kalka's camp. All of the prisoners were loaded up and taken a day's travel away to Katyn, where they spent the cold night in the middle of an empty field. They assumed they had been brought to Katyn to set up a new field repair station.

Before sunrise they were awakened and put to work digging a trench. SS guards had replaced the Wehrmacht soldiers. As the sky grew light Kalka saw another large group of Polish prisoners, obviously from another camp, digging a trench on the opposite side of the field. When twilight fell in the late afternoon, the Germans set up portable floodlights. Kalka assumed they would be digging into the night. Instead they were ordered to put down their shovels, line up along the edge of their trench, remove their shoes, and lower their trousers and underwear below their knees. Several closed trucks backed up to the trenches and stopped. Their rear gates dropped open. Bullets were spewing the length of the ditches by the time Kalka saw the machine guns. His world went black.

When he opened his eyes it was still black, and a tremendous weight was pressing him down. His left shoulder and chest hurt badly. He coughed up what he thought was phlegm but later realized was clotting blood. He finally worked his hands free. Gradually he pushed away the weight that was on top of him and stood up. He was standing on a carpet of dead bodies. He cut his right hand on the jagged trench wall in his attempts to climb out. The pain in his injured left arm was so intense that trying to raise it made him scream. He escaped his grave by slowly and painfully building a staircase of corpses. In the dark he walked along the edge of the trench, unable to locate the sounds of tortured breathing and cries for help. He was sickened with the realization that the trench was already partially filled in with earth. He vomited blood, then walked away.

Kalka walked on in a daze until he saw lights on the horizon and the silhouette of a farmhouse. He stumbled to the front door and knocked before collapsing. The door opened and an old peasant, a lamp in one hand, looked down at him. Kalka could only weep. The old man took him in.

The old farmer and his family bathed Kalka and crudely cared for his wounds, then for the next three days they hid him under the straw in their barn and brought him food. The old peasant contacted his son, who was a member of the Partisans and in touch with the Underground. Kalka left on the morning of the fourth day hidden on a farm wagon between two bales of hay.

He was eventually transferred to a cave in a forest that served as a nighttime hiding place for the Partisans and a refuge for their wounded. An army paramedic tending the wounded found that a bullet had pierced and exited Kalka's left lung. A second bullet had lodged in his left upper arm after breaking the bone. The medic removed the bullet, disinfected the wound, and put Kalka's left arm in a cast.

My face must finally have betrayed my growing skepticism of this fantastic story. Kalka suddenly interrupted his narrative to exclaim, "You don't believe me!" He pulled his shirt up to expose the bullet entry and exit scars on his left chest and back. I asked him to face me and inhale deeply. The left side of his chest expanded very little, which is characteristic of damaged lungs that have partially lost the capacity to expand.

He sensed that I was not fully convinced and began to bare his arm for me to examine. I said I had enough evidence. I really felt that nothing he could show me would erase my doubts.

The stool pigeon on the train had been a warning to me to mistrust my fellow prisoners, and so had some of Kalka's own descriptions about life in these camps. There were things about Kalka that just did not fit. Why, I wondered, had he picked me to confide in? Germany had told the whole world that the Russians were responsible for the Katyn massacre. They didn't want any living witnesses to the contrary. How could Kalka be so sure I wouldn't denounce him to get better treatment for myself? It also bothered me that his German fluency had seemed to improve markedly during his narration. And why had he given me his name as Peter Kalka when I had heard his countrymen refer to him earlier as Stashik or Lizinski?

I put this last question to Kalka. He explained that when he was arrested in France he had given his name as Stanislaw ("Stashik" is the diminutive) Lizinski because the Germans had his real name on their list of Polish officers massacred at Katyn.

"Why did you choose me as your confidant? I just got here yesterday, and you don't know anything about me."

"I have reached my breaking point," he said. "I know it is only a question of time before the Germans find out I am the only survivor of Katyn. I will be tortured and shot. My only hope is to escape using the garbage ditch, but I can't do it alone. I need two people to help me. One man will have to escape with me after we hide under the emptying garbage container and burrow under the garbage. The other man will shield us with the container when we jump into the ditch and empty it over us. I have chosen you because you speak German, which gives us a common language. And you are a doctor, which means I can trust you."

I was unenthusiastic about Kalka's plan to escape from Dachau. After my own brief escape fantasy, I foresaw too many risks. I told him we would talk more about it in the next few days. Then I wished him good night and went to bed.

I was still suspicious of Kalka. I would never trust him fully, but I didn't know then whether I was responding to something in his character or whether it was simply an inevitable consequence of the paranoia essential for survival in the concentration camp. Two incidents that were to occur in quick succession several days later forced me to ignore my suspicions and go along unhesitatingly with Kalka's plan.

We had barely finished our coffee and bread the next morning when the Blockführer, the SS officer responsible for our quarantine barrack, entered with several SS guards. He handed a list to the Blockältester. Ten prisoners were called off the list and separated from the rest of us before we were herded outside to begin our daily chores. It wasn't until the next day that we knew what had happened to these men.

On our way to the ditch with our first loaded container, Kalka advised me to notice how well hidden we would be while we emptied it. We would have time to dig ourselves down into the accumulated garbage and have a third person quickly cover any part of us that still stuck out. After we bent down behind the ditch wall, I realized that we really could not be seen by the Kapo and the SS guards! Kalka's basic plan seemed feasible, but it needed thorough thinking through and preparation.

More important, we needed a trustworthy third person. Kalka had no candidate. Our other major problem would be finding essential supplies. We needed civilian clothes. Our striped uniforms would be impossible to disguise. We would also need some food because of the difficulty, if not impossibility, of finding any en route. Finally, we had to know which way to go once we left the camp.

We continued to talk whenever we had the opportunity. Kalka had already worked out a plan to get us to the Frauenkirche, a church in Munich, where we would find a Catholic priest from Danzig whom he knew quite well. He was confident that this man of God would provide clothes and a refuge for us. The church would be the perfect hiding place until the search for us was called off.

To minimize our risks while we were still in the camp we had to avoid being seen together any more than would seem natural. Whoever had something to communicate to the other about our plan would make a sign or wait and go to the other person's bed during one of our mealtimes. For the present we had to concentrate on finding our third party.

Only five of the ten prisoners called out that morning were in the barrack when we returned from work. A young French Communist functionary from my transport was in such deep shock that he didn't even stir from his bed at suppertime. Dr. Aarons went to him, but the man's only sign of consciousness was the mumbled repetition of one phrase: "*Ce n'est pas possible!*" ("It's impossible!") When the Blockältester slapped his face and ordered him to get his food, the man stood up mechanically and moved with the glazed eyes of a sleepwalker. He didn't even react when the Stubendienst willfully missed his cup and spilled steaming coffee on his hand. During the night we heard him sobbing and talking to himself. It wasn't until the Blockführer arrived the next morning and ordered us to line up for the count that we realized he was missing. The SS found him in the lavatory hanging from a towel that had been fashioned into a noose.

We didn't know whether the Blockältester had murdered him so that he could sleep undisturbed, or his experience of the preceding day had so shattered him that he took his own life. Dr. Aarons, Dr. Ottinger, and I all felt that it was important to find out what kind of terror he could have suffered. There were no marks on his body —no wounds, no bruises—to indicate torture. None of us had been issued towels. The appearance of this towel with which he had apparently hanged himself was a mystery.

I quietly asked one of the remaining four what had happened yesterday. He wouldn't speak, but he raised his hand to his throat over and over again in a slashing motion. It made no sense. I turned away quickly as the Blockältester menacingly approached us. I saw Dr. Aarons trying to talk with another prisoner and shaking his head over and over. From the looks of it he wasn't learning anything either. Then Dr. Ottinger signaled that he wanted to talk to us, but we were herded into our unit formations and the opportunity was lost.

Kalka and I were at opposite ends of the cart, which made it impossible for us to communicate. I was furious that he was following my advice so obediently. During a moment's opportunity he stopped me, green and frightened, to whisper that he would come to the bunk below mine that evening. Out of nowhere the Kapo hit me and barked out, "*Nicht sprechen!*" That ended our conversation.

While we filled the garbage containers, I silently worked on our escape plans. Although the SS guards were supposed to count us every time we left and reentered the camp, I had observed that most of them didn't bother after the first of our four daily trips. After that the Kapo called out "thirty prisoners" and left us to dump the garbage, evidently confident that an escape was impossible.

It seemed wisest to start our escape with the last trip of the day because the guards wouldn't count us and the light would be growing dim. The Blockältester always counted us at dinner, but that wouldn't be until well after dark. By the time he realized that we were missing we would be nearing the Frauenkirche in Munich. I still couldn't figure out how to disguise our striped uniforms and shaven heads.

When I returned to the barrack I discovered Dr. Ottinger and Dr. Aarons in conversation in the washroom. Dr. Ottinger found out what had happened to the ten men the previous day. He described an exercise the SS called "*Hasen Jagd*" ("rabbit hunting"). It was the *Grand Guignol* come to life.

The men had supposedly been taken through the northern door into a courtyard and ordered to remove their shoes and run. Someone in a hidden spot started shooting into the courtyard, using the prisoners as moving targets. Each time one was hit the force of the bullet somersaulted his body forward. His face wore the terrified expression of a rabbit that has just been shot.

Although it recalled the sardonic words, "here are your little rabbits", with which we were handed over to our Blockältester, I couldn't believe the story. The author of this fantastic tale had obviously created a smoke screen. We had heard that survivors were warned they would die if they spoke a word of what had really happened to them. This man must have created the "rabbit hunt" story because he was afraid to tell the truth.

According to my naive logic, if the Germans really wanted to kill these prisoners they would simply have lined them up and shot them. Why such an elaborate scenario when a straightforward execution was so much easier?

As I returned to my bunk the Blockältester sent the under-Kapo for our food and shouted at us to line up for the evening count. I was surprised and disappointed not to see the four survivors of yesterday's mystery. My opportunity to question them was gone.

After our dinner was distributed I took my meager portion to Dr. Ottinger's bunk. Kalka joined us. I asked him what he knew of Dr. Ottinger's ridiculous rabbit hunt story, expecting him to laugh and pooh-pooh the whole thing. Instead he solemnly described the invidious torture designed to shatter the resistance of political prisoners who had refused to confess during interrogation.

The Gestapo were present at our registration interviews the day we had arrived to point out those marked for the rabbit hunt. All political prisoners were classified "NN"—*Nacht und Nebel* ("night and fog")—because they were supposed to disappear without a trace after they broke down and finally confessed. That was why our identification numbers were provisional. Why do the work of sewing a permanent number on a prisoner's jacket if he was about to die? Politicals slated for the rabbit hunt were thought to have important information and were known to be resistant to the more traditional forms of torture.

The actual rabbit hunt took place in the enclosed courtyard on the other side of the locked northern door. Each group of "rabbits" also contained several randomly chosen victims. The SS hunters knew which rabbits were and were not important. The dispensable ones were the moving targets, who were deliberately sacrificed to terrorize the others into talking. The occasional prisoner who withstood this pressure and still refused to confess was sent to the inner camp. The reprieve was very temporary.

I was stunned and speechless. But Dr. Ottinger's story had barely touched the surface of this terrifying and incomprehensible sadism.

Kalka and I went to his bunk to talk privately about who among us were likely to have been marked as important rabbits. I felt that Dr. Ottinger was possibly one. He was eminent professionally and on record as anti-German. He had never revealed any Resistance involvement to us, but if he had been in the Underground someone at Royallieu might have recognized and denounced him. I suddenly realized what should have been obvious to me from the start. My own name must certainly be on their list!

I awoke the next morning after an unusually restless sleep convinced that the SS would arrive shortly to call off the list of "rabbits" with my name on it. To my indescribable relief, the SS Blockführer came around only to receive the prisoner count.

Kalka wanted to start rehearsing our escape procedure. I still suspected that he might be a stool pigeon. If I was right, then I would be much less vulnerable if he took the first go at it. I was relieved that he easily agreed to hide first.

When we had emptied our first round of garbage I invited Dr. Ottinger to help us push at the back of our cart. I discussed our plan with him. Kalka didn't understand French, so I could speak freely. He approved of Kalka making the first dry run. If I got caught during my turn, I was to say that Kalka had disappeared and I was looking for him.

As we dumped our second load, Kalka jumped down into the garbage ditch and Dr. Ottinger and I lowered the container with the door open to conceal him. When we lifted it, only his heel remained visible. We covered it, added a precautionary layer of garbage on top, and put a shovelful of horse manure on the ground to mark the spot.

In thinking ahead to our actual escape I realized that, with six prisoners to each garbage cart, one missing could easily be camouflaged but two would be noticed at once. We needed to borrow someone from another team to cover our escape.

We left Kalka there for two hours before we returned to free him. His experience had been so unpleasant (the stench was unbearable, breathing was difficult, and he was painfully stiff from not moving) that we decided to postpone my dry run until the following day.

That evening Dr. Ottinger joined us to discuss solutions to the difficulty of breathing under the garbage and to deal with the muscular stiffness from remaining rigid for an extended period. For breathing he suggested something like the hollow tubes we used as children to swim underwater, but that still left us with the problem of where to find such a tube. Dr. Ottinger had the answer again—a *Schlag,* a "rubber hose". Now we had to get some! We considered bribing a Stubendienst, but we had nothing to give away. Our only alternative was stealing, but who had the courage to risk being caught sneaking into the Blockältester's room? By a sheer miracle of coincidence, the perfect opportunity presented itself the very next morning.

IO

The next morning during the count one of the Stubendiensts called out for doctors. My friends and I stepped forward. We were asked who spoke German, and I raised my hand. The three of us were led into the Blockältester's room. Kugelkopf, as we called him, lay on his bed perspiring profusely and doubled over in severe abdominal pain. Dr. Ottinger examined him, hastily diagnosed appendicitis, and advised him to go to the infirmary. The Blockältester insisted that we treat him in his room even if an operation were necessary.

We were at a loss to understand his refusal to go to the infirmary for proper surgery. We were unaware then that the infirmary had become a desperate battleground in the war between the politicals and the criminals, with the odds stacked one hundred percent against the criminal green triangles. The infirmary medical chief and his Kapo were political red triangles. Green triangle Prominents, such as our Blockältester, sought needed medical help only if a physician was in their own barrack. Another fact still to be learned was that even the death of a criminal Prominent was of no importance to the Germans. The only consequence would be removal of his number from the camp register and selection of another Prominent to take his place.

We feared for our lives if this murderer died in the barrack during our treatment. As a ploy to gain time, I told the Blockältester that we needed specialized instrumentation and sterile materials. He answered that he could get everything just as soon as we gave him a list of what we needed. He moaned and begged us to hurry.

As we started to draw up our list, Dr. Aarons suggested that we request several semirigid catheters. In response to our puzzled reactions, he explained that they would be better than rubber hoses as breathing tubes. Dr. Ottinger recited all of the pharmaceuticals and materials he would need, including iodine, then enumerated the necessary instruments. I translated everything into German and wrote it down.

Back at our bunks I asked Dr. Ottinger what he planned to do if Kugelkopf really turned out to have appendicitis.

"Operate, of course," he said. "A successful appendectomy on this Prominent will help me get out of this barrack and help you and Kalka escape. The catheters will make effective and scarcely visible breathing tubes. You can use the iodine to paint out the white stripes in your uniforms."

Roll call came then, and we were counted and lined up outside for work. We decided that it would be premature to share these sudden developments with Kalka because of our doubts about him. Before we could leave, one of the Stubendiensts called Dr. Ottinger from the line and disappeared with him into the barrack.

A pale and frightened Kalka approached me as we lifted our container onto its cart. "You went to the Blockältester to denounce me," he cried in a whisper. He would not let me explain, vehemently accusing us of signing his death sentence. He obviously did not trust us any more than we trusted him. Confronted with the pure terror in his eyes, I tried to reassure him with the news that, on the contrary, some very favorable things had happened that I would tell him about when we got to the ditch. I never got a chance to keep my promise.

One of our Stubendiensts was talking with the SS guards as we passed through the camp gate. He ran over to us and called out Dr. Aarons' and my numbers. Dr. Ottinger needed us. He was obviously going to operate. Seized by a dread apprehension, we turned to each other. Seeming to read each other's thoughts, we both muttered at the identical moment, "*Il a perdu son nord!*" ("He has lost his marbles!") What if Kugelkopf died? But we couldn't refuse to join Dr. Ottinger. Damned either way, we returned to our barrack.

We couldn't believe our eyes as we took in the scene before us! The Blockältester was lying on a table in his room, and the barrack's barber was shaving his pubic hair. Dr. Ottinger, completely unruffled, his gestures radiating total authority, was pushing the Stubendienst around and, in his customary operating room manner, thoroughly terrorizing everyone. You would have thought he was back in Paris at the Hôpital St. Louis.

"Thank God you're here," were his first words to us. "This idiot doesn't understand a word of my orders." I was too flabbergasted to point out that he was speaking French to people who didn't understand it. "Haas, have them bring hot water continuously. We will scrub. I'm sure he has a hot appendix."

I began to suspect that Dr. Ottinger's basic motivation in operating might well be an all-consuming need to perform again in front of an audience. How powerfully he must miss the operating theater. At this moment it seemed of little real importance to him that he might be placing us in jeopardy, but he was also helping us to escape and I couldn't refuse to assist him in achieving his goal. I asked if he felt Dr. Aaron's presence to be really necessary. Before he could reply Dr. Aarons said that we were all in this together, up to the end.

In front of my unbelieving eyes was every last item on our supply list, all anxiously guarded by the SS Blockführer. The three of us began to scrub. The door swung open and a high-ranking SS physician entered. The Blockschreiber shouted, "*Achtung!*" We stopped in the midst of scrubbing. He addressed himself directly to Dr. Ottinger, sensing that he was the captain of our team. "So you want to operate on this prisoner?" My heart jumped into my throat. Someone had denounced us! This was the end. But then he introduced himself.

"I am Hauptsturmführer Koenig, head of Dachau's medical division and our SS hospital. I personally approved your request for instruments and supplies. I know the excellent reputation of Dr. Ottinger. I want to watch him operate, and perhaps I will learn something from him."

I translated. Dr. Ottinger answered contemptuously that he was sure Dr. Koenig would learn something, and he was most welcome to observe. I translated again, judiciously omitting the first part.

Dr. Ottinger still seemed untouched by the extraordinary nature of our situation, continuing in his role of chief surgeon in the operating room at the Hôpital St. Louis. We finished scrubbing. Dr. Ottinger asked me to invite the SS physician to assist him. I stared—he had lost his hold on current reality, no longer a prisoner with a number on his chest but an authority figure in a real operating room in a real hospital. Despite my anxiety I asked if

Häftling ("prisoner") number such-and-such could have permission to address the Haupsturmführer. He nodded affirmatively. "The *Häftling Arzt* ["prisoner-physician"] would like to know if the Hauptsturmführer would like to assist him in the operation." "No, I just want to watch." Dr. Ottinger said, "Fine. Haas, you will assist me."

We tied on butcher aprons. Dr. Aarons poured iodine into a kidney dish while Dr. Ottinger and I pulled on rubber gloves. Holding a piece of gauze with his forceps, Dr. Ottinger soaked it in the iodine and painted the patient's abdominal area.

While he draped the sterile sheets he informed me of his plan to do a modified MacBurney incision, a demanding technique used on physical laborers and dancers that left a tiny three-inch scar and reduced postoperative weakness of the abdominal wall. I recalled then that Dr. Ottinger had been the physician for many of the dancers at the famous nightclubs in Paris. As he finished draping he muttered, "I'll show this *schleu* ("scum") what a surgeon can do under pressure!"

He signaled Dr. Aarons to inject the Evipan. The Blockältester began to count backward from one hundred and was asleep by the time he reached sixty-seven. Ether would be used to continue the anesthetic, using thick gauze in place of a mask. The SS physician, to my great surprise, secured Dr. Ottinger's mask for him. Mine was secured by Dr. Aarons before he stationed himself by the Blockältester's head to administer the ether.

Dr. Ottinger grasped his scalpel and made an incision of *only one inch!* I was so astonished that I forgot to clamp the severed vessels on my side. He bawled me out, irritably claiming that this always happened to him when he had an internist for an assistant. I began to perspire.

Dr. Ottinger carried on as if we were all medical students attending his surgery, describing each step in a kind of singsong chant. "I am now cutting the external oblique." He asked for retractors, then, "Put it in, Haas! Hold it, Haas! And here is the fascia, and here we are at the peritoneum. Clamp the vessels, Haas! You cut the peritoneum, and you see the cecum. Pull the retractors, Haas, so that I can see!" Using the forceps, he continued: "I retract the cecum, and here is the fat pad, which I am pushing to the left. And here is the dirty bastard." He held up the appendix and pointed out the black spot and sac of pus to the SS physician. "Ready to perforate."

A thud resounded as the Blockführer fainted and crashed to the floor. Dr. Ottinger didn't break his rhythm as he ordered, "Someone pour cold water on that idiot. Haas, clamp the appendix at the base!" He tied off the column of the appendix, secured it, then cut by the edge of the clamp. He pushed the fat pad back into place and closed the cecum, then the peritoneum. He put a small tube into the opening, closed the fascia, then the muscles, then the skin.

From the initial tiny incision to the final closing, the entire procedure had taken seven minutes. The SS physician was obviously impressed. He assured Dr. Ottinger that he would keep him in mind, then left as abruptly as he had entered.

I asked Dr. Ottinger what would have happened if the appendix, whose location is known to be variable, hadn't been in a common position and a different incision had been needed. He shrugged. "Sometimes, *mon petit,* you have to take a risk, and you can only pray that you are right. But I had to do so many almost invisible incisions on the cabaret dancers that I had something up my sleeve besides a prayer!"

I was so excited by this spectacular demonstration that I had forgotten to pocket the catheters before we walked out. As I turned back Dr. Aarons asked where I was going. I muttered, "Shit, I forgot the catheters!" He smiled—he hadn't forgotten them.

Dr. Ottinger remained with the Blockältester and the two of us spent the day loafing around the barrack. The other Blockältesters paid their sick calls around dinnertime and brought get-well gifts of food. Dr. Ottinger emerged momentarily to issue their invitation for us to join the celebration. It was a veritable feast in our present context, and on top of that we didn't have to go to evening roll call!

I wanted to see Kalka, who must be distraught with fear and worry. I hid bread and sausage under my jacket, added a small bottle of schnapps, and told everyone I had to go to the latrine. I made sure no one followed me.

Kalka was sitting in his bunk like a wide-eyed statue, mumbling to himself. I sat down and pulled out the bread and sausage. His pathetic face transformed almost instantly. By the time I added the schnapps he was wreathed in smiles. I whispered about the catheters and iodine and left him contentedly consuming his goodies.

The Blockältester, who was awake when I returned, asked me to convey his gratitude to Dr. Ottinger and Dr. Aarons and tell them that as soon as he was back on his feet he would have the three of us transferred out of this barrack. He knew that Dr. Koenig, the SS physician, liked to surround himself with physicians of renown.

Dr. Ottinger stayed with his patient. Dr. Aarons and I left for the night. We signaled Kalka to join us at our bunks and gave him a detailed rundown of what had happened since the morning. He decided to save the schnapps I had given him and give it to the Kapo on our big day. If the Kapo shared it with the SS guards, they would survey us even less carefully than they normally did. Dr. Aarons and I talked for a while longer after Kalka left us. Everything seemed to be working in our favor. For the first time since my arrival in Dachau four days before, I had a restful night's sleep.

"*Aufstehen! Aufstehen!*"

I was awakened by the Stubendienst's harsh barking. After breakfast the SS Blockführer appeared with his cohorts and handed a list to the Stubendienst, who called off fifteen numbers. The prisoners—ten from my group and five from the group already there when we arrived—were gathered in a corner. The rest of us were lined up outside and divided into our work units to begin our daily routine.

I had hidden a catheter under my shirt and told Kalka I was ready to do my dry run. At the ditch he and Dr. Aarons put the garbage container over me as I lay on my back. I felt myself being covered and heard Dr. Aarons whisper, "Relax, everything is okay. We'll be back soon."

Breathing through the catheter was quite easy, but the weight of the garbage paralyzed me. A scrap of something had lodged uncomfortably in a nostril. I couldn't remove it. I began to feel cold, particularly in my wounded toes. The piece of garbage in my nose tickled more and more.

An entire day seemed to go by. I could easily have dozed if I hadn't been so terrified of moving in my sleep and being discovered. I dreamed of Sonja and François and passed the time with elaborate fantasies of our reunion and future together. I watched François grow up and become a brilliant physician, upholding the family tradition. Sonja and I sat in front of a crackling fireplace at a ski resort, reminiscing about our war experiences.

Strange voices and sounds of falling garbage jarred my daydreams, and a voice whispered, "We're putting the garbage container over you. Move quickly!" I couldn't. My limbs were as stiff as boards and weighed a ton, and my joints were very painful. Somehow I managed to wriggle from under the garbage. Kalka and Dr. Aarons lifted the container, helped me to my feet and out of the ditch. I was reeking from the garbage, but I didn't even notice it. We had grown so used to the stench. I hid the catheter under my jacket and stuck my hands under my armpits to warm them.

On the way back I suggested that we do a series of dry runs to learn to adapt ourselves to these difficult conditions. We had to find cotton to stuff in our nose and ears to close them off. We also had to decide what to do in case it rained on our chosen day. Today's weather had been fine, yet I was still half frozen.

I was so chilled that lunch, the familiar tasteless and watery soup, was actually good. The flavor was no better, but it quickly warmed up everything except my feet. My old toe wounds still throbbed terribly. Rubbing my feet did not help much.

When Dr. Aarons and I returned to our barrack at the end of the day, Dr. Ottinger was waiting for us with real hot soup that had been brought in for the Blockältester. He felt we needed it much more than Kugelkopf did. When I told him about my dry run he suggested trying to get a bottle of schnapps to help ward off the cold. I mentioned the small bottle Kalka was keeping to give the Kapo, and Dr. Ottinger promised to steal a second bottle for us from the Blockältester's room. Dr. Aarons and I went off to share our real soup with Kalka.

Dr. Ottinger rejoined us shortly. He told us that he had witnessed a rabbit hunt that morning. He had watched it through a tiny hole in the Blockältester's room that opened onto the courtyard.

"If I hadn't witnessed it with my own eyes, I would never have believed that human beings could be capable of such cruelty and could remain so unmoved by the results. Seven of the fifteen men who were called out this morning were slaughtered by SS marksmen. I watched the SS soldiers shoot these men down, and then carefully examine the bloody corpses to see where their bullets had entered. Apparently they were comparing their predictions with their actual performance. They looked very pleased with themselves. The eight men who were still alive were taken away. I can't forget the terror on their faces."

I still could not quite believe this was really happening.

The next morning it was to be my turn for the rabbit hunt.

We were just leaving for our work assignments when the SS Blockführer strode up and ordered, "Everybody back inside. Shut the barrack doors!" We had already learned that this order inevitably preceded an unpleasant surprise.

We stood in front of our bunks for at least an hour while the Blockführer paced the room and occasionally glanced at his watch. I was very tired and my toes were hurting, but I dared not sit down. Finally the Blockführer ordered the Stubendiensts to get a small table and four chairs. Dr. Ottinger helped the Blockältester out to one of the chairs, then went to stand by his bunk. Dr. Ottinger's face was very pale. He mumbled the word "selection" as he passed me. I was thoroughly puzzled. The word made no sense in this context.

The long silence in the barrack continued unbroken. Even the Blockältester whispered.

At last the SS Blockführer spoke again. "Is everything ready?" The Blockschreiber and Stubendiensts chorused, "*Jawohl.*" The Blockführer left and I heard the ominous sound of the door being locked behind him.

Dr. Ottinger walked over to me. His whispered words were almost inaudible. "The Blockältester told me there's a new transport of prisoners coming in. Room has to be made in our quarantine barrack. They're going to step up the rabbit hunts and transfer the less important political prisoners into the main camp. If you, Kalka, Dr. Aarons, and I stay in this barrack, we're kaput."

We heard the lock turning in the door. Dr. Ottinger quickly stepped back to his place.

The Blockführer had brought another SS officer and two briefcase-carrying Gestapo agents from the *Politische Abteilung.* They appeared to be in no hurry and chatted with the Blockältester about his recent appendix surgery. Suddenly the Blockführer shouted, "Where is this surgeon?" He ordered Dr. Ottinger to go to the Blockältester's room and close the door.

The SS officer called off a long list of numbers. The rest of us stood in place as the thirty people eventually called were ordered to step aside. The Gestapo agents took a list of their numbers and left, locking the barrack door behind them.

The Stubendiensts chose prisoners to fetch breakfast. Not until the food was distributed were we finally allowed to sit down. Dr. Ottinger told us that the thirty prisoners were going to be transferred. No one knew where or why. He had also been told of his own imminent transfer as surgeon to the HKB, the prisoners' infirmary. The SS physician had kept his word.

But Dr. Ottinger had some immediately disturbing news. A special rabbit hunt had been scheduled for this afternoon. He handed me a roll of bandages and some gauze from his pocket "for when you get out of here." I hid them under my pillow. We spent the remainder of the morning locked in the barrack.

At lunch I felt out of focus, despondent, threatened. Despite my gnawing hunger it was difficult to swallow my food. I didn't know what was wrong with me. Dr. Ottinger brought me some more news that temporarily banished my malaise. There were civilian overcoats hanging in the Blockältester's closet. He was going to try and steal two for Kalka and me.

I knew the risks to which Dr. Ottinger was exposing himself and asked him if he was fully aware of the dangers.

"Of course I am," he said, "but if you have the courage to risk arrest and death, then I should have the courage to help you. If you're caught with the coats, I'll tell the SS that the coats had been thank-you presents from the Blockältester for your participation in his surgery."

I doubted that anyone would accept this lie in view of our Blockältester's miserly character, but I accepted Dr. Ottinger's help without further question. He was proud and strong-willed, and it would be futile to try and discourage him.

I asked if he knew why our barrack door was still locked. According to what the Blockältester had told him, we were waiting for a van to pick up the thirty transfers. He had barely finished his sentence when we heard an approaching engine stop outside. The door was unlocked and opened, and the thirty prisoners were herded out and into the van. Then the door was locked again.

Twilight was approaching when the Blockführer reappeared with the two Gestapo men and handed another list to the Stubendienst. He called out the first number. A man stepped forward. He called out the second number. It was mine.

My number! My feet rooted to the floor, and the repeating phrase "they are calling *your* number" ricocheted through my head. I stepped forward. Dr. Aarons had also stepped forward. Most of the prisoners in our growing group were from our transport. I no longer recall what was in my mind at that point, but I vividly remember staring at Dr. Aarons' terror-stricken eyes.

Twelve of us were called. The remaining prisoners were herded out of the barrack. I heard the door lock. The Stubendiensts said we were merely going to the *Brausebad* to shower. I was surprised not to see Dr. Ottinger anywhere around.

The two Gestapo officials looked us over disdainfully. To my astonishment, my anxiety suddenly melted away and I became calm and quiet. They opened that fateful rear door and drove us out into the courtyard. It ran the length of the barrack and was almost as wide. Under the last rays of the sun we were lined up in single file about ten feet apart. One of the Gestapo men asked who among us spoke German. Everyone turned and looked at me. I raised my hand.

Two SS generals came out of our barrack. I saw the gun handles exposed in their open holsters. The Gestapo man told me to translate what he was about to say, which was that we were a bunch of lazy dogs who needed some exercise to improve our health, and that we should be very honored that two SS generals had come to watch our exercise program. He gave each of us a safety pin and a square piece of white paper with numbered concentric circles on it, like those we had used for target practice in the Resistance. We were ordered to pin our target to the back of the person standing in front of us, so that it was positioned behind his heart.

Dr. Aarons was in front of me. As I struggled with the safety pin I told him I hoped the bullet would hit my heart and finish me quickly. I tried to pin the paper directly over his heart so that his death might also be fast and relatively painless. The SS generals had been watching us intently. One of them ordered us to run. I was trembling so much that I couldn't move, but I was shoved forward by the prisoner behind me. I started running.

Suddenly a shot rang out. A man near me flipped in the air from the force of the bullet and fell to the ground, dead. One of the SS generals laughed. The gun in his hand was smoking. We all ran like a group of crazed headless chickens, everyone trying desperately to hide behind someone else. In this nightmare panorama I saw Dr. Aarons holding his bleeding left arm.

I was driven by pure terror, blood pounding in my head and my breath a hoarse rattle filling my ears. My feet began to hurt so badly that I could barely lift them, and my lungs and chest began to burn. I couldn't get enough air. I started gasping. A prisoner near me was shot. I saw the terrified rabbit look on his face, and then I stumbled and everything went black. I heard distant voices. "I've been shot" was my last thought as I floated off into space.

I regained consciousness as water was thrown in my face. I stood up. I looked around at the courtyard strewn with lifeless bodies. A sharp kick sent me over to the group by the barrack door. We were driven back inside. I dropped onto my bunk and began to vomit and cry at the same time. Dr. Ottinger appeared with a sedative injection from the supplies that still remained from the operation. He told me that Dr. Aarons had been grazed by a bullet but was okay. I blacked out again.

When I came to, the evening meal was being distributed. I had no appetite and gave my food to Dr. Aarons, who sat next to me. He asked me to clean and bandage his wound, then told me to wait until the lights were out for the night so no one would know we had any first aid supplies.

That evening Dr. Ottinger brought us sulfa powder and iodine, more gauze and bandages, and the news that he had stolen the two coats. He had hidden a heavy black one under my pillow and given a green loden coat to Kalka. Dr. Ottinger had anxiously witnessed the rabbit hunt from his peephole in the Blockältester's room. He could not understand how anyone held on to his sanity after living through it. We thought back to the young Communist who had hanged himself our second day in Dachau!

All I could think of was how to speed up our escape plans and do it during the next few days. Kalka suggested the coming Saturday because camp supervision would be somewhat relaxed. I thought Saturday was a poor choice. When we reached the town of Dachau there would be fewer people in the streets than on weekdays, so it would be more difficult to be inconspicuous. I wanted to enter the town as the workday ended and blend into the crowd of workers leaving the factory. Kalka thought we could still blend in easily on a Saturday. Someone suggested avoiding the issue and waiting until darkness gave us a natural cover. Dr. Aarons felt that would be unwise. Although we would be less visible at night we wouldn't be able to see what was happening around us. And if there was an air raid, where would we hide? Dr. Aarons also thought we should try to get some German money so we could take a bus directly to Munich and so avoid being seen on the roads.

At that point the Stubendienst called "*licht aus*" ("lights out"). Dr. Ottinger returned to the Blockältester's room. Kalka went back to his bunk. The lights, except for the two small ones at either end of the barrack, went out.

I whispered to Dr. Aarons to take his jacket off, and I slipped my supplies out from under my pillow. Sitting on his bunk in the near darkness I examined his wound as best I could. Luckily, it didn't require stitches. I cleaned the area, sprayed it with sulfa powder, and bandaged it. I whispered "good night" and was just about to return to my bed when the barrack door burst open.

The Blockführer and another SS man lurched in, dead drunk, waving their guns in the air and laughing. I bolted for my bunk in panic. The lights went on before I got there. The Blockführer spotted me instantly. "*Häftling,* come here!" I jumped to the floor. He held his revolver to my neck.

"What are you doing out of bed after lights out?"

"I was coming back from the latrine."

"What are you holding in your hand?"

I had no answer. In my fright I had completely forgotten that I was clutching the gauze and bandages.

"Give that to me!" he commanded. I stretched out my hand and he grabbed them from me, yelling, "These supplies were stolen from the SS pharmacy! What are you doing with them?"

I couldn't say anything about bandaging Dr. Aarons' wound because I was afraid I would include him in my troubles instead of getting myself off the hook.

Dr. Ottinger came out of the Blockältester's room. The SS Blockführer demanded to know how the gauze and bandages had come into my possession. Dr. Ottinger calmly answered in French that he had given them to me to bandage Dr. Aarons' gunshot wound. I was afraid that translating the truth would kill us all. Speaking quickly in French I asked Dr. Ottinger to let me handle the Blockführer as I thought best. He agreed, but he thought the Germans were too drunk to be predictable.

I told the SS Blockführer that during the Blockältester's operation Dr. Ottinger had noticed me putting the gauze in my pocket because there was no place to lay it down when I suddenly needed both hands free. The Blockführer was shouting at me, "Our German soldiers are dying on the front because they have no first aid materials, and you steal such vital supplies for your own use! I'm going to teach you a lesson you'll never forget!"

I thought this outburst ended the incident and turned to go back to my bed. "Don't move!" he screamed, and slapped my face. He and his companion began whispering. I caught the word "*Strick*" ("rope"). They began to laugh. It wasn't a good-natured sound. The SS man left. The Blockführer stared at me like a snake trying to hypnotize his victim and ordered the Stubendiensts to wake all the prisoners and have them stand by their beds. I hadn't the slightest idea what was going on. A great iron hand seemed to close over my heart. Panic threatened to overwhelm me.

The drunken SS man reappeared with a length of rope in his hands. My skin prickled. Were they going to beat me with it or tie me up? The SS man jumped up and fastened the rope to the ceiling beam. The Blockführer was screaming again, "This Häftling has stolen war materials from the German Reich. He is guilty of sabotage! His punishment will set an example for the rest of you." He stopped to swig from the bottle in his coat pocket, passed it to the SS man and wiped his mouth on his coat sleeve. "Therefore he will be hanged!"

I heard the words. I could not believe he meant me. I froze, time ceased, everything around me became distant and dreamlike. My mind stopped translating. I watched his lips move, but the rest of his words made no sense whatsoever. I felt only the impersonal shock of a passerby glancing into a stranger's window at an unexpectedly horrible event that was beyond his ability to stop.

A Stubendienst brought in a stool. I realized that Dr. Ottinger was looking at me with terror in his eyes. I was half pushed, half lifted onto the stool, vaguely wondering why my hands hadn't been tied. I remembered a toy I had made for François, a clown fastened with string that you pull on to make its arms and legs jump up and down. "These sadists want to see me dance," I thought.

The Blockführer ordered me to put on my necktie. I was too paralyzed with fear to understand what he wanted of me. A Stubendienst finally jumped onto the table and placed the noose around my neck.

I began to shake. I closed my eyes to shut out reality despite my intense sensation that it wasn't real. The Blockführer spoke. I could not hear him. I squeezed my eyes tighter, ready to collapse. I gasped and opened my eyes and realized that I had just fallen onto the floor! Laughter exploded around me. My relief, spilling out in hysterical laughter, added to the cacophony. I began to realize that I was not the almost-victim of an accidentally aborted execution. The SS had deliberately staged this for their own amusement. Paroxysms of sobs choked off my laughter.

Dr. Ottinger and Dr. Aarons picked me up and gently carried me to my bunk. One cradled my head in his arms and the other did his utmost to calm me. I was too shattered to utter even a word. I trembled, overwhelmed by a sadness without end. By sheer caprice I was still alive. If their drunken mood had tipped just slightly in the opposite direction, I would have been irrevocably dead now. It was beyond my comprehension.

My two friends spent the entire night with me, Dr. Ottinger quietly talking and Dr. Aarons holding my hand. I watched the dawn herald the end of my first week in Dachau.

Dr. Ottinger wanted me to remain in the barrack next day, but I refused. I was afraid to be alone, afraid that I might give in to the temptation to commit suicide and finish what the SS soldiers had begun the night before. Dr. Ottinger discovered superficial rope burns on my neck. He expressed amazement that I hadn't begged for my life at any point the previous night. I told him it was only because none of it had seemed real until it was over. He shook his head slowly and sadly.

Dr. Aarons volunteered to help fetch breakfast, hoping to pick up some news. I lay on my bunk, only half listening to what was going on around me. I could barely make it to my feet when the Stubendienst ordered us to line up for the first count of the day. I was still alive, but I felt like a corpse who would never see sunshine or his friends' faces again.

Dr. Aarons returned with breakfast. He had overheard the kitchen Kapo and our Stubendienst talking about our barrack. It was about to be cleaned out. Some of us were going to be transferred to the inner camp. The rest of us would be going to the bunkers, the "interrogation dungeons." It was much too clear to me where I was destined after the rabbit hunt and my near hanging last night.

We called Kalka over right away and told him what was about to happen. He got very panicky and wanted to carry out our escape immediately because he was convinced that he was going to the bunkers. I agreed that we could not wait, but I wanted to use today to gather all of our supplies and bury them in the garbage ditch in readiness. We needed this time to prepare. Our escape already entailed enough risks that we could not anticipate. It was foolhardy to risk increasing our dangers by doing something stupid and avoidable simply because we had been too rash. Dr. Ottinger was also in favor of waiting another day. It would give him time to steal some money he had discovered in a matchbox on the Blockältester's night table. That decided it. Today we would get ready. Tomorrow we would escape from Dachau.

After breakfast we tied our civilian coats around our waists and hid our catheters under our shirts. There were camp identifying patches on the coats, but we could rip them off later. Dr. Aarons put the iodine in his pocket, I put the gauze and tape in my pockets, and Kalka carried the schnapps in his pocket. Dr. Ottinger was still tending the Blockältester full time. We had no time to find a replacement for him.

Our preparations went smoothly. We hid our supplies at one edge of the ditch and covered them with horse manure, knowing that no one would try to dig through it. When we returned to the barrack Dr. Ottinger gave each of us five marks. He was being transferred to head the HKB, the prisoners' infirmary, first thing in the morning. He did not want to call further attention to us by spending too much time in our barrack. As he said this brief good-bye to us his eyes were filled with warm affection. He promised Dr. Aarons that he would do everything possible to get him transferred to the HKB as well. He wished us "*Merde,*" "good luck" in French slang, and turned and walked away without shaking hands.

Dr. Aarons' words echoed my thoughts. "He's not so tough after all! I always took him to be a cold and calloused professional. It's become obvious to me now that he surrounds himself with an aura of arrogant detachment to hide his true feelings." He then added an important piece of news for us. "One of the workers here is an old comrade of mine from the Maquis. He will arrange things with his work unit so he can join us and take Dr. Ottinger's place."

That night I was dead tired but much too excited and frightened to close my eyes. I stared at the ceiling beam and thought about tomorrow. If we were caught in the garbage ditch it would mean certain death. But so would remaining in the camp. The Gestapo knew that I had been trained in London and worked with the French Resistance. They had sent me to Dachau and kept me alive only to get information out of me. I would be interrogated again, tortured and, whether they broke me or not, I would finally be murdered. My days here were numbered, and the number was very small. I still had misgivings about Kalka, but now I had no choice. I had to go through with our escape plan.

As I lay there my weakened bowels began to bother me. I had to control them because I was terrified of being caught out of bed if I went to the latrine. It dawned on me that this might be a real problem for me tomorrow and might even incapacitate me. Why hadn't I realized this earlier and gotten myself some constipating medication through the Blockältester's surgery? I could have kicked myself.

What would we do if the priest Kalka knew was not at the Frauenkirche when we got there? The more I worried, the more guilty and anxious I began to feel about those I was leaving behind, particularly Dr. Aarons, who was physically very frail. Fears, rational and irrational, besieged me, gradually turning my excitement to panic. I called Dr. Aarons, but he was asleep.

I tried to focus on the reality that escape was my only alternative. I told myself over and over that nothing could go wrong. Our plans were as good as they could be. My thoughts grew fuzzy and confused. I closed my eyes. Exhaustion and the sleeplessness of the previous night finally overtook me.

The next morning I hoped in vain to see Dr. Ottinger in the washroom. Kalka, Dr. Aarons, and I volunteered to carry breakfast back to the barrack. On our way to the kitchen we rehearsed our plans again. At each dumping Kalka and I would take turns concealing ourselves in the ditch. At the last dumping of the day, both of us would hide. Dr. Aarons and his Maquis comrade would make sure that we were both well covered by the garbage.

The day was gray and overcast. It was impossible to say how much of my gloom reflected the weather and how much my own forebodings. I was dismayed that this cloud cover might mean no moonlight that night. It would reduce our risk of being seen but increase the difficulties of finding our way.

We were called to our work units after breakfast. Our routine was the same as always, loading the container, pushing it to the gate where our SS guards joined us, and proceeding to the ditch. I hid during this first transport, then it was Kalka's turn, and we continued to alternate.

Finally, Kalka jumped into the ditch and I joined him. As I placed my catheter in my mouth I felt the garbage covering me. My heart was pounding in my throat. The squeaking noise of the wagon receded farther and farther into the distance. Then there was only silence.

I let a little time go by before pushing the garbage away from my face and opening my eyes to the dusky twilight. I cleaned off my mouth and said, "Psst!" Kalka answered. We dug ourselves out and retrieved our belongings from under the horse manure. Then we ripped the patch of striped fabric off the back of each coat. We decided to put off painting our striped berets and pants until we had put a little distance between ourselves and the camp. Kalka took out his small bottle of schnapps and we each took a warming swallow.

Kalka suggested that we crawl away from the ditch so that if we were spotted from a distance, we could be mistaken for stray cats or dogs foraging for food. I didn't want to. I had heard that the German food shortages were so acute in rural areas that stray pets often ended up on the dinner table. I wanted to find the railroad tracks and follow them to Munich. Kalka vetoed that idea. The German railroad workers were among the most fanatic of Nazis. We should make our way to the neighboring village of Dachau and find the bus that went to Munich.

Our discussion had taken us to the end of the garbage ditch. We stood up and started to clean ourselves off. Despite all our efforts the stench of rotting garbage clung to us. Night was falling quickly. It had gotten perceptibly colder. We took another swallow of schnapps before walking briskly for the next twenty minutes. We kept off the road and didn't see anyone.

As we approached the outskirts of the village in the dark of early evening the rural silence was torn by sirens. In the distance we saw the glow of exploding bombs and searchlights raking the sky for attacking American planes.

"Now," Kalka said, "at least we know where Munich is!"

We neared the road and saw people running, all in the same direction, most likely to a community bomb shelter. We sat down in a ditch off the roadside and began to paint our pants and berets right away. We did not want to run into a patrol with our stripes clearly advertising where we had just come from.

The American airplanes swarmed overhead, black silhouettes in tight formation against the translucent navy sky. The scene held a power and beauty that sent shivers down my spine and covered me with goosebumps. We were surprised not to see any German planes giving chase. We could hear the distinctive pom-pom sounds of antiaircraft guns, and there were distant white smoke puffs from the flack. A single airplane suddenly fell away from the formation and traced a slow arc across the sky. As we stared there was a sudden burst of light. The plane spiraled down faster and faster. We watched the sky, mesmerized. Another plane was hit, then another, and another, and another. . . . We were amazed to see the others continue without the slightest tendency to break formation, as if the destruction of their comrades didn't in any way concern them.

The flak grew more intense. A bright light shot up toward the sky and we heard the muffled explosion of bombs. The planes, having unloaded their bombs on Munich, were coming our way again. Their droning was fading into the distance when the sirens sounded the all-clear signal.

Then the night was quiet. A flickering light somewhere in the distance was the only reminder of the deathly chaos that had reigned for the past half hour. Our pants and berets were painted by now. We left our shelter for the road and walked in the direction in which the airplanes had flown—toward Munich.

We found ourselves in the center of the small village of Dachau. The few people who were out in the street were all looking toward the distant flickering lights. By listening carefully we could make out faraway firetruck sirens. In Kalka's experience, bus stations were usually located in the center of a village. We scanned the central area carefully and discovered not only the bus station but an approaching bus. Its headlights were camouflaged to hide all but a very faint blue light. The inside of the bus was in total darkness. We climbed in and asked the driver if he went to the Frauenkirche in Munich. He nodded, and we took our seats in the dark.

The ticket conductor came over and asked with a nasty chuckle if we had been hiding in a latrine during the raid, or perhaps, he snickered, we had simply shit in our pants? I had become so used to the stench of the garbage ditch that I didn't realize how offensive we smelled. I ignored the man's jibes and asked for two tickets, holding up my five marks. He handed me the tickets and some change.

After the bus left Dachau the conductor began to talk indignantly about the barbaric American pilots killing innocent German civilians. He talked on steadfastly, and we arrived in Munich without once having to answer him, although I had to restrain myself from asking him what he thought of the famous concentration camp so near his village.

When we finally arrived at our destination the conductor stopped talking just long enough to point it out. "There's the Frauenkirche. If you want to live in a free world, go inside and pray for Germany's victory!" I invited him to join us, laughing inwardly at my private joke. I even gave him a swallow of schnapps before we got off the bus and headed toward the church.

The old stone church faced a town square that must have been filled with people during the day. We went around to the side and pushed through an unlocked door and went directly to the confessional booth. Kalka pushed the button to summon the priest. We waited. For twenty minutes we sat almost motionless and exchanged an occasional whispered word. Kalka pushed the button again. We heard approaching footsteps. I made out a black-clothed figure in the dim light. Kalka questioned the man as to his name. The voice on the other side of the grill let out a joyous whoop, and he and Kalka launched into a rapid exchange in Polish. Kalka seemed completely unmindful of the cramped confines of the narrow booth. I finally interrupted him to suggest that we continue this conversation elsewhere.

As we stepped out of the confessional the priest kissed each of us warmly on both cheeks. He was slender and fair-haired, with the high cheekbones and almond-shaped blue eyes of the Polish aristocrat. An air of cool reserve was softened by a kindness that was obvious in his gentle eyes. The priest led us through his private quarters and down to the church cellar to join two Polish escapees from a nearby forced labor camp. He told us to get some sleep. In the morning he would give us a nourishing meal and begin all the preparations for our flight. We had come to the right place for help.

We could not disguise the fact that we had come from a jail or a concentration camp but, ever wary, we withheld our names from the two other escapees. Kalka became involved in conversation with them. I sat off in a corner. Although they were speaking in Polish, Kalka's gestures soon indicated that he was telling them the story of my "hanging" in the camp. He came over and embraced me affectionately. One of the other men stood up and brought me a piece of his sausage. It was his only way of expressing his sympathy and his desire to comfort me. I was deeply touched by his gesture. I recalled all the gifts of food brought to my bedside at Royallieu. I was again struck by the profound symbolic and emotional qualities food takes on when men have been stripped bare of virtually everything else.

I spread a blanket on the floor and lay down. Kalka pulled his over and lay down beside me. The chill of the stone floor was softened by the almost palpable warmth in Kalka's voice as he spoke of our having reached safety and of the strong brotherhood between us. Still the nugget of mistrust in me refused to melt.

I pulled my knees up to my chest to try and keep warm. There I lay, Albert Haas, the Jew. I had just escaped from the Germans and found asylum in the company of a group of Poles. Virulent anti-Semitism had for centuries been given as welcome a home in Poland as in Germany and Austria. Were these three men in the cellar Jew haters? Was the priest? I couldn't risk finding out. The Germans had not discovered that I was Jewish. I prayed that these men would not either.

In the dark Kalka reached for my hand and squeezed it, saying we would never separate until we were both completely free. He talked about what our lives would be like after the war and how I would be reunited with my family and all those dear to me. His voice gradually lulled me into a sense of security and warmth, and I slowly sank into a deep and dreamless sleep.

A few hours later I awoke, very cold, and dimly aware of someone covering me with a blanket. Through the haze of my sleepiness and the cellar's murky light a nun's habit materialized from the shadows. I gazed up at a woman's face in the semidarkness. Her finger at her lips gestured silence. I didn't move a muscle. The black dress merged back into the darkness. I fell asleep.

II

The aroma of freshly brewed coffee woke me. In the cellar light I saw Kalka sitting at a table with the two Polish escapees and the priest. A nun was pouring hot coffee. She must have been the woman who had covered me during the night. I wished everyone good morning and took an empty chair. Breakfast, to my astonishment and delight, was real coffee, fresh *Bauernbrot* ("farmer's bread"), and a dish of *saindoux* ("pork lard"). I spread lard on a thick piece of bread and salted it. Each marvelous bite was to be savored slowly, turned over and over in my mouth before being swallowed. The others stared and then began to laugh in delight at my obvious pleasure.

I took my time relishing this breakfast. The food was delicious, and it was my first meal as a free man in four months. When I finally finished, I took the priest aside to ask if he had heard anything about our escape, but it was too early in the day for the Underground fighters from the village of Dachau to have arrived in Munich with any news. He didn't think anyone had ever tried to escape from Dachau before. That was probably why we had found it relatively easy so far.

"Now," he said, "you have to be patient. It will take me time to get you clothing and false identification papers through the Underground. You also need time to cover up your emaciated bodies and grow back the strip of hair shaved down the center of your heads. Meanwhile, you are safe in my church. And if the church itself is ever searched, you have the catacomb to hide in. Even though such a search is highly unlikely," he continued, "we have to take all possible precautions. I want you to keep the cellar free of all signs of habitation. You will have to practice disappearing at a moment's notice, leaving no trace of your presence. That is why I cannot let you have mattresses or even straw pallets. Sleeping on the floor isn't comfortable, but it's safer."

The priest took us through a well-hidden entrance down to the catacomb itself. We entered a long, musty stone corridor. Tombs were recessed in its shadowy walls, catafalques lined the center. We could dislodge a stone from the base of a catafalque and hide underneath it. But we had to be careful not to rattle the skeletons inside and give ourselves away.

Because of my weakened bowels I was concerned about provisions for our toilet needs. The priest said that he would bring us upstairs to his quarters when there was no danger. We would have cans filled with lysofoam or carbolic acid for use in the privacy of the crypt when it wasn't safe to go up. They could be emptied upstairs when it was safe. There was drinking water available to us in the cellar. At night, when circumstances permitted, we could use the priest's bathtub upstairs.

He gave us a chessboard and some books, and said he would give us writing materials. But our letters must not contain anything negative about the Germans, disclose our whereabouts, or request an answer. Our return address would be the number of a forced labor camp. The letters would be mailed from inside this camp by Underground fighters.

The priest took our measurements for our new clothes. He suggested that I top off my new outfit with a beret and that Kalka wear a typical Polish hat, a soft visored cap. He told us both to start growing moustaches. For the time being we should get as much sleep as we could and, when awake, remember not to talk above a whisper.

Finally, the priest said that he would come down and pray with us each morning, and he offered his services for confession. I was less than thrilled by this announcement. I sensed danger. "Now let us pray together," he began. He looked at me before he started. "I have not asked your religion. My only concern is that we are praying to God. Whether you're kneeling or standing, a prayer is still a prayer."

The others knelt. I caught surprise in the priest's eyes as I joined them on my knees. When he had finished praying in Latin, he blessed us and asked us to remain on our knees and pray for the Jewish people who had perished because of their loyalty to their religion. To my astonishment he recited the *Kaddish,* the Jewish prayer for the dead. I was so overcome with emotion that, when he finished, I rushed up and kissed his hands. My gesture was so spontaneous and natural that it was done before I even realized it might give me away to my Polish comrades. What, in fact, happened is that all three followed me and did the same thing.

I am now sure that the priest knew I was Jewish, yet he never alluded to it. I was the one who acknowledged it the day I went to see him with a private request. In the confessional booth he recited the *Kaddish* and I repeated it after him. I also recited the *Shema Yisroël,* as I had in Fresnes. I was ashamed that I could not recite these two basic prayers of my religion without help. At that moment I made history. I was the first Jew to recite the *Kaddish* and the *Shema* in a confessional booth, prompted by a Polish priest!

On our second morning in the cellar the priest brought me letter-writing materials. I had a formidable task trying to explain my situation to Mme. Billard in Nice between the lines of my letter. I began by saying, "I just missed seeing my grandfather." I had to hope that she would realize he was dead. My letter went on: "I am making my way toward Montpellier, where my Uncle Jacques is living. I am hoping to find a train connection and arrive home for the holiday." *(I had escaped and was planning to make my way back to France by Christmas.)* "I have not heard from Sonja since she went to work in a different factory from mine." *(I had no idea of her fate since we were arrested and separated.)* "I would very much like to have a book of François Villon's poems because I would love to refresh my memory of him. Please tell the librarian how much I miss this particular book." *(I deeply miss my son* [*François*], *who I know is safe with my family* [*in Villon*]. *Please let them know.)* "It's pointless for you to answer my letter because I will soon be transferred to another place, one from which I will be able to help Ramoine with the fine mechanics. I will write again as soon as I can." *(I will have valuable information for the Resistance if I can get back home.)*

I closed the letter with love to everyone, sealed the envelope, and gave it to the priest.

That afternoon he brought some of our new clothes—a suit for Kalka, military boots and thick khaki-colored socks for me. When the remaining items arrived he would bring a photographer for our false identification and working papers. He examined our hands for some clue in choosing our new occupations. None of us had the heavily calloused hands of manual laborers, but they were easily rough enough to belong to skilled mechanics or lathe workers.

After the priest left Kalka wanted to play a game of chess, but before we had even made our opening moves he began to worry morosely about his future. He was confident that I would survive, but what would happen to him? He had not seen his wife and little daughter since the collapse of the Polish army, and he was afraid he would never see them again. He made me memorize his address in Warsaw, and I had to promise to visit his family after the war.

"Don't worry so much. You'll be back home soon," I reassured him. "And when I come to Warsaw to visit I'll bring French perfume and a silk scarf for your wife and a doll for your little girl."

Kalka was far from comforted. He shouted angrily at me, "If you're so damned sure I'll be okay, why didn't you mention any present for *me?*"

"Because I was saving you for last," I said, and I quoted a well-known German proverb to the effect that sugar always sinks to the bottom of the cup and makes the last sip of coffee the sweetest. "I'm going to bring you a bottle of vintage champagne and some pâté de foie gras with truffles. We'll sit down together and eat and drink, and talk about old times and our escape from Dachau."

Kalka looked at me earnestly and asked if I really thought we had a fighting chance of getting out of this alive. I couldn't talk about my doubts, since he was partially responsible for them. I put as much enthusiasm in my voice as I could muster and told him, "When we were buried under the garbage in Dachau did you dare to think, even for one moment, that before morning we'd be in this Munich church, far away from the SS guards and their lackeys, sleeping comfortably in a protected cellar, eating well, playing chess in the afternoons?"

Kalka cheered up and exclaimed, "Wait till you see how beautiful my little girl is! She must be fourteen years old by now."

He continued to talk about his family and their high-ranking careers in the Polish army. He was his parents' first son, and his authoritarian father had demanded that he follow in the family's military footsteps. Kalka's passions were music and animals, but at the age of twelve he was sent to military school. He graduated as a lieutenant and went on to the cavalry, where he soon married his colonel's daughter, Maria, a haughty aristocrat. The marriage, so typical of many young military men, had been one of convenience.

Kalka's reminiscing led me to think and talk about my own marriage. It stood in such stark contrast to Kalka's loveless union. Sonja's imagination and intelligence had enriched my life to such an extent that I would now feel like a shell without her.

Our talk moved on. I wanted to know Kalka's thoughts as to why the Polish people were, on the whole, so anti-Semitic. He reminded me that anti-Semitism was now as virulent in France as it had been for many centuries in Poland and the Baltic countries. Never had the government of an occupied country cooperated as zealously with the conqueror as France was doing in hunting down her Jewish citizens to help Hitler.

Kalka spoke of a Jewish friend of his from the army, a man who had called himself a Polish national of the Hebrew religion. Kalka had never understood how a man could see himself as a Pole first and a Jew second; never, that is, until he saw his friend executed at Katyn, not as a Jew but as a Polish patriot.

Our conversation was cut short by the priest's arrival. With him was a middle-aged man carrying a camera who had come to take our identification photographs for our new papers. The priest handed me a package containing a black suit, a collarless peasant shirt, and a separate collar that was fastened on with a collar stud. I got my hat and shoes and tried everything on. The photographer stood us in front of a dark and unidentifiable wall. Our pictures were taken with our hats on to hide our shaved heads. The photographer gave the exposed roll of film to the priest and wished us *Hals und Beinbruch*—the German good-luck equivalent of "break a leg."

"Why," I asked the priest, "is the photographer walking around the streets with his camera empty? I know the Gestapo frequently confiscate civilians' cameras. If a Gestapo agent takes his camera and finds it empty, he'll be very suspicious that something is going on."

"We're not that naive!" the priest answered. "Today is market day. The photographer always comes here on market day, and he always carries an undeveloped roll of film in his pocket that was exposed on a previous market day. We've taken precautions all the way around. No one will be able to trace any of the materials used to make your papers, and none of the people involved know each other. They only know me." The priest never divulged any more information than that about his Underground organization.

Now that the chain of events had been set in motion for the next phase of our escape, the priest talked to us about what would happen once we left the church. Since Kalka and I were both going to France, we would make our way together. Public transport was to be avoided if at all possible. We would usually use private vehicles, a peasant's wagon being the ideal. In rural areas we would spend our nights in a friendly farmhouse. In cities and large towns we could easily disappear into the masses.

We were instructed that two guides would always accompany us from one destination to the next. One would covertly lead the way while the other would unobtrusively follow from behind. The leader would never enter a house with us but would stop in front of it. He would give us the all-clear sign by lighting a cigarette, then crushing it out. A new set of guides would pick us up when we left the house. They would identify themselves by standing in front of our hiding place and lighting a cigarette. A cigarette quickly crushed out meant "danger, stay inside." Continuing to smoke meant "the coast is clear."

Only a few escapees with guides had been caught so far. Nothing had been learned of their fate. I wondered out loud what plans had been made for the two Poles hiding with us. The priest politely indicated that it was none of my business, and I apologized.

He left us and returned a short time later with our lunch and some sacramental wine. It had been so long since my last glass of wine! The warm glow left in its wake almost made me forget our precarious situation.

After lunch the priest offered us another special treat. One of us could take a hot bath. We drew lots. Kalka won. I was beside myself with jealousy. The stench of the garbage ditch would have to stay with me for a while longer.

While Kalka was upstairs scrubbing Dachau out of his pores, I looked through a copy of the *Voelkischer Beobachter*, the German daily newspaper, which the priest had left us. The Germans were retreating along the Russian front, which they euphemistically called a "strategic improvement of their situation." The editorials were filled with simplistic and obvious lies. Any normal ten-year-old could see through them with ease, yet the German public swallowed them whole! I had just put the paper down when the priest returned with a sweet-smelling Kalka and announced that there was still enough hot water left for me.

He took me up to his bathroom and gave me a small piece of soap with the admonition not to let it sit in the water. Soap was very hard to come by. I lowered myself into the tub, feeling the warm, clean water caress my skin and enfold my body. It was sensuous, timeless, spiritual, exquisite—an ultimate pleasure for which there are no words. I drowsed in this luxurious cocoon until I was startled awake by a knock on the door and a request to finish. I had been in the tub for a whole hour.

The water had grown cold. I soaped and rinsed quickly, then savored the pleasure of drying myself with a large, soft bath towel. I put on my new clothes. I felt truly like a new man, happy, optimistic, and full of energy.

Kalka and I, both feeling immeasurably refreshed, tackled a game of chess. The priest spent the next hour conferring privately with the two Polish escapees, who were to start on their journey the next morning. The nun brought down a late supper for everyone—sauerkraut, sausage, and dark beer. It was my first beer in a very long time. Ersatz beer had never tasted so good!

The priest then asked us all to go to sleep early so that the two Poles could have a good night's rest before their morning departure. After we bedded down I tried to imagine myself in their place, but whether it was someone's rhythmic snoring in the dark or the effects of the wine and beer, I was asleep in an instant.

The next thing I knew it was 7:00 A.M. and the priest was waking me for breakfast and prayer. The others were already up and dressed, and deep in conversation. I fervently hoped that Kalka had heeded the priest's warnings not to reveal who we were, where we had escaped from and where we were headed. After breakfast, the nun came down and asked the two Poles to follow her. We exchanged farewells, wished them luck, and they went upstairs.

Kalka and I retreated into our separate thoughts until he broke the silence. "I hope everything goes okay for them." I asked him point-blank if he had revealed anything about ourselves to them. He said, "All I gave them was my family's address and a letter to my wife." Kalka remained somber and subdued for the rest of the morning. I grew increasingly edgy.

When the nun appeared with our lunch I didn't dare ask how things had gone. She volunteered nothing. The priest came down shortly afterward and reported that the guides and their charges had departed without incident. He had no news yet of what had happened to them since they left the church.

He thought we might like something interesting to read and offered a choice from among the French, Latin, and German books in his library. I was partial to Zola and Hugo. Kalka expressed no preference.

The priest returned from his library with two of Hugo's novels —*Les Misérables* and *The Hunchback of Notre Dame.* He also brought me a book on Chinese philosophy. In view of our circumstances he felt that I would appreciate the Chinese philosophers who advocated living one day at a time and the futility of worrying about tomorrow. I was fascinated with the apparent similarity between Sonja's and my attitudes and those espoused by these Oriental sages.

For Kalka the priest brought *Quo Vadis,* a Polish novel I had read as a teenager. The apt title—"Where Are You Going?"—called forth an ironic chuckle as I thought of the overwhelming uncertainty obscuring our own future.

I began one of Hugo's books but could not concentrate on anything other than the fate of the two Polish escapees. What had happened to them? Had they gotten away? I was very restless. I got up to stretch my legs. My walking about disturbed Kalka. I sat down again and opened the book on Chinese philosophy, but it didn't hold my interest.

The time continued to pass slowly until the priest finally arrived with the cheerful news that the two Poles had safely reached their first destination. We had dinner soon after, then loafed around. Neither of us was in much of a conversational mood. In the camp, where we were forbidden to speak to each other, we had wanted to talk all the time. Here we were free, yet we felt too anxious to talk at all.

The priest enlivened—if not cheered—our evening with, at long last, some news from Dachau. Our absence had been discovered at suppertime. Everyone in our barrack, including the Blockältester and Dr. Aarons, had to spend that entire night standing outside in the freezing cold. The SS had turned the entire camp upside down, but it had never occurred to them to interrogate the men in the garbage collection units. Several prisoners, including two Frenchmen, had been "taken away." The priest was sure the Germans would be hotly on the lookout for us for a while. We should remain in the church long enough for the furor to die down. Perhaps our escape had been a lucky break for our two Polish comrades. The hunt for us might divert attention from them and enable them to get away.

I was very pleased when I saw myself in the shaving mirror the next morning. My moustache was beginning to look like a moustache, even though I didn't like its reddish color, and my telltale bald stripe was beginning to disappear. The nun brought our breakfast and told us that the priest had been called out. She would stay to pray with us.

After breakfast I asked her, a little hesitantly, if she might be able to find a toothbrush for me because I hadn't been able to brush my teeth for quite a while. The roots of some of the teeth broken during my interrogation had become infected. She promised to find something I could use. At that moment the priest returned, bringing with him our false papers and a *musette* (a kind of "shoulder bag") for each of us. Kalka and I played chess until lunch. Then we read for a while before returning to the chessboard.

This was to be our routine for many days, interrupted only by occasional baths and frequent instruction sessions with the priest. Like small schoolchildren we had to repeat his instructions out loud, over and over and over again.

"Never tell even your guides who you are and where you have escaped from."

"What those in the escape organization don't know they can never confess if they are arrested."

"Never start a conversation with the inhabitants of a house in which you are being hidden."

"Be wary if the inhabitants become inquisitive. This may indicate stool pigeons who have infiltrated the organization."

"Be particularly wary in a house where there are children. They can be far more dangerous than adults, especially if they belong to the Hitler *Jugend.* They'll give away their parents as well."

"Keep your wits about you, and behave as naturally as possible."

12

One morning when the nun came down to wake us I was struck by her uncharacteristic agitation. She had bad news. The two Poles had been caught. The priest was out trying to locate their guides and get them safely into hiding in case the escapees confessed under torture.

As this news sank in, my stomach began to cramp. I told the nun I had to run to the bathroom upstairs. She warned me to go slowly because there were early worshipers in the church. I had no hope of controlling my bowels for that distance if I had to walk unobtrusively. I ran into the catacombs with barely a moment to spare.

While I was relieving myself unfamiliar voices began to filter in from the cellar. My God! Had the Poles told the Gestapo about the church? I finished as quickly as I could and approached the door, frightened and cautious. I gradually recognized one of the voices as that of the priest. The other was definitely a stranger. I vainly tried to determine whether the voice was friendly. I heard the priest ask for me. The nun said something I couldn't make out. The priest laughed. She had only been describing my intestinal response to the bad news. I was weak-kneed with relief.

I emerged from the catacomb to see the priest in conversation with a middle-aged man. A younger man was helping Kalka pack his belongings in his musette. The priest caught sight of me and suggested that I sit down for a few minutes. I was mortified and mumbled that last night's dinner must have been too much for my weakened system. At the same time I was angry with myself for feeling compelled to excuse what was surely a natural reaction in these circumstances. What would have been so terrible about openly admitting to these people that I was so badly scared I had almost shit in my pants?

The priest recounted what had happened to the Polish escapees. "SS men in the street spotted them and began tailing them. Following procedure, their guides gave a danger signal to the escapees and then unobtrusively retreated to a safe observation distance. They watched the SS men grab the Poles. One of them summoned a car with a whistle. The Poles were hustled inside and taken away. The entire incident was carried off so smoothly that no one in the street except the guides seemed to realize what had happened.

"This means the Germans may learn of the church and search it. The two of you have to be on your way immediately. There isn't a moment to lose. The two men here with me are your guides."

My head was spinning. This was not the leave-taking I had imagined.

The nun handed Kalka and me some bread and sausage. Our guides helped us pack while the priest rechecked our papers and reiterated his stern warning to observe our instructions meticulously. He expressed his hope that we would meet again after the war. Then he prayed, gave us his blessings, and wished us luck.

I wanted the nun to know of my gratitude and affection. Words seemed too shallow and impersonal to express what I felt. I was hesitant about physical contact, but I bravely decided to kiss her. At the last moment I compromised and switched from her cheek to her hand, and added a few words. Kalka did the same, then there was nothing more to be done. I followed my guide out of the church to the square, my heart pounding.

It was Munich's market day. My guide, who was the older of the two men, walked over to a vendor and innocently inquired about his prices while I appeared to admire some richly worked slippers at another stand. From the corner of my eye I was following Kalka as he left the church and came toward me. My guide began to move away from the market stalls. Kalka's guide nodded for us to join them.

Kalka and I walked along like a couple of friends chatting about trivial matters. We left the Kirchenplatz behind us and kept going. Suddenly I felt overwhelmed by the dizzying realization of my freedom. The fresh air in my lungs was like alcohol on an empty stomach. The younger guide caught up to us for a moment to direct us to the bus stop nearby. There had been no time to arrange for a peasant's wagon. We must take the bus to Augsburg.

At the bus stop a man in civilian dress asked to see our papers. I handed mine over, trying to appear nonchalant as he scrutinized my face and picture for a few endless seconds. Trepidation had knotted my insides by the time I heard his approving "*Alles ist in Ordnung.*" Kalka's papers were also approved, and we boarded the bus. Our guides sat apart, one in the front and one in the rear. We had to wait while the bus filled with passengers. The door closed and the bus started. We turned into Bayernstrasse—I caught the name on a street sign. I saw a sign indicating fifty-nine kilometers (thirty-six miles) to Augsburg. We were on our way.

Kalka's spirits were high, which always made him garrulous. I listened vaguely, absently nodding yes or no. My mind was preoccupied with my return home. What if my Resistance comrades supposed I had been released from Dachau because I agreed to become a double agent? It certainly had happened in the past. How would I deal with such suspicions if they arose? I had to put these anxious fears out of my mind in order to keep my spirits up. I daydreamed about family and food. I began with breakfast—the croissants my sister baked, slathered with sweet butter and homemade preserves, served with freshly brewed café au lait. I tasted her croissants for the next thirty miles.

Kalka nudged me from my culinary dreams. Our lead guide had gotten up. We gathered our packs at the next stop and got off behind him, followed by the younger guide.

It was close to noon, and we were in the outskirts of Augsburg. Our guides led us to a roadside tavern and indicated that we should enter first. Evidently they had to check that we were not being followed. We went in and sat down. They joined us at our table a moment later and informed us that their job had been to deliver us to the tavern owner. He was part of the escape network and would take charge of us from here. They wished us *Hals und Beinbruch* and disappeared out the door.

A short, chubby man in his thirties brought us each a beer, telling us as he set them down that our new guides would be arriving very soon. We felt terribly alone and apprehensive. I stared fixedly at my beer. My face felt very somber. Kalka constantly glanced toward the front door.

The owner returned to our table. "You fellows had better be patient and try to act naturally even though there isn't anyone else here. It's lunchtime. Somebody could come in at any moment."

Attempting to look more natural, I took the bread and sausage out of my musette and started to eat. The appetite I'd had during the bus ride had deserted me. I could scarcely swallow because of something that seemed to be sticking in my throat, though I knew it was only an illusion. The sensation abruptly recalled a case of *globus hystericus* from my student days.

A pretty young woman had been hospitalized complaining that a piece of food had stuck in her throat during dinner the previous evening following a terrible fight with her mother. Our examination disclosed only a slight redness of the throat. At that time medical students were supposed to be seen and never heard, but I spoke up and diagnosed abrasion of the tissue from a breadcrust or piece of bone. The resident disdainfully explained that the woman's symptom was purely psychological, merely the result of the angry fight with her mother. I laughed to myself, certain that he was wrong. Her condition could not have been caused by emotions alone.

I was smiling at the recollection of my own naiveté. While Kalka testily asked me what I had found so amusing, two men entered the tavern and ordered white wine at the counter. The owner served them immediately but never once looked up at us. Obviously these men were regular customers, not our guides. One of the men left.

The other man waited a few minutes, then he came over to our table and sat down. He began to whisper instructions. "My companion and I are going to take you to your next destination. We're going by trolley. We'll sit near you, but we have to keep our distance on the street. We'll stop and light cigarettes in front of the house you're supposed to hide in for the afternoon. If we keep smoking, you keep walking. If the coast is clear, we'll stamp our cigarettes out right away. You knock on the front door and tell whoever opens it that 'we are the volunteers.' The answer should be 'welcome to Augsburg.' "

We gathered our things and began looking for the owner. Our new guide insisted there was no time for thank-you's and goodbye's. We must leave at once.

We took the trolley for a short distance, then walked for a few blocks. One of our guides instructed us to drop behind them when they crossed the street. They stopped by a house in the middle of the next block. They each lit a cigarette and then threw them on the sidewalk, stamped on them, and walked away.

Kalka and I crossed the street and approached the house apprehensively. I knocked. After a few anxious minutes a pleasant-looking young woman opened the door. I gave our password. She welcomed us to Augsburg. We entered the house. Before closing the door she stuck her head out for a moment and appeared to survey the street, then turned to us and smiled.

We found ourselves in a tiny hallway. Several men's overcoats hung from the mirrored Victorian coat stand against the wall. A military cap was perched on the little table below the mirror. Our hostess ushered us into her living room and introduced us to the three people who were sitting there, her husband, brother-in-law, and sixteen-year-old son.

Attempts at conversation were awkward and led nowhere. The hostess served us hot coffee and *Babka,* a typical German coffee cake. Kalka attempted to break the uncomfortable silence by complimenting her cake and coffee (which, in truth, were awful). Silence closed in again. The four of them sat and stared at us. When I could endure it no longer, I decided to ask a harmless question. "How long will we be staying here?"

The boy answered, "Not long. Just until the car comes."

My heart skipped and began to pound. The priest had not said a word about traveling by car. There were so few cars in the street nowadays that any automobile would automatically draw a lot of attention to us.

My face obviously reflected my suspicions. Our hostess immediately began to reassure me. She related the "incredible piece of luck" that a local civil servant, who also belonged to the escape network, had been ordered to travel to Strasbourg by car. We would go to France with him as his interpreters. He had only to enter our names on the official document.

My apprehension rapidly increased. This sudden and dangerous change in procedure, plus the need to disclose our names, meant that something was very wrong. I turned to Kalka. He was deep in conversation with the young boy. In stunned disbelief I heard him naming Katyn and Dachau.

One of the adults exclaimed, "Ah! So you are the two who escaped from Dachau!"

I tried to save us. "No, no, we are not from Dachau. We are

volunteer workers. We have to return home at once for family reasons. It takes too long to get official authorization so we're doing it this way."

Kalka interrupted me, "Oh, doctor, these people are our friends. We don't have to pretend with them!"

"*Ferme ta guele!*" ("Shut up!") I hissed at him. I spoke in French through the reflexes of fear, desperate to warn Kalka in words that our German hosts culd not understand. "*Docteur, ne t'inquietes pas!*" ("Doctor, don't worry!") he answered me, leaving me stunned and even more frightened. I thought Kalka couldn't speak French. Obviously he had lied to me about that. Was he a spy after all, and had I walked into a trap? Or was it simply that he had become so drunk with freedom that he had lost all sense of our danger? I sat in horror as he talked on and on in eager response to the chorus of "oh's" and "ah's" of the others. But some last vestige of caution or sentiment kept him from naming the various people who had helped us.

Making an excited gesture, Kalka accidentally knocked his coffee cup onto the floor and the boy kneeled down to pick it up. His sweater rode up over his belt in back, exposing the brown shirt of the Hitler Jugend. I recalled the military cap in the foyer. I turned to ice.

A car pulled up outside. The sequence of events that followed was played out with dizzying speed. Someone knocked on the door. The boy ran out of the room. Our hostess followed him, saying, "Here is your car." She returned with two men who barked, "We are from the *Geheimstaatspolizei* (Gestapo)! Don't move or we'll shoot you where you stand!"

For an instant I thought of escape, but in that same instant I realized it was hopeless. Still, later I would be furious with myself for not having attempted some kind of resistance.

As the Gestapo men marched us out of the house the boy was standing by the black Citroën. He had pulled off his sweater and was proudly showing off his Hitler Jugend uniform. I saw the swastika on his military cap, and several decorations dangling from his jacket. No doubt he would get another one for our arrest. He hooted derisively at us, "You think we Germans are so stupid."

He helped us into the car with a kick.

I glanced at Kalka, a face of pale terror staring into space. One of the Gestapo men sat to Kalka's right and the other, in the front seat, turned and wordlessly fixed us with a contemptuous stare.

Events had moved so swiftly that the present was not yet real. Only five minutes before I was sitting in a comfortable living room, a relatively free man excited by my dreams of the future. Yet despite my dazed sense that this was not really happening to me, one corner of my mind continued to grapple with the terrifying events. The detachment I always seemed to experience during imminent danger allowed me to wonder cooly whether we had been betrayed in the tavern or if it had happened even before we left the church that morning.

The Citroën pulled up in front of the Augsburg police station. A tall gate swung open and the car entered an inner courtyard. Kalka and I got out. Just before we were separated I whispered, "Have courage! Deny everything!" He walked away, leaden and ashen-faced, with the mechanical gait of a zombie.

I was taken to a ground floor cell. An exposed toilet stood in one corner. The window was of normal height but completely boarded over. The only furniture was a small bed with a bare mattress. My wooden cell door looked strangely out of place, as if it belonged in someone's apartment.

The time passed in agony because of the silence bearing down on me. I tried pacing for a while—five steps forward, five steps back, five forward, five back, five forward, five back. I began to feel dizzy and lay down on the mattress.

What was going to happen to me? I was sure that our capture alone would not satisfy the Gestapo. They would be determined to find out who had helped us, and Kalka and I had not even prepared a plausible story that we could rely on. What a careless, stupid oversight. What most frightened me was not torture but the possibility of breaking and revealing how we had been helped by the priest and the nun. Kalka might talk even if I didn't. All I could do was try to invent a believable story for the interrogation that I knew would come, and hope for the best.

I grew very thirsty and got up and stood with my ear against the door. The only thing that greeted me was that same deafening silence. I knocked, but no one came. I pounded with my fists and my feet and gave up only when my fists were on the verge of bleeding. I returned to the bed and lay there, staring at the ceiling and torturing myself with our failure.

We had been so confident in our carefully laid plans, but obviously we had made a mistake somewhere. I tried to review this twisted and frightening day. What had we done wrong? Why hadn't my intuition warned me sooner of danger in that house? Was I too trusting, or had they played their roles too well? My morbid reverie was interrupted by footsteps.

13

Two SS guards and a civilian opened the door and entered my cell. One of the guards handcuffed my wrists behind my back. I was taken out to the courtyard and pushed into the back of a waiting Citroen. Two SS guards sat on either side of me. A Gestapo officer sat in front beside the driver. He gave the order to start and the car pulled out of the courtyard. I asked where Kalka was. My question went unanswered.

It wasn't long before we entered another courtyard. It was night and too dark for me to have any idea where they had taken me. Not until I left did I realize that I had been in Gestapo headquarters.

I was shoved out of the car into a big white building that looked like an armory or a monastery. I was marched up two flights of steps and brought into a very large and brightly lit room. At first I thought it was a gymnasium because I could see ropes dangling from the ceiling, a wooden horse near one corner, dumbbells racked on the walls. I also noticed an electrical generator and what seemed to be a high-powered floodlight. A man in civilian clothes with a pleasant face was sitting behind a desk. I was shoved into the chair facing him.

The man said, "Sit down, please," which threw me momentarily off balance since I was already seated. He offered me a cigarette. I accepted. It would buy me some time to organize my thoughts since you can't talk while inhaling. He took a cigarette for himself and put one in my mouth. He took out a lighter and lit his cigarette. As he leaned over the desk toward me his hand suddenly jerked forward and the flame from his lighter burned my lips.

"Oh, how clumsy, please forgive me," he said.

Then, as though it hadn't happened, he told the guard standing next to me, "Handcuff him in front so he can smoke his cigarette."

He placed an ashtray by my handcuffed hands and asked me if I was ready to talk. How had we escaped from Dachau? Who helped us there? Who helped us get to Augsburg?

I told him that two people had helped me in Dachau. I named the young Frenchman who had hanged himself and a prisoner who had died in my rabbit hunt. I knew that it would be impossible to withhold everything. I told him about stealing some of my materials from the supplies delivered for the Blockältester's operation. I included the story of my near hanging after I was caught with the gauze and bandages. He already seemed to know about this because he nodded in agreement as I talked.

Then he cut me off. "That is a beautiful story, but save it for your return to Dachau. We know Allied agents have been parachuted into this area. We also know that you managed to avoid civilian patrols and that you had money for bus tickets to Munich. Obviously you had help as soon as you were out of the garbage ditch. Who was it?"

I replied that no one had helped us. We were able to make our way without being discovered because we had the cover of both darkness and the air raid. He looked at me skeptically. How come we hadn't run into any civil-defense patrols? I said we had hidden ourselves in a roadside ditch when they passed us by. Then we had made our way to the bus station and taken the bus to Munich.

"How did you know where the bus station was?"

I said we had simply gone under the assumption that there had to be a bus station in the town, and we had walked until we found it.

"Who gave you the money for your tickets?"

"I stole it from the Blockältester's room during his appendicitis operation."

"What did you do when you got to Munich?"

"We walked around for a while, then hid in a public garden."

I started to put out my cigarette in the ashtray. I was suddenly speechless with excruciating pain on the back of my right hand. I looked at it. My interrogator had crushed out his cigarette in my flesh.

Again, the sarcastic apology, "I'm so sorry. Pardon my clumsiness. I was so absorbed by your fairy tale I didn't notice your hand in the ashtray." His tone scarcely changed as he went on, "That's what happens to people who lie so well. Why don't you just tell me who hid you in Munich? Who provided your clothes? Who gave you your false papers?"

The disarmingly casual manner of my interrogator and the deadly cat-and-mouse game he was playing with me disrupted my ability to think quickly and literally gave me the shivers. My mind tried to keep up with his questions. I regretted even more that I had never worked out a cover story, however farfetched, so I would have something to recite and hold on to now. Had the priest been careless? Had he prepared such a story and simply forgotten to tell it to us?

I decided to try the moral appeal that had worked with the Gestapo during my torture in Nice.

"I can't tell you names and addresses that I don't know," I said. "But even if I knew them I could not knowingly put someone's life in jeopardy."

My Gestapo interrogator looked at me with what seemed a kindly expression, and replied, "Look, we're not in the Middle Ages. Why should you want to suffer for the sake of your honor? You won't be allowed to leave this room until I have your confession. Your friend Kalka is also here. If your story and his differ in any way, it will be a very sad day for you. You are going to be executed as soon as you return to Dachau. But it's up to you whether we torture you or leave you in peace before you return there to die." This was all stated in a voice as relaxed and natural as if I were being asked to help him solve an interesting but unimportant problem.

As usual, my bowels began to respond to my fear. By now I was also hungry, thirsty, and exhausted. An SS guard entered and put a few sheets of paper in front of my interrogator. He glanced through them and casually announced that it was the transcript of Kalka's confession. He would give me a few hours to think things over. A Gestapo man came in and took me out, down a corridor, up three flights of steps, and down another corridor lined with heavy doors. At an open door he pushed me into a small, well-lit cell and locked me in.

I became aware of moaning and sobbing very close by. Was it Kalka? I yelled, "Peter, Peter" at the top of my lungs. No one answered. I tried again. I finally gave up and lay down, falling into an exhausted sleep. It wasn't long before my interrogator woke me up. He wanted my decision. Would I corroborate Kalka's confession or not? I stood up and stretched my hands out for the handcuffs. Was I really the same man who had begun his trip home to France only that morning?

I was taken to the gymnasium again and pushed back into the seat opposite my pleasant-faced interrogator. My handcuffs were removed.

"What was the priest's name? Who else helped you in Munich? Who took your pictures? Who brought the clothing?"

I tried to let the questions guide my answers so that I wouldn't contradict myself. The Gestapo obviously had some accurate information, but I wondered why they made no mention of the nun. Had Kalka omitted mention of her, or was their information from someone who had never been inside the church cellar and wasn't aware of her participation?

I told my interrogator that I didn't know the priest's name. Two men had come to the cellar, one with a camera and another who actually took the pictures. An SS guard kicked me in the ribs. My interrogator asked again for the priest's name. I answered truthfully that the priest had never told me. The SS guard hit me on the head. I decided to say I thought his name ended with a "-witz" or a "-ski". "Does he have an accent?" I replied that my German was not good enough to judge.

The calmly impersonal and detached speaking voice of my interrogator was a jarring contrast to what was actually happening. "Stand up," he softly ordered. I was sent to the wall by a kick in my stomach. He commanded me to stand on one leg and stretch out my arms. He placed a heavy dumbbell in each of my hands and repeated, "Does the priest have an accent?" I repeated that I didn't know. "What does he look like?" I described an average-looking man.

The questions did not stop. My hands and arms became heavier and heavier and my leg was growing unbearably tired. I tried to get some support by leaning slightly against the wall. One of the dumbbells fell. An SS guard kicked me in my stomach. I fell backward onto the floor. It was a momentary relief from standing on one leg, although stomach spasms made me want to vomit. I was so terrified of their anger that I swallowed it.

"Stand up, you dog!"

As soon as I got to my knees I was kicked and beaten until I fell to the floor again. They made me stand up and then beat me down several more times. My stomach and back were beginning to hurt badly. I could feel my face and lips swelling. I was picked up off the floor and put in the chair, but I couldn't hold myself up and fell to the floor. They threw cold water on my face.

The interrogation continued in a merry-go-round of questioning and beating, questioning and beating, questioning and beating. Then everything suddenly stopped and I was carried away and dropped on the bare bed in my cell. I could finally let myself throw up, but since I had had nothing substantial since my stop in the tavern, all I had were dry heaves. I lay on the bed, weak and feverish, in pain from even the slightest movement.

In a little while an SS guard took me back to the interrogation room. A woman in an SS uniform was typing what turned out to be my "confession." My interrogator read it to me, a detailed description of the priest, which I had never given, and the names of the camera carrier and the photographer, which I had never known. Were they gotten from Kalka or a spy who had infiltrated the Frauenkirche escape organization?

He placed the transcript in front of me for my signature. I said indignantly that I couldn't honestly endorse it. "I know," he said. "Actually it's your comrade's confession. But we don't want you to suffer his awful fate, nor do we want you to have to repeat your unpleasant experience all over again at Dachau." I still refused to sign.

My pleasant-faced interrogator suddenly became enraged. He ordered the SS guard to "bring in the horse and the singing machine." The typist appeared to be enjoying the spectacle.

The guards pulled down my pants and underwear, then slung me over a wooden horse and tied my dangling hands and legs together underneath it. Again I refused to sign. In retrospect I realize how ridiculous it was that I was more embarrassed by my naked, fully exposed rear end than I was afraid of what they were going to do to me.

Rubber-gloved hands separated my buttocks and something hard and cold was pushed up my anus. I didn't know that it was an electrode. A searing shock tore through my innards and tied my intestines in previously unknown knots of agony. I screamed. The odor of feces and burning flesh reached my nose.

When I looked around me I was on the floor being doused with cold water. I had fainted. A voice said, "He's had enough. He'll talk now." I still smelled burnt flesh and feces. I was trembling from excruciating abdominal pain. I lost consciousness.

Pain brought me back to awareness. An SS guard was kicking me in the head and screaming, "Pig, clean yourself up!" Through the foggy slits of my eyelids I saw something dangling over my head. I reached for it. It was a towel.

I dried myself off, but I couldn't clean off the excrement. The SS guard took me to a bathroom to wash. The pain was so intense when I touched my anus that I did a very poor job. Half naked, I was taken back to the room and the interrogator pushed the transcript in front of me. I refused to sign. The two SS guards picked me up and threw me over the horse.

I couldn't take it again. Silently I begged the forgiveness from everyone mentioned in the confession. In utter resignation I called out, "I'll sign."

As soon as it was done, the interrogator asked me, "Was your resistance really worth it?" What could I say to him? With that kindly expression still eerily softening his features, he ordered the guards to take me back to my cell and informed me that I would be returned to Dachau the next day. He had ordered coffee and food to be brought to my cell because he "didn't want to be accused of being a poor host and neglecting his guests."

The food was waiting for me. Hot ersatz coffee, a piece of dark bread, and some steaming cabbage and potato soup. But my spasmodic and knotted insides left me with very little desire to eat. My need to lie down was overwhelming, but if I did I would fall asleep and the coffee and soup would be cold when I woke up. So I ate, using the bread as a spoon. The meal was immeasurably better than I had anticipated, but whether it was really delicious or just that I hadn't eaten in so long, I don't know. The old proverb does say that "the best cook is hunger."

I felt somewhat better after eating. I tried to anticipate what would happen to me when I returned to Dachau. My last coherent thought before falling asleep was the realization that nothing was to be gained from torturing myself with terrible fantasies. Those old Chinese philosophers were right. One day at a time was enough, and tomorrow was out of my hands.

I woke up and threw up my dinner. Then I must have slept for quite a while, certainly long enough for my bladder and bowels to fill, because I had to empty them. I was using the toilet in the corner when a key turned in my cell door and it opened to reveal my interrogator and an SS lieutenant. I was still crouching with my pants at my ankles as my interrogator introduced me.

"Here is our special guest."

The lieutenant said, "Get dressed and come with me. The reception committee at Dachau is waiting for you—with anticipation."

As we left, the door next to mine stood open on an empty cell. Fervently and foolishly, I expected to be reunited with Kalka downstairs. A khaki-colored Mercedes was waiting to take me back to Dachau. I was bitterly disappointed that Kalka was nowhere to be seen.

I was sandwiched between two SS guards in the back seat. The lieutenant climbed in front. As we left Gestapo headquarters I couldn't restrain myself from asking where my friend was. The lieutenant drew his hand across his throat.

"*Kaput, fini!*"

During the hour's drive I wondered in terror what the return to Dachau would hold in store for me, and what had really happened to Kalka? The gates of Dachau came into view and the car stopped just inside them. The car doors were opened. I was dragged to the guardhouse and hit in the mouth with a rifle barrel. My lips began to swell again and I tasted blood. Everyone—and there were many—who came in contact with me hit me, usually with their hands or feet, sometimes with gun barrels. Not a single word was uttered. The silence was unbroken except for heavy breathing and an occasional grunt of effort or pleasure.

When the word "*Halt*" rang out I was on the floor trying to protect my face and head. My battered face was swelling so badly that my eyes had shut completely. I was blind. "Leave something for the others," the authoritarian voice continued. "We don't want to deprive them of some amusement. Bring in the board." The beating stopped as suddenly as it had begun.

Someone pulled me up and put a rope around my neck. I was dragged across the floor by my arms, then through the door and down the one step to the ground. I tried to breathe deeply of the fresh air, but such intense pain stabbed through my chest that I could only take small half breaths. I was incapable of sitting up. I was lifted up and placed on my knees on top of something wooden. My hands were tied behind me, then my body was tied into an upright position. It wasn't until an order to move was issued that I realized I had been hoisted onto a wooden cart. I had no idea where I was being taken.

I could feel a breeze on my face when the cart finally stopped, so I concluded that we must have arrived at a large, open place.

Was it the Assembly Place? I felt the ropes supporting me being loosened, then I was lifted off the cart with my hands still tied behind me. I could not hear a sound.

My arms were raised up behind me and the rope that bound my wrists was used to suspend me from something a little higher than my head. Still not a word was uttered. When I stood flat on my feet my shoulders felt as if they were tearing out of their sockets. The only way to soften my agony was to balance on the tips of my toes and lean forward.

The silence surrounding me, which had become, if possible, even more intense, was ruptured by the same sharp voice. "This Häftling dared to escape from Dachau. Now that we have captured him, his punishment will set a fine example for the rest of you. Until we decide just what his punishment will be, he will hang here without food or water while the rest of you parade by him to see what will happen to anyone else who tries to escape. There is only one way out of here, and that is through the chimney!"

The voice uttered a command to march. An infinite number of wooden shoes began clattering on the cobblestones. Each footstep of these unseen thousands exploded in my brain. It was becoming impossible to stay on the tips of my toes. My arms were being torn out of my shoulder sockets. I wanted to scream out that I couldn't stay erect any longer! Finally, no amount of willpower could keep my legs from collapsing. I hung, suspended by the rope around my wrists, as the inside of my head grew as dark as the world around me and I sank again into total blackness.

The numbing mantle of unconsciousness receded under a wave of cold. My shivering body ached all over. I was lying on a damp, cold floor. Gradually I opened my eyes, but in the darkness I could see nothing, not the barest clue as to where I was. The slightest attempts to move my arms shot incredible pain through my shoulders. I tried moving my hands, but my wrists hurt so terribly that I could use only my fingers to explore the stone surface on which I lay. My efforts exhausted me.

I discovered that I could manage to roll sideways along the floor. I rolled until I came to a wall. I wanted to use it to lean my back against, but the capacity to sit up was beyond me. Lying on my back I used my feet to swivel myself until my head was against the wall, then used my feet to try and push myself, slowly, up the wall. Eventually, after many exhausting failures, I managed to raise myself to a sitting position. My head was throbbing and I was so out of breath that I had to rest. The pain remained excruciating even when I stopped moving, but it was a little duller now and somewhat easier to bear.

The silence around me was deep and insulating. I tried to focus my senses on outside noises. I began to hear voices outside, but after a while I realized that my burning ears were only playing tricks on me. Because of the inactivity, the cold penetrated into my marrow. I began to shiver again. I realized, as I wondered how long I had been in this place, that I was barefoot.

I pulled my legs up to my chin and began the long and exhausting struggle to push myself up the wall. I tried placing my hands lightly against the wall to steady myself. The least pressure on my shoulders made me gasp from the pain. I finally made it to my feet.

I rested again before cautiously moving sideways along the wall. After five or six steps I encountered a corner and continued my blindman's walk along the adjoining wall. Fifteen or so small steps brought me to the next corner, then ten steps along this third wall brought me to a door. Since I couldn't use my hands to explore it, I lowered my head and used my nose to probe for a keyhole or Judas window. My thorough examination uncovered no opening or knob. All I felt was rough wood. I hit the door with my forehead hoping to draw the attention of whoever was on the other side, but the thick door seemed incapable of transmitting sounds.

Exhausted, in pain, and with my feet half frozen, I finished exploring the walls, which I dared not leave, and then slid carefully down to the floor and pulled my knees up to my chin. I wanted to bring my icy toes close enough so that I could rub them, not realizing the pain from moving my arms would cut off my breath. My toes were too numb to wiggle, so I rubbed them gingerly against my pants legs until my hips hurt and I was exhausted from this new effort.

After settling into a relatively comfortable sitting position I wondered what had happened to me after I had blacked out. Had Dr. Aarons and Dr. Ottinger paraded past me while I was on display? The cold started bothering me again and I began to stiffen up. I had to keep moving! I stretched out my legs and managed to use my buttocks to "walk" sideways along the wall. The worst chills were gone by the time I reached a corner. Moving my arms at the elbow became less painful, and the pain from rotating my wrists also grew more tolerable. As I pored over the past events, I couldn't figure out why I was still alive. Why had I been hung up by my arms instead of by my neck? Did they still hope to force me to talk about my Resistance work? That would mean more torture, and I didn't think I could stand it.

I noticed a yellowish rectangular frame across the room. It was some moments before I realized I was looking at light shining around the contours of the door. Footsteps eventually emerged from the silence, followed by the sound of a key in a lock. I was blinded by the light suddenly pouring through the open door.

When I finally succeeded in keeping my eyes partly open, the man standing in the brilliantly lit doorway introduced himself as Hugo, the bunker Kapo. So, I thought, I'm in the interrogation dungeon.

"I'm under orders to take you to the Politische Abteilung."

I asked him when I had been brought here.

"Last night," he answered. "A French doctor came by to see you then. Your shoulders were dislocated. The doctor reset them and gave me a piece of bread and some pills for you."

Hugo handed them to me and offered to bring me some coffee to help wash it all down. I had already noticed the green triangle on his pocket and was very puzzled by his kindness to me. Hugo must have sensed my confusion.

"I'm a green triangle," he said, "but I'm not a real criminal. I didn't kill anyone. I'm in Dachau only because I embezzled some money from the company I worked for. My sympathy goes out to those who are about to die. I try to make their last moments as comfortable as possible."

My thoughts echoed his words: "to those who are about to die . . . about to die . . . about to die." How much time did I have left? My immediate needs still demanded attention. I showed Hugo that I had nothing to cover my freezing feet. He promised to give me a pair of shoes as soon as we returned from my interrogation. I told him that I needed a toilet. He raised his hand and I followed his pointing finger to an open can provided for that very purpose. It had been too dark for me to see it. "You could have smelled it," he said.

I asked Hugo if Kalka had also been brought here to the bunker.

"No," he told me, "you were the only one brought back alive. I've heard two different stories about how they killed him. The nicer one is that they shot him immediately. The other is that they found out he had survived the Katyn massacre, and they hung him from a meat hook and gouged out his eyes to make him name other surviving witnesses."

Kalka wasn't here, that was obvious, but it didn't fully dispel my suspicions. He could be a very much alive spy now, comfortably on his way to another job. But if those rumors were true, particularly the second one, poor Kalka!

I used the can, ate the bread, and took my pills. Then I followed Hugo up the stairs to a room in which a number of men were waiting. Most of them were from the SS, including a few in the security branch, and two were obviously Gestapo officers. One of the SS shouted "*Achtung!*" as the camp commandant entered.

A pudgy, ordinary-looking man stopped six feet in front of me and stared at me impassively. His middle-aged burgher's face, with its red cheeks and potato nose, would have been impossible to pick out in a German crowd. But his SS uniform was a standout, covered with gold braid and military decorations. He carried a riding crop. Finally he took his eyes off me and turned to his audience.

"So this is one of the little birds who wanted to fly away and leave us heartbroken!" The commandant turned back to me. "And you have told my people that no one here helped you?"

I answered affirmatively.

"I can't believe it," he said, "but since the other one said the same thing, I guess I'll have to. But before I decide whether I want to hang you or let you rot in the bunker, you are going to show me just how you did it."

I was taken outside, where a garbage container waited on a cart manned by four prisoners. The commandant and his entourage followed as I helped push the cart to the ditch. I showed them how I had hidden under the garbage container and dug myself farther under before it was lifted off. The commandant's single comment was, "Terrific!" He seemed convinced and satisfied.

Back in his office the commandant announced his decision. "Since your escape was fifty percent successful and your partner is already dead, it isn't very satisfying to us if you just simply die. We have places where dogs like you can serve as an example. . . . Take him away!"

An SS guard roughly led me to the bunker Kapo downstairs. Hugo seemed very relieved to see me return.

I was put into a somewhat larger cell. Now I had a bed and an electric bulb as well as the standard toilet can. I was galled at the commandant's clearly sadistic amusement in taunting me by describing my escape as "fifty percent" successful—until I was captured. I didn't have any energy left for contemplating my future. At the moment I was simply glad to find myself still alive.

Hugo brought me a message from Dr. Ottinger. He would visit me as soon as he could. He was now one of the camp Prominents and had asked Hugo to treat me especially well. According to Hugo, it was only through his influence that I hadn't been left hanging from the post until I died. The grapevine had it that he had tried to persuade the SS to put me in an upcoming transport to another camp.

Hugo did his best to make me comfortable. Later in the afternoon he brought me a hunk of good bread and a bowl of soup from the SS kitchen, a sheer delight of thick potato soup flavored with smoked meat. I found a morsel of meat at the bottom and rolled it around in my mouth for quite a while before swallowing it. When Hugo came for my empty bowl, he left me a newspaper. What a lucky guy I was after all! I had survived my capture, I had just eaten a wonderful meal, and now I was sitting like a gentleman in his easy chair reading the evening paper! My satisfaction deepened as I read of the "tactical retreats" the Germans were carrying out on all fronts. I even began to feel a glimmer of hope for my own future.

Two such days passed in relative peace and comfort, punctuated by Hugo's solitary visits. On the third morning, Dr. Ottinger surprised me with a sudden visit. I took in his smiling face and the "Prominent" armband bearing his medical chief title. He ran to me and embraced me. What pleasure that moment held!

My first question was about Dr. Aarons. I learned that our friend had been sent to the inner camp and promoted to the comfortable position of chief of lice control, visiting each barrack to examine the prisoners' shirts for lice. Since lice spread diseases that could infect the Germans as well as the inmates, louse infestation was a serious crime that often required "disinfection" in the gas chamber. It did not matter that the host died along with the parasite. Dr. Aarons was given many edible bribes to ignore damning evidence. Louse control didn't utilize his professional skills, but it performed the more vital functions of keeping him fed and alive.

Dr. Ottinger told me that I was to be sent to Flossenburg, a camp at the edge of the Bohemian Forest near the Czechoslovakian border. He had used all of his influence to get me this transfer out of Dachau. I was very nonplussed.

"Why did you do it? At least here I have friends!"

"Here your friends can't help you. You are going to be killed. At Flossenburg at least you will have a chance to survive. Nothing can be worse than this hell hole."

Dr. Ottinger handed me some cigarettes and a few analgesics after he examined my shoulders. Then he wished me *merde* and left in his typically abrupt manner.

My morale was uplifted by this unhindered contact with a friend, but I was torn by uncertainty and apprehension. I could not understand why the Germans would put so much effort into capturing and torturing me, and then give in so easily to Dr. Ottinger's request. I did not have to wait long for my answer. The Germans were playing a fiendish joke on us both. And they had by no means finished their game.

Hugo's next visit provided me with a pleasant meal, a clean striped uniform, and a pair of shoes with two pieces of cloth to use as socks. What a pleasure to shed my blood-stained, filthy clothes even for a new outfit that was two sizes too big. My night was relatively quiet. Hugo woke me in the morning with hot coffee and bread he had stolen from the SS kitchen. He warned me to be ready to leave when he returned in a short while, and disappeared before I could ask him if he knew where I was going.

When he came to get me, I put my question to him. He confirmed Dr. Ottinger's information. I was going to Flossenburg in a transport of sixty prisoners. I asked him what he knew about Flossenburg. Hugo was silent. He grew increasingly uncomfortable as I insisted on an answer. Finally he shouted at me to leave him alone and follow him to the Assembly Place to join the other prisoners.

Several mutilated corpses lay in a heap outside the bunker. Hugo pointed to the most battered of the three—chunks torn from the flesh, eyeballs dangling from blood-caked sockets. He said it was Kalka. The body was torn and bloodied beyond any possibility of recognition.

I joined two rows of prisoners who were standing in the Assembly Place. We waited in line until two open trucks arrived. A small portion of bread and ersatz sausage was handed to each of us, and we climbed in. We were driven to the railroad station and transferred to two cattle cars. The doors were sealed. We listened to the sounds of the engine pulling out of the station. The noise of the wheels soon indicated that we were moving at a fair speed. With only thirty men to a car and straw on the floor we were fairly comfortable, but there was little talk. The sporadic fragments of conversation that reached my ears were mostly in a Slovakian language. Occasionally I picked up a few sounds of Dutch or Flemish. At one point I called out, "Are there any Frenchmen here?" There was no answer. I lay down and tried to sleep.

I was half asleep when I realized that someone was touching me. I asked, "What is it?" No one spoke. The hand was silently withdrawn. Someone had probably hoped to steal my meager food supply. I realized that I was hungry and started to nibble, consuming everything. I reproached myself for my stupidity. Who knew how long it would be before I encountered food again. My consolation was that I could sleep peacefully since I no longer had to fear that my food would be stolen. I slept until the grinding of the brakes woke me. The door opened. We got out to empty our toilet cans. Then we were lined up and counted, and herded back into the freight car to continue our journey.

I was painfully cold without a blanket, and I curled up to warm myself. The next I knew someone was shaking me urgently. He anxiously explained, half in German and half in Dutch, that the sound of my breath rattling in my chest had made him think I might be dying. I tried to combine my German with the bit of Flemish I had picked up in Antwerp in an attempt at conversation. It turned out that he spoke a little French. We quickly became quite animated as we wove our four languages back and forth.

He was a teacher and a member of the Dutch Underground who had been manning an illegal radio station to receive and transmit messages and anti-Nazi propaganda. He was identified and tailed by the Germans, unwittingly leading them directly to his radio station. He was in the midst of sending a message when they began pounding on his door, so he was able to transmit what was happening before destroying his code book.

His resistance won him a bullet wound and a typical SS interrogation. He confessed nothing. During his transport out of Holland the Underground had liberated him, but he was recaptured only a mile and a half from the scene and taken to Gestapo headquarters in Amsterdam. He had arrived at Dachau yesterday, most probably as a candidate for the rabbit hunt, but there was no room in the camp and he spent the night in the Bunker and was on his way again. (In Flossenburg I would see him brutally killed.) I told him the bare minimum about myself, talking mostly about the quarantine barrack, the rabbit hunt, and my ill-fated escape.

"It's interesting," he mused. "Most of the Dutchmen in this transport have histories of escape and recapture."

He offered me a bite of chocolate that he had hidden away, which I accepted with greedy pleasure. Just as we decided to sleep, the brakes began screeching and the train ground to a noisy halt. We obeyed the shouted orders to jump out. I looked up at the name painted on a sign—"FLOSSENBURG."

FLOSSENBURG, BUNA, LAURAHUTTE, AND GUSEN I

15

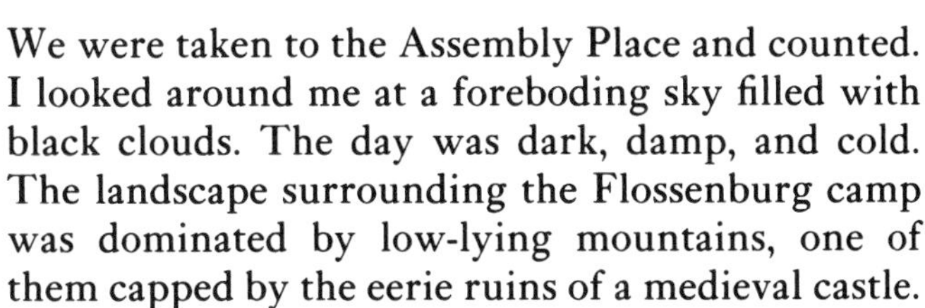

We were taken to the Assembly Place and counted. I looked around me at a foreboding sky filled with black clouds. The day was dark, damp, and cold. The landscape surrounding the Flossenburg camp was dominated by low-lying mountains, one of them capped by the eerie ruins of a medieval castle. The setting was straight out of a horror movie. Standing behind the supervising SS officer and watching us was a group of well-fed and cleanly dressed Kapos.

The SS officer asked for the engineers among us to step forward. About ten men did so. A Kapo was instructed to take them to his barrack and give them new clothes and a bowl of factory soup (flavored hot water provided by the factory for which the prisoners of Flossenburg labored). Then the SS officer called out the industrial designers, then the machinists, then the electricians, then the lathe workers, each time assigning the group to a Kapo who took them away.

Finally only one other prisoner and I remained. I told myself that physicians are always needed. To my knowledge I was the only one in this transport. The SS officer silently looked us over, then said to one of the Kapos, "Take this filth away!"

We were marched along a winding road to a cluster of three barracks that were separated from the camp proper by barbed wire. We entered the enclosure and stood in front of one of the barracks until we were ordered inside.

The interior was totally empty except for one long table in the center with a bench along each side. There was no glass in the windows. Forty or so men were already waiting. We were all ordered to line up by the Blockältester, a fat but powerful-looking man in his mid-forties. The raw brutality etched into the features below his shaved head made him seem more animal than human.

A Prominent whose face was frustratingly familiar entered the barrack. I watched him sit down at one end of the table, open a book, and raise his arm to signal the start of registration. He was the Lagerschreiber. He began to register us one by one in a small book, obviously not the main camp registry.

When my turn came the Lagerschreiber repeated, without looking up, "Name and occupation?" As I replied, he hesitated before completing the entry. I had the distinct impression that we had known each other in happier times, but I racked my brains fruitlessly trying to place him. He wrote a few words on a small piece of paper that he pushed over to me. I had already read it upside down as he wrote it, so I left it on the table, saying, "It's okay, I've read it." He called out, "Next!"

The note had said, "*I know you from Nice.*"

The Lagerschreiber went away, and the Blockältester took over. He fished a piece of chalk from his pocket, handing it to the first prisoner in line with the order, "Make your bed!" We were all bewildered. With a nasty smile the Blockältester explained, "Each prisoner must have a place to sleep, but we have no bunks or beds here. So you have to draw two parallel lines on the floor three feet apart, and this will mark your bed. After you make your bed you will give the chalk to your neighbor, who will then make his bed, and so on down the line."

We drew our identical beds on the floor and waited for our straw and blankets, but there was nothing to wait for, no blankets to protect us from the cold night air coming through the empty windows, no straw to keep our undernourished bodies off the cold, damp floor. None of us could survive for long under these conditions. I understood now why Hugo had been so awkwardly silent about Flossenburg. I had been sent here to be exterminated. It would be slower and far more cruel than shooting, hanging, or gassing me at Dachau.

After we "made our beds" the Blockältester chose five prisoners to fetch dinner. The rest of us had to stand at attention by the table. Most of us were shivering from the cold by the time the soup arrived. The kettle was set down by a pile of rusty bowls encrusted with the rotting leavings from previous meals. At last I would be getting something hot to eat. We were ordered to place our bowls on the table and remain at attention while the watery soup was ladled out. But we were not permitted to taste it yet. We had to sit motionlessly with our hands flat on the table.

We sat and stared at the steaming liquid. Our noses began to run from the cold. I lifted my hand to wipe my nose before it dripped into my soup. A sharp pain in my back stopped me. One of the Steubendiensts barked, "You did not ask permission to move!" My request for permission to wipe my nose was also answered with a stick across my back, and the Steubendienst shouted, "Who authorized you to speak?" I raised my hand, as we did when we were schoolchildren. He hit it with his stick as he exclaimed, "Who gave you permission to move?" It finally sank in that I wasn't going to be allowed to wipe my nose, which was now dripping precariously near my bowl of cooling soup.

Finally the order came. "*Essen*" ("Eat"). I wiped my nose surreptitiously and picked up my bowl. I had barely gotten the liquid to my lips when we were commanded to jump under the table. In the ensuing chaos some of the inmates dropped their soup bowls on the floor. A few seconds later we were told to sit with our hands on the table again, then again came the order to eat. Before anyone could get a mouthful of soup the command to crawl under the table was repeated. This senseless exercise went on for about an hour. Each new order, needless to say, was accompanied by kicking and beating.

By now most of the soup was on the table and floor. What finally brought this sadistic game to an end was the order to clean up. Night had fallen by the time we finished mopping up with our jackets. We had to spend more time sitting motionlessly at the table before eventually lining up and going to the latrines. Then we had to stand at attention by our "beds" before the order came to lie down and sleep.

We were all too hungry, cold, and exhausted to survive this treatment for long even if our tormenters didn't murder us.

I tried to make some sense out of what was happening to me. The Germans obviously intended the men in this barrack to die. No one had shown any interest in interrogating me before I died. So why hadn't they simply executed me at Dachau instead of transporting me here to Flossenburg to die? And here, as at Dachau, I was in a barrack that was registered in a separate book from the main camp registry. I came to the numbing, inescapable conclusion that, for all intents and purposes, all of us here were already dead. Since we had never officially entered the camp, we did not exist. We were clearly expected to disappear, quite soon (within a week, I found out the next day), without a trace. We were now open targets for murder.

I wondered why the Nazis had bothered to register us at all and what they planned to do with these separate registration books. They never did anything without a reason.

I tried not to think of the damp stone floor beneath me and concentrated intently on the Lagerschreiber who said he knew me. Suddenly it all fell together! He was a Polish refugee I had met at La Jetté, the gambling casino on the Promenade des Anglais in Nice. I used to go to the casino occasionally, sometimes to pass Resistance information to the *chemin de fer* croupier, and sometimes just to play at the gambling tables or attend the performances of the Paris Opera at the casino's theater.

One evening Sonja and I had arrived quite early for the opera. I took the opportunity to play *chemin de fer,* my favorite game. As I took a place at the table I gave the croupier an unobtrusive sign that I had come solely for my own pleasure. As I placed my bet, a stranger opposite me called out that he wanted to play with me against the bank. I accepted and we won four hands in a row. When I rejoined Sonja for the start of the opera I was three thousand francs richer. I gave her two thousand francs to buy a couturier sports coat she had been admiring. In fact, it was the coat she was wearing on the day we were arrested.

During the first intermission I returned to the gambling tables to try my luck at roulette. I won my first two bets, lost one, then won five consecutive plays. I was so flabbergasted by my extraordinary winning streak that I was unaware of the crowd gathering around me until the bell rang to signal the end of intermission. My *chemin de fer* partner, who was in the crowd, came over and introduced himself (a Polish name I no longer remember). He had copied all of my roulette plays and thanked me for his winnings. I explained that I had to return to the theater, and bade him good night.

Sonja wanted to celebrate our small fortune with a *gueleton* ("feast"). We chose the Chanteclair, one of the best restaurants in the area and one where you didn't have to hand in your ration stamps if you tipped the maitre d' adequately. We ordered oysters and a bottle of Haute Sauterne, Sonja's favorite wine. Then fish soup, to be followed by chicken à la crème, green beans, and rissolé potatoes. Suddenly Sonja complained that a stranger at a nearby table was trying to flirt with her. I told her to relax and ignore it. When the wine for our first course arrived, the sommelier informed us that it had been sent with the compliments of the gentleman at the nearby table. Now it was Sonja's turn to tell me to relax and enjoy the wine. I looked around to catch a glimpse of this intrusive gentleman and found myself looking at my partner from the gambling tables. He was alone, and seemed delighted when I sent over an invitation for him to join us. I teased Sonja that it obviously wasn't her beauty that had magnetized him, but the money he had won with me!

We met fairly often after that, usually gambling together at the casino. After the Germans occupied the south of France in the late fall of 1942 and turned the casino into a military fortification, I saw him only a few more times. The last time I had encountered him on the street he had run away from me. To find him here and now in Flossenburg was incredible. How had he recognized me so easily despite the marks I bore from my recent beatings?

That first night in Flossenburg my mind was so actively full of memories and conjecture, and my body was so cold, I thought I would never fall asleep. As I lay there shivering, I became aware of the unusual closeness of the bodies on either side of me. Being huddled together gradually helped us to conserve our body heat. I no longer felt cold, and finally I slept.

It was still dark when we were awakened. Four prisoners went to get our breakfast while two more were ordered to distribute last night's dirty bowls. A very unpleasant odor had invaded the barrack, despite the open windows, from two comrades who had accidentally emptied their bowels in their sleep. The Blockältester chased them outside to stand at attention while the rest of us had our tasteless luke-warm coffee and thin slice of ersatz bread. This time, at least, we were allowed to eat.

After breakfast we were lined up for the latrines and washroom, with the two unfortunate prisoners relegated to the end of the line. In the latrine there was nothing—not a solitary piece of dirty paper—with which to clean ourselves. The two dirtied prisoners were pushed forward by our SS guards and ordered to take off their soiled pants and wash off the excrement with their tongues.

The older man refused and was beaten almost to death. The young man obeyed. The prisoners with more sensitive stomachs vomited their breakfast.

After the old man's body was carried back to the barrack, I examined him and informed the Blockältester that he was still alive. I was smacked on the head and told not to worry, he would be dead soon.

A small group of us, including the young man, were given shovels that we carried, along with the half-dead old man, deep into the surrounding forest. Presently one of the six SS soldiers guarding us ordered us to stop and dig a grave for the old man. Digging in the frozen ground was so difficult and slow that the old man had time to regain consciousness. When we had dug down about four feet, the SS guards ordered us to throw him in. No one moved.

An SS soldier pointed his gun at the young man and two others, and they rolled the old man into the hole. I saw his bewildered and frightened eyes just before he rolled over the edge. I heard the faint crying of a human voice beneath the sounds made by the frozen earth as it was shoveled back into the grave. Finally there was only silence.

We had to stamp down the earth to make the new grave virtually invisible. Then the young man and four others were ordered to dig another ditch. When it was sufficiently deep, the other four rejoined us and the young man was ordered to jump in and lie flat on the bottom. He refused to enter the ditch. He fell to his knees and began to beg for his life, dropping his head forward in abject supplication. One of the SS soldiers picked up a shovel and whacked the back of his neck with the sharp side. His head dropped off his shoulders, hanging from his body by a few shreds of muscle, the gaping wound spraying us with blood as the twitching body fell to the ground. All of us, including the SS soldier, froze in mesmerized horror as the eyes and mouth in the severed head snapped open and shut, open and shut, open and shut, the head gasping for breath, desperately clinging to life as the body was gripped in the rhythmic spasms of death. An SS guard kicked the body away from him into the ditch and we finished the second grave.

Returning to the barrack I was haunted by the staring eyes of that old man buried alive and the horrible severed head dangling from the young man's body. No one knows how many such bodies may still lie entombed in that Bohemian forest. No one has ever thought to dig there and investigate.

Back at the barrack we lined up and stood at attention again until the Lagerschreiber arrived. He went in to the Blockältester's room and stayed there for half an hour. When he came out he pointed to me and another prisoner and motioned us to walk to the door. Three other prisoners and several SS guards were waiting outside. Not a word was spoken. The guards marched the five of us out of the barbed-wire enclosure and up to the Brausebad where we showered, were shaved from tip to toe, and given clean uniforms. Then we were taken to another barrack and told to wait—for what, we still didn't know.

Eventually the Lagerschreiber reappeared and motioned for me to join him in a private room. As soon as the door was shut he embraced me with an outpouring of emotion.

"Thank God I recognized you!" he said. "You are in a special disciplinary camp that few people know about. All prisoners sent here are dead, one way or another, within five to six days. Your only hope is to get out at once. Because I'm in charge of all paperwork in the camp, I have been able to put you in a transport of specialists leaving tomorrow for Buna. Buna is one of Auschwitz's satellites, a factory camp run by I. G. Farben. Conditions there aren't too bad. And if the SS guards who run the place don't start digging into your records, you'll have a chance to stay alive. Thank God I recognized you," he repeated.

I told the Lagerschreiber how long it took me to search my memory before I finally remembered who he was. I was glad it hadn't taken him so long to place my face.

"Your face!" he exclaimed. "You've been so badly beaten that I wouldn't have known you in a million years if your face was all I had to go on. It was your voice I recognized."

I asked him why, seeing his compassion toward me, he had run away the last time I had seen him in Nice. He explained that I had been wearing my Todt organization uniform, and he had thought I was a German spy. Not until he heard of my arrest did he realize that I was in the Resistance. At that time he was a Polish pilot trying to get to England, but he was caught by the Germans and sent to Flossenburg. With a combination of cleverness and a lot of luck, he had managed to save himself and become one of the Prominents.

He gave me some food, knowing that I had hardly eaten since my arrival, and advised me to keep silent and avoid doing anything that might draw attention to myself. The transport was going to leave early in the morning. He showed me to my bunk, and I slept until "*Aufstehen*" woke me.

I jumped up and stood in line for my coffee and bread, my last breakfast before leaving for another unknown destination. The Lagerschreiber came over to shake hands just before I left to be herded, once again, into a waiting cattle car.

Compared with the brutally harsh and debilitating conditions of my previous transports, this voyage was almost pleasant. The German war machine must really have needed our "imported" skills! The basic arrangements in the cattle car were standard—a garbage can for a toilet, some bread and sausage for food. But there were only about thirty of us in the wagon. We had ample straw to sit on, two blankets apiece, and room to stretch our legs. It became obvious that orders had been given to treat us less harshly. The apelike Kapo who accompanied us beat us with restraint, never once drawing blood.

Two of the men near me were French engineers. One had worked near Cannes in a small aviation factory in La Boca, Maurice Chevalier's home town. The other was a chemical engineer from Rhône-Poulenc. There was a Jewish chemist who had worked at I. G. Farben before Hitler's rise, then escaped to France only to be captured and brought back to Germany. He tried to joke about it. "Why did I ever bother to leave I. G. Farben, since all it has done is bring me back to work for them again!"

After eight or twelve hours we stopped at a small station. We were herded out, lined up and counted, and marched for about half an hour to a hangarlike building where we passed through the rituals of washing, barbering, and shaving again. Then we were each given a permanent number and registered—name, profession, number—in the official camp book. At last, I felt, I *existed!*

I called that official camp registry our birthday book because it signified, for each of us, our official rebirth into concentration camp existence. I had metamorphosed from Dr. Albert Haas of France to Häftling Arzt #1762222 of Auschwitz. As with all prisoners entering Auschwitz or one of its satellites, my number was tattooed on my left forearm.

I had finally become a "respectable" member of concentration camp society. This did not mean that I could no longer be hanged, gassed, electrocuted, shot, or beaten to death, but it meant that if any of these ends should befall me, the camp commandant would have to report my death. Now, discounting the unforeseen, I had a reasonable chance of surviving the next several weeks. It was no mean achievement in my circumstances.

We were given new striped uniforms before being taken by open truck to our final destination at Auschwitz *Drei,* popularly known as Buna, the rubber factory and forced labor camp near the Polish town of Monovitz. An SS soldier hurried me out of the truck with the full force of his gun barrel. As I fell to the ground I heard a crack and felt a sharp pain in my leg. One of the French engineers helped me get to our barrack.

I requested permission to talk to our Blockältester, Alphonse, and told him I had probably broken my leg. He laughed.

"Where do you think you are? This is a labor camp. Prisoners who need time out for medical treatment are automatically given over to the gas chamber squad."

I replied that I was a physician. I had been sent here to Buna because my skills were urgently needed and a broken leg certainly didn't interfere with my ability to practice medicine. Alphonse laughed again.

"We have so many extra physicians in this camp that doctors are doing manual labor in the rubber factory. But," he reflected, "my father was a physician too. I'll see what I can do for you."

I later learned that Alphonse's father was Jewish. The entire family had been killed by the Germans. Alphonse, orphaned by the Germans, had been shuttled around their concentration camps for the past twelve years.

Alphonse supplied me with the two wooden slats and string that enabled me to improvise an ankle support, and warned me not to go to the infirmary sick call. All new arrivals to Buna had to see the Häftling medical chief during their first few days anyway. I recognized the doctor's name, a well-known hematologist from Strasbourg. Alphonse described him as a very decent man who would go out of his way to help a fellow physician. In the meantime, I could be the barrack Stubendienst so I wouldn't have to be on my feet.

Even with the makeshift splint my ankle was too painful to bear my weight. It was so immensely swollen that for once I was thankful for the clumsy but roomy *sabots* ("wooden shoes") we had to wear. I used my Stubendienst's broom as a crutch to limp around on one leg. Whenever a stranger entered our barrack I was able to pretend that I was sweeping the floor.

On my fifth day at Buna, the recent new arrivals were sent to the HKB, the prisoners' infirmary. I introduced myself to the medical chief and told him about my leg. An X-ray confirmed a broken fibula. I was stunned by the presence of X-ray equipment in light of Dr. Ottinger's description of the primitive conditions at Dachau's HKB. I learned that I. G. Farben, the factory owner, wanted its infirmary properly equipped. The SS had appropriated equipment from German-occupied territories, charged I. G. Farben an alleged purchase price, and pocketed the entire sum.

The medical chief put my leg in a light cast to give me support and some mobility, and promised to do his best to get me transferred to a smaller work camp, an *Ausser Kommando* (an "outside work unit"). Until then I would continue as Stubendienst in my barrack. I thanked him for his kindness.

He was true to his word. I remained in Buna only a few more days. My transfer was to Laurahutte, about thirty miles away, a new Ausser Kommando in the ring of Auschwitz' satellites. It was attached to the Hermann Goering cannon factory in the little town of Laurahutte. The camp's only physician was Jewish. The camp commandant was greatly relieved to be able to replace the man with an Aryan and demote the Jew to the position of assistant medical chief. My masquerade certainly had its ironic overtones. I was now to be the Aryan medical chief of Laurahutte!

16

The next morning a truck transported a small group of us to Laurahutte, a tiny work camp situated in the center of town. Prisoners entered the camp through a gate on one of the town's main residential streets. The guardhouse stood just inside the gate. To the right was the cannon factory, forming one of the long sides of the rectangular layout. The factory had its own street entrance for the free workers, and was attached to the SS quarters and offices.

There was a makeshift infirmary in the prisoners' quarters, an immense hangarlike building. Although a high electrified fence effectively isolated us from the outside world, we could see the inhabitants of the town looking inquisitively into the camp from their apartment windows. This continuous scrutiny actually gave us a sense of relative security. We knew that impulsive beatings and killings could not take place publicly or during daylight, and would therefore be much less frequent than they would be if we had our privacy.

For the first few weeks we were only a few hundred prisoners in the camp, all involved in cleaning up the scrap metals and other debris from the courtyard where the factory had regularly dumped its garbage before the prisoners' camp was added on. The medical chief was Dr. Robert Bloch, a Parisian internist in his fifties. I had convinced the camp commandant that his experience, which was far greater than mine, outweighed his yellow star. Robert put in only half a day of menial work and then prepared the HKB for the patient visitation that would be taking place between the end of the work day and the start of dinner. I had to put in a full day of labor and then treat all of the minor cuts and injuries during patient visitation. It was incongruous to see Robert, a brilliant specialist, spending his time and expertise treating headaches, diarrhea, and chest colds.

When an order arrived that our camp was to be enlarged to hold fifteen hundred inmates, plans were drawn up for two large buildings to augment the prisoners' quarters and a smaller one for our infirmary. This latter was to be built from prefabricated plates that we would manufacture at the camp. I was put into an unskilled Kommando to make cement. This involved lugging one-hundred-pound sacks of cement from the delivery truck to the construction site, fetching gravel in a wheelbarrow, then mixing everything with water. I never figured out why I was forced to carry the cement sacks on my back, one at a time, instead of carting several at once in my wheelbarrow. Mixing the cement was the one truly unpleasant part of my job. My hands were not used to that kind of rough manual labor. By the end of my first day the handle of the mixing shovel had given me huge blisters that soon ruptured. To my immense relief, they healed without any infection setting in. In the evening, after working in the cement Kommando all day, I had to attend to an ever-growing number of sick prisoners.

The SS Sanitätsdienst in Laurahutte was a Silesian-born Pole named Vloka, a Volksdeutscher in his fifties who had been assigned to his home territory, where Auschwitz and its satellite camps were located. Vloka was small and thin. The wrinkles at the corners of his eyes emphasized his perpetual squint. He had the air of a cunning old fox.

Vloka was really not a bad fellow. In fact, I would even call him a basically good person, but his kindness always had a price. He gave us special food if Robert and I behaved well, and he would be especially generous with these food payments if we also managed to obtain gold teeth or foreign currency from any of the prisoners.

Before long Vloka decided to formalize this aspect of our relationship and asked me to become his black-market middleman. I accepted the offer and carried out my responsibilities honestly and to the benefit of all concerned. The prisoners received health-sustaining food for their otherwise useless valuables, Vloka reaped pure profit from the currency and gold given him for the food and medication that he stole, and my services bought the infirmary staff welcome peace, a modest degree of independence, and some extra food and medication. I had to be a very aggressive bargainer, a living example of the Nazi stereotype of the Jew trying to exploit the Christian. Vloka never suspected that I really was Jewish, and I saw no danger of being unmasked. He bargained twice as ruthlessly as I did!

After our first formal black-market transaction I asked Vloka for a favor. I wanted to work full time in the infirmary. My physically demanding days with the cement Kommando and the lengthening evening hours in the still-growing infirmary were exhausting me. A few days later Vloka announced that the best he could do for me was a transfer to the kitchen as a part-time cook's helper. His price, of course, was stealing food from the kitchen for his black-market trades. I accepted without hesitation. To this day I still love working in the kitchen, although now I do it as a hobby for which I can choose my supplies and pay my own price.

Laurahutte was commanded by SS Hauptscharführer ("staff sergeant") Quackenacker. Contrary to SS rules, he had been only a noncommissioned officer before taking over this command. Unlike the brutes who typically rose to a position of command in the concentration camps, his aristocratic demeanor harked back to the top-ranking career officers of a civilized era. Even at our initial meeting these differences were immediately apparent. The very first thing I noticed was his hair, which he wore parted instead of in the military crewcut adopted by the SS. Tall, slender, and exceptionally neat, Commandant Quackenacker looked every inch the refined aristocrat from his shiny boots to his manicured and polished nails to his classically shaped head. When he addressed me, his manner was distant yet respectful. What struck me most forcefully after that first meeting was the absence of any display of anger. Of course, my standards for evaluating human behavior had been drastically modified by my own recent experiences. The relatively positive light in which Quackenacker appears must be assessed with this in mind. Although in my experience Commandant Quackenacker did not possess the brutality that was obvious in many of the SS camp commandants who escaped punishment after the war, he was hanged for atrocities committed at Laurahutte.

Our commandant was a strict disciplinarian. The slightest infraction was always punished, and the punishment was always more severe than the infraction warranted. Yet even the most serious punishments were never intended to incapacitate the prisoner.

Commandant Quackenacker claimed that he evaluated each infraction individually, but I soon realized that the final punishment really depended on how long it took him to reach his conclusion. If he felt that he needed time to reflect, the sentence would be twenty-five lashes on the buttocks administered in public by a Kapo who always put his back into it to show his cooperative spirit. But if Quackenacker made an immediate decision, he would carry out the sentence himself. He always put on a glove before hitting the prisoner, never hit below the belt, and never lost his composure.

Although the exact number of blows depended on the prisoner's crime, Quackenacker always concluded with a knockout punch to the chin. The prisoner would be given a rest period that was commensurate with the length of time he remained unconscious. On days when Quackenacker was in the factory, the crime of speaking to a free worker was committed with unusual frequency. It was a surefire guarantee for an on-the-spot Quackenacker knockout punch and the rest of the day off from work!

Quackenacker was so proud of his knockout punch that the longer a prisoner remained out, the greater was the commandant's immediate fondness for him. Visitors to the camp were always treated to a view of the prisoner who, at that point, held the knockout time record. This led to some impressive demonstrations of inmates' abilities to fake unconsciousness for longer and longer durations.

A nasty unofficial competition eventually developed between the SS guards and Quackenacker. The goal of the SS guards was to beat their commandant's knockout record. Their technique was to provoke an inmate into committing an infraction, knock him down, and kick him hard in the head to make sure he would be unconscious for a good while. The real losers were, as always, the inmates. Those who couldn't restrain themselves from responding to the provocation began appearing in the infirmary with symptoms of brain damage.

When Quackenacker wasn't at the factory he was in his office, so he had virtually nothing to do with the operation of the infirmary. I normally saw him only at roll calls, and on those occasions when he himself needed medical attention. Before long he had entrusted his health care completely to us—Robert, myself, and a Jewish Yugoslavian prisoner-physician who eventually joined us. He always addressed the three of us respectfully with the formal *Sie* ("you") instead of the familiar *du* ("thou"), going against the routine practiced in all the other camps. Yet I was the only one actually permitted to touch him, since I was ostensibly not Jewish.

Four executions took place in Laurahutte during the seven or so months I was there. Four such deaths was an extraordinarily low number in this time period. The first hanging I witnessed was of a seventeen-year-old Russian boy who had tried to escape after stealing bread and a knife from the factory foreman. The boy had allegedly attempted to defend himself with the knife and wounded one of his SS captors. Hangings were always on a Sunday so they wouldn't interfere with camp work, because the entire camp population had to sit in audience. All of our civilian neighbors were home to take ringside seats at their apartment windows, watching as the stool under the boy's feet was kicked away and two prisoners each grabbed a leg and pulled down to tighten the rope around his neck. The young life ended after a few spastic movements.

I looked up at the audience's reaction and saw a priest framed in an open window just across the street. I expected the priest to make the sign of the cross for the departing soul, but instead he turned away. A moment later he was back in the window devouring a piece of cake. I succeeded in catching his eye. I made the sign of the cross with my arms and gestured my question, "Where are your last blessings?"

He gave me a look of disgust and disappeared from the window.

The cadaver was cut down and put in the cellar of the new infirmary. I searched for a Christian man of God among the prisoners and found only a rabbi. I promised the rabbi two extra rations of bread and one week of supplementary soup if he would hold a wake in the cellar for the dead boy. Although all religious activity was strictly forbidden in the concentration camps, the rabbi agreed to do it. I hospitalized him for an alleged malady. My medical colleagues thought I was crazy to take such a risk, yet they cooperated with me and we held our wake in the cellar.

In a far distant epoch a pagan Roman soldier had guarded the grave of Jesus Christ for one shekel of Jewish money. I gave an

Orthodox rabbi some extra bread and soup to hold the wake for a Christian boy's soul. I cannot know exactly what was in the heart of the ancient Jews who had secured the Roman's services. But for me it was an impulsively rebellious act in the face of our helplessness in an epoch so heartless that even men of God had no compassion to spare for the sufferings of their fellow men.

Much later I found out that we were thought to be specifically a Jewish camp, and that the priest in question had come expressly to witness the hanging of a Jew. Perhaps we had that reputation because we were a satellite of Auschwitz. Or perhaps it was calculated misinformation spread by the German authorities who feared that the partisans would attempt to liberate the camp if it became known that we comprised all nationalities, Polish gentiles included. Many of the Polish and Russian partisans were so implacably anti-Semitic that they would never risk their lives for Jewish prisoners.

The next hanging I witnessed was of an aviation engineer who had temporarily disrupted the factory by dropping a piece of metal in the machinery he worked at. No one ever learned whether it had actually been an accident. He was hanged after brutal torture at the hands of the Gestapo. Our civilian neighbors were again in the audience, but this time I didn't see the priest among them. I did not call the rabbi to perform any religious ritual. I had come to feel too hopeless to be able to derive any sense of meaning from such an obviously useless gesture.

When the new HKB was ready, my part-time position in the kitchen was terminated. I moved into quarters in the new building. I had my own room, a door that I could close, and a real bed fitted with clean sheets and a homemade Hungarian quilt. Many of the Hungarian prisoners transported to the camp had packed up their beloved quilts to take with them. The SS guards had stolen their quilts as soon as the prisoners reached the camp.

The new HKB facilities included two large inpatient dormitories, consultation and treatment rooms, and a surgical treatment operating room. My responsibility was to look after inmates who had been hurt in accidents in the factory. Visitation and treatment began right after dinner and continued until the last patient was tended to, often lasting beyond midnight. During the day I attended to those whose injuries had required their hospitalization.

The windows of our operating room faced the third floor of an apartment building across the street, giving free visual access to anyone who stood at their window. I was always too busy to notice whether anyone was sufficiently interested to watch us. Eventually I discovered someone who often spent her time doing just that. This one contact with the outside world was a unique and charming experience in a life that varied between my bitterly tolerable existence at Laurahutte and the inner circles of hell.

My unsuspected intermediary with my outside contact was the green triangle Kapo of the mechanics Kommando. Like Hugo, the kind Kapo at the Dachau bunker, Hans was a mild-tempered, tolerant German thief. There the resemblance ended. Hugo had been arrested for high-class thievery—embezzling. Hans had been a common pickpocket for whom Hitler's justice had decreed amputation of the fingers on his right hand. Hans and I began our friendship in the HKB. One of his finger stumps was not healing properly. We saw each other regularly as I tried, one after the other, all the medications we had. Hans's wound did not respond. I wanted to try cod liver oil and asked Hans to keep his eye out for some.

One morning Hans came to the HKB and very secretively asked to see me in private. He had something very special for me. I laughed at his unnecessary caution in giving me cod liver oil. He shook his head and smiled. "That is not your surprise!" Robert Bloch took over my bandaging chores, and Hans and I went to my room and closed the door.

He took a small package from under his shirt, explaining that a French volunteer worker had smuggled it into the camp during the night shift and given it to him to deliver to the doctor who treated the injured prisoners at night. I opened it. The contents were incredible: a bar of chocolate, a packet of tobacco, cigarette papers, and a handwritten letter.

The tobacco's value, in particular, has to be understood in terms of what concentration camp inmates were willing to sacrifice for it. So many men died from starvation because, too often, they traded part of their already inadequate food ration for one thin cigarette rolled in newspaper.

The letter, written in German, was from a young woman named Maria Cielack. She wrote that she had been observing my work from her apartment across the street from my operating room. She admired my unselfish efforts in helping my comrades despite my own tragic situation. She hoped the few small items in the package would help me to bear up. If I wished to write to her, the French worker would be a reliable intermediary.

The reality in front of my eyes was practically impossible to believe. These two strangers were willing to risk ending up in a concentration camp themselves for the sake of trying to soften the harshness of an unknown inmate's life. I was deeply moved by Maria Cielack's gesture.

That night, when the treatment period started, I looked up at the windows across the street. I wanted to give my newfound friend a sign that I had received her letter and gifts. All of her windows were dark. For one uncertain moment I thought I saw a shadowy silhouette, but because of the strong lights in the treatment room I couldn't see a thing outside.

How could I make Maria Cielack understand that I could not see her? I walked over to the light switch. To the startled looks of the other prisoners, I turned off the lights for just a few seconds. The next time I looked up I saw a light go on in a third floor room and a woman approach the window. I distinctly saw her hand move to her face and throw a kiss through the air. Then the light went out.

The following day I wrote a thank-you note and gave it to Hans. I told her how touched I was and warned her of the risk she was taking if she continued to identify herself in her letters. Two days later I received a package of cookies and a note begging me to share them with my colleagues. In a separate unsigned letter, Maria Cielack described herself as a widow living with her mother. Her husband and father had been killed fighting the Germans, and her brother was in a German prisoner-of-war camp. She was running the family shoe store and shoe repair shop.

In my next letter I asked if she could send me a photograph of herself. Each day I waited expectantly for her answer, and each night I looked across at her window. She would always turn on the light to announce her presence, as if to say, "I am here. I am real. You are not dreaming." Hans, at long last, brought a package from her that included a photograph of a pretty young woman with large brown eyes and long brown hair.

Maria Cielack and I corresponded regularly for the rest of my stay at Laurahutte. The more I wrote, the longer my letters became. I told her all about my family, my uncertainties and fears about what had happened to my wife, and how wonderful it was to have a free human being to whom I could unburden myself.

She was giving me so much, and I had nothing to offer. I wanted to send her a present, some concrete token of my profound gratitude. A fellow prisoner, a former commercial artist, made special lampshades for the Prominents and the SS soldiers to earn himself extra soup. He painted Parisian scenes on the inside of the shades and incorporated a mechanism that made the lampshade revolve. I commissioned him to do a lampshade for Maria. When you turned on the light you saw a little bit of Paris with a detailed portrait of a few blocks in the Eiffel Tower district, and the opening notes of a current popular song called "*Maria, Si Je Vois Tes Yeux*" ("Maria, If I See Your Eyes"). I sent a note with my gift to make sure she knew it had been created by a real artist, and not by myself! She sent me a packet of tobacco for him.

17

In the late winter and early spring of 1944 we began to get a very clear indication of what was happening on the military front. Incoming prisoners brought us news of the Russian counteroffensives. As the Germans retreated before the advancing Russians, they evacuated the concentration camps in the areas from which they were fleeing. Survivors from Maidanek and the other extermination camps in Russia and northern Poland were being distributed among the camps closer to home, including Auschwitz and its satellites. Quite a few of them were sent to Laurahutte. These inmates had managed to survive either because their specialties were important to the troubled German war effort, or because they had ingeniously managed to hang onto the valuables they had brought with them from home.

Our black market was flourishing as never before, thanks to Russia's military progress. The supply of gold teeth, foreign currency, and diamonds streaming into our camp became barter for more food and less strenuous labor, both essential to our unending fight for survival.

The HKB was finally beginning to look like a real infirmary because of our burgeoning black market. Gone were the days when all we had were unsterile strips of paper for bandages, unsterile cellulose paper for cotton, primitive equipment, and little more than aspirins in our medicine cabinet. Now we had sterile gauze, real cotton, and a variety of essential prescription drugs that Vloka obtained from the local pharmacy. The Hungarian transport, which included many physicians, had given the camp an impressive store of the most sophisticated surgical instruments available. Through Vloka, these also found their way into our infirmary. My black market efforts enabled us to provide a level of medical care that vastly reduced the death rate. There were only twenty-five deaths during my stay at Laurahutte—not counting the four executions, of course.

I was somewhat puzzled when, as my payment one day, Vloka brought instruments for pelvic examination and surgery. With a smile I asked him if the homosexual Prominents thought they were going to need obstetric and curettage services. He looked at me silently, measuring me, then asked if I could interrupt a pregnancy. I continued to joke, but he was very serious. "It's for an abortion," he said. He added, "It would be a favor for Quackenacker."

I was stunned. Aborting any Aryan birth was a crime punishable by death in Nazi Germany. In fact unmarried Aryan women and German soldiers were encouraged to have illegitimate babies, who were delivered at a special hospital in Munich and raised by the German state.

"I don't know anything about gynecologic medicine," I told Vloka very truthfully. I named Bela Robert, a highly capable gynecologist whom he could find in the camp factory. Bela, a Yugoslavian Jew, was a factory messenger and first-aid dispenser. I had failed the several times I had asked Vloka to transfer him to the HKB to help with our heavy patient load. This abortion might give me leverage weighty enough to change Vloka's attitude and help the Yugoslavian gynecologist.

Vloka continued to press me. I repeated my request for Bela Robert's transfer to the HKB. "But he's Jewish," Vloka replied. "That makes it absolutely out of the question. The pregnant woman is the wife, or mistress, of a high-ranking SS officer." I insisted that I lacked the necessary skills to perform an abortion, and I would not endanger the life of another human being by performing an operation for which I was not qualified. We ended our talk with the dilemma unresolved.

That Sunday Commandant Quackenacker sent a messenger to bring me to his office at once. I wondered whether he was going to press me directly about the abortion. On the way over, his messenger told me that I had gotten a postcard from my wife, but this weasel was such a notorious stool pigeon that I didn't believe him. Maybe this had nothing to do with the abortion. Perhaps my records had caught up with me and some tricky interrogation trap had been set up.

At the commandant's office a Gestapo agent was examining a card lying on Quackenacker's desk. Quackenacker immediately demanded to know when I had written to my wife.

"I haven't written to her. I don't even know where she is."

He snatched up the card from his desk. "Then how do you account for this, which arrived addressed to you?"

I looked at him in total bewilderment. "I don't know."

Quackenacker replaced the card on his desk and the Gestapo agent began to question me about Sonja. As I answered him I stole a glance at the card. The handwriting was remarkably like Sonja's.

Quackenacker ordered me to remain in his office while he stepped outside to talk with the Gestapo agent privately. I felt sure he was giving me privacy to read my card, but I could not shake the reflexive suspicions of my months in the concentration camps. I remained where I was and slowly began to read the card upside down.

The opening words were "*Piciny papajo.*" The card really was from Sonja! Since François' birth, she had begun all of her letters to me with this phonetically spelled Hungarian phrase for "dear little father." I was so beside myself with joy that I picked up her card. It was difficult to read through my tears. Consequences be damned! This joy was worth the severest punishment.

Quackenacker was suddenly behind me. Did I recognize the handwriting? Could it have been a code meant for someone else? I was caught red-handed. I couldn't deny having read the card. I assured the commandant that the card was genuinely from Sonja. The opening and closing phrases were the phonetic Hungarian I had made up for her.

To my astonishment, Quackenacker let me go without another word. I had expected him to accuse me of smuggling a letter out of the camp, but apparently he was convinced that I could not have known Sonja's whereabouts. My excitement over word from Sonja kept me awake far into the night. By Monday morning I had decided to risk asking Vloka's help in transmitting an answer to her at Auschwitz.

Vloka's next visit was three long days away. In readiness I had collected some gold teeth, a brooch with a small diamond, about twenty-five Swiss francs, and one thousand *pengö* (a Hungarian currency). I gave Vloka the merchandise to examine, and after seeing the greedy pleasure in his eyes I was ready to begin haggling for however much margarine, bread, and sausage it was worth. Normally I was a hard bargainer, but this once I did something that I am still not very proud of. I longed so desperately to reach Sonja that I offered to give Vloka a particularly good bargain if he would carry out my favor. He unhesitatingly agreed to this soft deal but said he needed one more thing from me to clinch it. I had to say yes now, and when he returned with the food he would tell me what I had agreed to.

To my surprise, Vloka brought more food than I had expected, plus a big piece of cake that his wife had baked. Something was very fishy. Vloka asked me how many prisoners were sick in the HKB. I described how overworked Bloch and I were. He again asked me for the name of the gynecologist who worked in the factory, adding that he was going to ask the commandant to transfer him to the HKB immediately.

Vloka chose that moment to reveal the final part of our bargain, the something extra he wanted from me. I was to play a role in the abortion, the final act of a story I found hard to believe despite the twisted moral code by which the Nazis lived.

The commandant of one of the larger camps had become involved in an incestuous relationship with his sixteen-year-old daughter, and the girl had become pregnant. The gynecologist would do the abortion, but because he was a Jew no one must know he had touched the girl. It would have to appear to her as if I had performed the actual surgery. I asked Vloka why the commandant hadn't simply ordered an SS physician to do the abortion. He regarded me with amusement. "Do you know what any SS physician would do with such an opportunity for blackmail?" This could only mean that the expectant father was in the highest echelons of the SS organization.

Vloka informed me that if I still refused to get involved, he would not have the gynecologist transferred out of the factory. I agreed to arrange the abortion if Vloka would agree not only to get my note to Sonja but to deliver her answer to me. He agreed, and our deal was closed.

I wrote my letter to Sonja that same day. It was short and optimistic, containing just the bare necessities so that it could be delivered as quickly as possible. I told her of my boundless joy when I recognized her handwriting and that her card, which I could now recite from memory, had become my bedtime prayer. I said that my last news from Mme. Billard was that François was safe with Jan (my mother's maiden name). I spoke of how lonely I was without her. I asked her to answer my letter and give her reply to Vloka's messenger.

Bela Robert, the gynecologist, was transferred to the HKB the following day. I wasted no time in telling him the price for his transfer. He agreed to do the curettage. Bela looked over the available instrumentation and supplies. The only equipment he needed was stirrups to hold his patient's feet and keep her legs apart, so we designed two simple leg supports for a long table that was easy to convert for our purpose. Since the girl was not to know who actually performed the curettage, we needed anesthesia. I suggested that we begin with Evipan and switch to ether once she was asleep. Bela wanted a root from the local pharmacist, a preparation used by midwives in middle and central Europe to dilate the cervix. We also needed disinfectant and some means of sterilizing our equipment. Last but not least, Bela wanted the girl to come in for an examination so that he could determine the best date for the procedure.

I discussed all of these needs with Vloka. Nothing presented a problem except the preliminary examination. It was out of the question. The girl could come only once, and only on the day when there was the least chance of anyone seeing her. This meant a Sunday, since the factory was closed and the inmates spent the day in their barracks. Vloka would take care of having the leg supports made. The supplies he could get through the local pharmacy. He would provide me with a sterilizing drum for our instruments and arrange with the local city hospital for me to bring it in—under escort, of course.

Just before leaving, Vloka slapped an envelope into my hand. In the bold strokes of Sonja's handwriting I read: "*Albert Haas, Häftling Arzt, Nebenlager Laurahutte.*" I told Bloch that I had a bad headache and was going to my room to try and sleep. I slipped a scalpel into my pocket as I left.

Once in my room I took out the letter and stared at my name. It seemed like a dream! My hands shook as I read my address over and over. I was almost afraid to read the contents, afraid it would all disappear if I actually opened the envelope. My anticipation finally won out over my sense of dread. I picked up my scalpel and inserted it under the envelope flap—slowly and carefully, as if it were the most delicate operation—not wanting to risk scratching a single word.

At last the envelope was open. I unfolded two sheets of legal-sized paper. My heart was beating in my throat. My eyes jumped hungrily to the opening "*Piciny papajo.*" I read:

"*You can't imagine my pleasure and happiness knowing you are not far from me. You wonder how I knew where to find you? I was put to work in the Politische Abteilung because of my command of languages. My work here gives me access to the prisoner files for all of the Auschwitz camps. Every day I checked the new arrivals for your name. Then one day I found you in a transport to Buna, and then to Laurahutte. I couldn't restrain myself from putting a postcard to you with the SS mail to your camp. Now I go to the card file every morning to say hello to you and kiss your card!*

"*In my daily search for your name I came across an Arpad Haas. I wasn't at all sure he was your cousin Arpad, since the last we knew he had been living in Moscow, and this Arpad was arrested in Montpellier, France. When I contacted him through a Kapo to ask if he had a cousin named Albert, the answer was, 'Of course'! Arpad had been in touch with your family until his arrest. Mme. Raffia* [Mme. Billard] *wrote to them from Nice that she thought you had been executed after your arrest, since there had been no word from you.*

"*Arpad brought news about our baby.*" My heart almost stopped. Had François been found and brought to Auschwitz, which I had long ago learned was a dreaded extermination camp?

"*François is safe in Villon. He had a middle ear infection, which healed without the need for surgery. The village priest took him to a local hospital for treatment.*"

Sonja ended her letter with the hope that she would see me soon and asked me "not to forget Beethoven." This was a coded reference to the clandestine Free French radio broadcasts. I knew she was really saying, "*Ici la France Libre*" ("Here is Free France") and "*On les aura, les boches*" ("We'll get the Germans!"). The first was the phrase that always followed the musical opening, and the second was the sign-off used at the end of the broadcast.

I read her letter many times and fell asleep with it on my chest.

Luckily it was Robert Bloch who came looking for me, although I had a moment of panic when I first felt someone shaking me awake. Since he had seen my letter from Sonja, I could no longer keep our correspondence a secret. Actually, I was so keyed up that I was very glad to be able to share my feelings with a friend. Bloch was very moved. I showed him the letter. When he finished reading it he said, "Your wife is quite a gal!" He sighed deeply, wondering what had happened to his own wife. He hoped the fact that she wasn't Jewish was protecting her. "I'm very happy for you," were his parting words.

I walked on air for two days and concentrated on my answer to Sonja. I worked at devising innocent-looking code words that made sense in the superficial context of my letter and also communicated to her all the things I wanted to say but would be dangerous to refer to directly. If my efforts were successful, Sonja would know everything that had happened to me since we were separated.

I also sent a note to Maria Cielack, through the usual route, to tell her about my imminent excursion to the city hospital. I asked for her business address in case I succeeded in bribing my SS guards to take me to see her.

The next day Vloka brought all the supplies that Bela had requested, plus the sterilizing drum that he had arranged for me to bring to the hospital the following morning. For the day's black market transaction I had several foreign currencies and two dental bridges—one gold, the other *wipla,* a German dental alloy that resembled platinum. I withheld ten dollars from the currency to use as a bribe the next day, but I more than made up for this deficit by convincing Vloka that the *wipla* bridge was really platinum.

In the event that I managed to meet Maria Cielack, I wrote a letter to France that she could mail for me. It was actually a letter within a letter. The first letter was addressed to a nurse with whom I had worked in Nice, asking her to send the enclosed letter to Mme. Billard at the *poste restante* with a request that she send it on to my mother and then mail an acknowledgment to Maria Cielack in Laurahutte that this had been done.

At eight the next morning my two SS "escorts" and I were ready to visit the city hospital. The sound of Laurahutte's heavy metal gates closing behind me literally made me dizzy with the illusion that I was finally free. I spent the walk to the hospital thinking about the best way to try and bribe my guards into taking me to Maria Cielack's shoe store.

When we arrived at the hospital one of my guards asked the other if he could go to his home for just a short while. The other guard agreed because he trusted me not to escape. The two of us gave the sterilizing drum to the nurse in charge and sat down on a bench in the hall to wait. People passing by stared as if we were visitors from another planet—a tall, thin man in an odd striped uniform and an SS guard with his gun held at the ready. A few minutes passed in silence. I decided it was time to try.

"You know," I said, "there is a woman here in town who would pay you well if you would take me to visit her. She runs a shoe-repair shop, and no one would be suspicious at seeing us there." I waited for his answer. He thought about it, then asked the nurse when the sterilization would be finished. "Not for at least two hours," she told him. "Okay," he said to me, "let's go. But you'd better promise that there won't be any monkey business!" I gave him Maria Cielack's address.

The store looked empty as we approached. Perhaps she was frightened of our being seen together. My guard knocked on the door. It opened, and there stood the lovely young woman I had seen in the photograph. Words can't begin to describe how I felt at being so near a woman after so long! Wordlessly, I took her hand and kissed it. Her skin was so smooth and smelled so good. I, Albert Haas, who was never at a loss for something to say, was as mute as an adolescent on his first date. The SS man shifted uncomfortably from foot to foot, seeming to sense my personal awkwardness and the strangeness of the situation—this beautiful woman before me and his gun at my back.

Maria, speaking German, invited us into a back room that contained a table and chairs, with some large sacks standing in a corner. We sat down while she set out wine, bread, and cold cuts. I found my tongue and wondered where the shop workers were. She explained that the shop was open only three days a week and would normally be closed that day. The SS man was watching us with hawkeyes, and I could barely swallow my food. Suddenly Maria spoke up and asked if he would do her a great favor and leave us alone. To my astonishment he agreed, remarking that this back room held no escape exits. Maria gave him the bottle of wine for company as he left.

She sat down and pulled her chair very close to me and tenderly kissed my cheek. Then she sat back and asked me how I had been caught. As I recounted my story I suddenly remembered the envelope that I had prepared for the nurse in Nice. I asked Maria if she would mail it for me, explaining that it was simply a letter to my sick mother to let her know that Sonja and I were still alive. She promised to do it and send me the answer as soon as it arrived.

Sitting next to her I felt so excited, so ill at ease, and so unsure of myself. Her closeness reminded me that, despite all the torture I had suffered, I was still a man capable of experiencing the special physical and emotional pleasures of a woman's company. I leaned closer to her, magnetized by her presence, and looked into her eyes, at her lips. I lost my self-control. I took her in my arms and kissed her. My awareness of her warm and fragrant body blotted out everything else. Agony and ecstasy intermingled. I felt so alive! Slowly she freed herself from my arms, kissed my eyes and said, "The next time you come here I will hide you and help you to escape. I will organize and prepare for it and send you a note when everything is ready." She quietly confided that she had already helped several forced laborers to escape.

My guard knocked. It was over. We had to leave. She gave a small package of goodies to the SS man and one to me. I slipped the almost forgotten ten dollars into her hand, answering her inquisitive look with a remark that it was for the food. She checked the street to see that we wouldn't draw attention as we left, then let us out.

We returned to the hospital and picked up the sterilizing drum before the other guard rejoined us. Back at the camp I informed Bela that the instruments were sterilized and ready.

The next morning Vloka appeared during the roll call. It was barely 6:00 A.M. He took me out of the lineup and over to the surgery. We worked out the arrangements for the abortion for the coming Sunday, ensuring that the girl would not be spotted entering or leaving. All the barracks would be temporarily closed, and all the windows in the barracks and the infirmary would be painted over.

Word of a plan to lock the barracks leaked out almost immediately, creating a rumble of apprehension among the inmates. The Hungarian prisoners who had come from other camps knew that mass gassing was always preceded by a daytime locking of the barracks. On the day the windows were painted, these inmates reacted with a panic that quickly infected the entire camp as they aroused terror with their accounts of mass exterminations they had seen. Some had even witnessed occasions on which there were too many victims for the gas supply. There was only enough gas to stun, not kill. When this happened, the bodies of the dazed victims—some still conscious, including those of children—were piled into a huge pit, soaked with gasoline, and torched. The witnesses could not forget the screams, and their tortured memories created frenzied chaos in the camp.

I tried to calm my fellow inmates' fears. I reminded them that we were only in a tiny work camp. They looked at me with closed faces. I pointed out that there was no gas chamber here at Laurahutte. They felt that I was betraying them. They were certain that the SS guards were about to drop gas canisters into their barracks. On Saturday morning two evacuees from the death camps committed suicide. Prisoners began to mob the HKB, promising me money and gold for sanctuary. I kept insisting they were not in any danger. They refused to believe me. They begged me to give them poison so they would not be burned alive.

The situation had gotten totally out of hand. We had to come up with something to calm these terrorized people. We discussed alternatives and finally decided on Bloch's solution. We made up packets of aspirin and laxatives, disguised them as poison, and handed them out to everyone. The lampshade artist collaborated with us by making a bunch of little paper bags decorated with a skull and crossbones and the words "Warning! Poison!" written on them. We made the frightened men promise to take this poison only if they were forcibly removed from their barrack. The panic subsided.

On Saturday afternoon the leg supports were delivered. Vloka informed me that I would be bringing the girl in from the SS kitchen as soon as she arrived. I tended to my patients during the afternoon, silently wondering what she would be like. I imagined an arrogant person full of contempt for us, her inferiors. As the daughter of an SS commander, she would be thoroughly infected with the Nazi propaganda of the super race.

At 10:00 A.M. Sunday, Vloka came to the HKB to take me to the SS kitchen. When we arrived I looked around for a young woman disguised in a soldier's uniform. The room was empty. A door opened suddenly, and standing before me was a beautiful blue-eyed, black-haired young woman wearing a simple and tasteful dress. She looked about twenty, not sixteen. She stretched out her hand. "*Guten Tag, Herr Doktor.* I am happy to meet you." I was thoroughly unprepared for this lovely, polite and surprisingly unpretentious creature.

I began to perspire from nervousness and had no idea how to start a conversation with her. For the first time in my professional experience it was the patient who put the physician at ease. She unaffectedly began to ask me questions about myself—how long had I been incarcerated, and why. I remarked that I had no idea why I had been arrested, since I was innocent of the charges brought against me, and I almost laughed out loud as I realized that I was saying what every criminal says about his arrest. My young patient expressed the hope that I would soon regain my freedom, then I could tell the world that the ugly rumors spread about the German concentration camps were untrue. She had learned from her father that all of the prisoners were well treated. Even Jewish families were permitted to stay together. I stared in disbelief at her easy acceptance of her father's lies. She chattered away in a manner that was typical of young people her age. If she was at all nervous, it didn't become obvious to me until I realized that she was talking compulsively about everything except the abortion. She was anxious, after all.

She fell silent as we entered the surgical room. Her face tightened when she saw the table with the leg supports. I felt so lost, so helpless. In retrospect it brings a smile to my face to remember how I stood there, face to face with a beautiful young girl, too embarrassed to ask her to undress! We stood like two statues. I, the physician who had always found words with which to soothe my patients, was mute. I finally broke the silence with, "Would you like to have a glass of water?"

I had to explain to her what would be happening. Before I began, I asked if I could call her Tina. She laughed and agreed to this temporary name. I couldn't know her real name, of course, but if she remained nameless for me the situation would be distressingly impersonal. Why did I choose Tina as her pseudonym? No real reason. The name just popped into my head, prompted perhaps by Maurice Chevalier's popular song "Titine."

I explained to Tina that I would anesthetize her so that she wouldn't feel any pain. I would remove her clothes after she was asleep if she felt uncomfortable about doing it beforehand. She was still much more at ease than I was, volunteering to undress once I had given her a sheet and turned my back. I decided to leave for a few minutes and advised her to empty her bladder and bowels while I was gone. She surprised me with her response that, on the advice of a friend, she had already taken a laxative and was thoroughly cleaned out.

I found Bela and Vloka standing near the door and told them that I would inject only enough Evipan to produce a slight anesthesia before calling the gynecologist in to take over. I returned to my patient to find her already on the table, covered by the sheet. She was really lovely, with her thick black hair and deep blue eyes. I shook my head at the irony of her Semitic looks as I prepared the syringe, the ether, and the ether mask.

As I opened the sterilizing drum and laid out the materials and instruments, she followed my movements with a curious gaze. I asked her to hold out her left arm so that I could tourniquet it and inject the anesthetic. As she moved her arm the sheet slipped down, exposing breasts that were round and firm and very exciting. My fingers began to tremble, and I couldn't take my eyes off her breasts. I was horribly embarrassed. I tried to look down at her arm, but I couldn't take my eyes away from that beautiful sight. She must have sensed my feelings because she made a move to pull the sheet back up. Finally I managed to cover her again, pulling the sheet up to her neck. Her vein was prominent, and the injection went easily despite my unsteady hands. She began to count backward from one hundred. By the time she had reached eighty, she was asleep.

I went out and told Bela Robert that the patient was ready for him. To my dismay, Vloka insisted on coming in with us. He insisted that he was dying of curiosity to see a curettage performed, but I was sure he was far more interested in the girl's anatomy than in the procedure. Since he had no sterile coat, I pointed out that if an infection developed from contamination he would be just as responsible as we and we'd all hold hands in the gas chamber. To get him off my back once and for all I promised, stupidly, that I would arrange other abortions for him since we now had the setup. He could attend as many of those as he wanted.

Vloka agreed to wait outside while Bela joined me in the operating room.

Bela slid the girl down to the end of the table, and positioned and secured her legs while I began to administer ether through the mask. He removed the abortifacient root from its glass tube and inserted the dilator in the cervix. Bela stopped and whistled in astonishment. I was afraid that something had gone wrong.

"Can you imagine," he said, "she's full of sperm. She had intercourse this very morning!"

He cleaned away the sperm and examined her, pronouncing that she was between ten and twelve weeks pregnant and just right for the abortion. Then he inserted the root into the cervix, took off his gloves and told me to stop the anesthesia and call him back in about four hours.

I freed the girl's legs and stretched her out on the table. Just as I finished covering her body she opened her eyes. She was disoriented for a moment, and then burst into sobs. She calmed down in a few minutes and wanted to know if it was all over. I told her we had to wait for about four hours before I could complete the procedure. She said she felt very drowsy and wanted to sleep off the effects of the anesthetic. I offered to stay to make sure that she didn't accidentally fall off the table during her sleep. She answered no when I asked if she was hungry or thirsty, but was silent when I asked if she felt any discomfort. I looked up to find her fast asleep again.

As I sat watching over this enigmatic patient, I pondered her bizarre situation. This girl was far from the spoiled SS commandant's daughter that I had anticipated. I couldn't conceive of her father actually raping her, and I couldn't comprehend what would prompt this apparently intelligent and well-brought-up girl not only to commit incest but to do so apparently without the slightest trace of guilt.

I had been lost in my thoughts for a while when I realized that she was speaking to me. I looked up to find her open eyes scrutinizing me. She must have been observing me for quite a while. She asked what I had been thinking about, since I seemed so far away. I felt very reluctant to share my thoughts with her, but my mouth opened and I heard my voice asking, "What makes a beautiful young girl have intercourse with her father? Did he rape you?"

She made a funny little noise in her throat. "Sort of," she said.

I suddenly realized what a risk I was taking. If her father learned about my curiosity the lives of everyone in the HKB would be in jeopardy. I told her to forget my questions and explained why. She was amazed that I could even think she might tell her father.

"Just ask," she said. "I have nothing to hide."

The biggest puzzle to me was her mother's attitude. She asked if I really thought this could have happened if her mother were still alive, and then launched into her story.

Her mother died when she was eleven. Her father seemed unable to pull himself together. All of his time at home was spent shut up in his room. It was a very unhappy home, and she began to have difficulty sleeping. One particularly anxious and sleepless night after she had turned twelve, she went to her father's room for comfort. She could hear him pacing behind his door, so she knocked and asked if she could sleep with him because she was afraid. She got into his bed and fell asleep curled up in his arms. She woke suddenly to feel a sharp internal pain. Her thighs were spread open and her father lay on top of her making strange sounds. She was too terrified to voice her fear. She submitted to several more rapes that night and finally fell asleep. She woke up in her own bed and thought that she had dreamed the entire experience until she stood up and became aware of pain between her legs and saw blood on her thighs. Her father was gone and didn't come home for several days. During that time she slept in his bed each night, very unhappy and feeling that she was responsible for her father's absence.

The night her father returned, she woke up while he was undressing and experienced a strong sexual desire for him. When he got into bed she pulled up her nightgown and seduced him, and since then she had slept with him nightly. He had always taken contraceptive precautions, but on this past New Year's Eve he came home a little drunk from a party. She had always enjoyed having intercourse with him and missed him achingly when he was away.

I asked her if she loved him. She assured me that she did. "He is very good-looking, an excellent lover, and a gentle person. I'm only sad that I can't bear him a child, because he would be such a good father!"

She had completely relinquished her identity as his daughter.

They lived together fully as man and wife. I would have suggested that she see a psychiatrist, but the suggestion might have cost me my life. I sat in confused silence trying to sort out my thoughts and feelings. The intensity of my initial shock had dissipated. Perhaps knowing the ancient Oedipus legend somehow made this twentieth-century story a little more acceptable, if not understandable.

A sudden uncomfortable silence caused me to look at her. Her eyes were closed. She looked no more than twelve years old. I realized that tears were streaming down her cheeks. I had no handkerchief, only a piece of gauze. As I tried to dry her tears with it, she opened her eyes. With a pitifully sad sound in her voice she said, "I am sorry."

It was time to continue the abortion. As I prepared her for what was to come, the sheet accidentally slipped off her again. I made a strong effort to push away the impression that her naked body made on me and tried to view her only as an unhappy child. She seemed embarrassed too, anxiously grasping a corner of the sheet and pulling it over her body. She covered her eyes with her right forearm. After what I had suffered at German hands, I was bewildered by the compassion I was feeling for the daughter of a concentration camp commandant.

I injected the Evipan, and when she fell asleep I called Bela from his room. He finished the curettage in about three-quarters of an hour. When he left I gave her an injection of Rubiasol to prevent infection. When she woke up I explained that I had given her an injection of medication that would color her urine red and that she was not to be frightened by it. I made up a heavy pad of gauze-covered cotton and placed it in her panties to absorb the blood. She rested for about an hour and a half, and finally asked for something to eat. I asked her to dress while I looked for food in the kitchen, where Vloka found bread and margarine and some kind of tea for her.

When I returned she was dressed, sitting in the chair with her head on the table. She had become dizzy after putting on her clothes. She slowly began to eat and drink but suddenly stopped to ask me what the fetus had looked like. "Was it a boy or a girl?" Since I hadn't really seen it, I hedged and told her that it wasn't developed enough for me to be able to tell.

When she began to feel a little stronger I told Vloka that she was ready to leave. He instructed me to take her back through the SS kitchen. She didn't say a word until we reached the outside door. Then she thanked me and said she hoped I would soon be reunited with my family. Then she disappeared.

Returning to the surgery, I was still preoccupied with Tina. Why had she spoken so frankly to me of her relationship with her father? Had she powerful feelings that had to be shared and no one else with whom she could share them? Or was it an expression of sympathy for me, a way of saying that everyone is a prisoner of his past? Or had she temporarily lowered her defenses because of the loosening effects of the anesthetic? And might she not regret it afterward? Whatever her reason for confiding in me, this child's deep emotional disturbance was obviously the result of her father's knowing and purposeful behavior.

An unexpected quiet in the courtyard interrupted my thoughts. The prisoners should have been outside by now, but the courtyard was virtually deserted. I went over to the barracks to find out what was wrong. The Blockältester informed me that an unexplained epidemic of dysentery had suddenly broken out among a large number of the prisoners. They could now be seen in a long line at the latrine. The Blockältester was thoroughly mystified. So was I, until it hit me! Vloka had ordered lunch to be held up until the girl had left and he had given the "all clear." But he forgot to inform the prisoners that their lunch was going to be delayed. So when lunchtime came and their soup didn't arrive, they took it as absolute proof that a mass gassing really was about to take place. They had all taken their "poison" and those who had been given the laxatives all had the runs.

This unanticipated turn of events had unfortunate consequences. Although the prisoners soon realized I had been truthful in assuring them that no one was going to be harmed, they also knew that I had lied when I agreed to give them poison. They interpreted my actions as a cruel joke, and an unbreachable credibility gap existed between us for the remainder of my stay in Laurahutte. It was painful for me to lose my comrades' confidence.

Vloka brought us a whole parcel of medications to express his thanks, a mere fraction, I am sure, of the reward he must have received for his role in the abortion. Even so, the parcel was a gold mine. As I sorted out the contents, which included cod liver oil, it was obvious from the original labels still glued on all of the containers that they had once sat on the shelves of a Jewish pharmacy in Hungary. What, I wondered, had become of the pharmacist who had stocked these medications to help the people in his community?

I did not dare ask Vloka how Tina was doing, but I felt sure that the girl's hand was involved in our unexpected bounty. A small part of me wondered if my sense of her involvement was an illusion that sprang from my need to believe that someone out there still felt compassion for us. But if it was an illusion, I didn't want to tamper with it. To lose that belief was to lose my sanity.

About a month later, Vloka brought me an envelope containing a thank-you note and a small handkerchief with "Tina" embroidered in one corner. The little handkerchief, a touching reminder of the gauze that was all I had had to dry up her tears, was moving evidence of the girl's gratitude, trust, and compassion.

18

By the late winter of 1944 the Russian armies had begun advancing along the entire eastern front of the war. The concentration camps that were in their path had been abandoned by the retreating Germans, the inmates exterminated or evacuated elsewhere. While these evacuations were in progress our black market had had a seemingly endless source of riches to draw on for trade. We had begun to take this regular supply of new arrivals—and their currency and other valuables—for granted. But by late spring, when the Russians reached Warsaw, the evacuations stopped. We saw no more new faces in Laurahutte. Our treasure trove began to dry up.

The depressing day quickly arrived when I had to tell Vloka to take back the food and medications he brought us. I had nothing to offer in exchange.

"You can still have these essential supplies," Vloka said, grinning at me. "Just let me attend the next abortion."

I was thunderstruck. I had never dreamed that he would take me up on my hasty promise. Vloka, unhappily, was determined to hold me to it and exploit it to the full. I found myself backed into a corner. All I could do was insist on certain conditions. Each time, Vloka would have to arrange for one of us to take the instruments to the hospital for sterilization and pay us generously with food, bandages, gauze, antiseptics, and whatever oral medications he could lay his hands on. He agreed.

I was still very reluctant to arrange any more abortions for him. I tried to stall by saying that I must talk it over with my colleagues. I informed Bloch and Bela about this unexpected and depressing turn of events. They saw no way out. We had no real choice but to give in to his blackmail.

Vloka brought in our next patient after 9:00 P.M. on the arranged date. The abortion had to take place the same night, and Vloka eagerly watched the entire procedure. I hoped the experience would satisfy him, but it served only to whet his appetite. He probably also pocketed a large fee for his services.

We had no alternative but to stay in business. Our only consolation was the benefits we were able to gain for our comrades. We began to call Vloka's abortion clinic our angel factory. In addition to a reliable supply of good medication and food, we were able to rotate the seriously run-down prisoners through the HKB for enough bedrest and enriched meals to build up their strength. If not ultimately lifesaving, at least it postponed their transfer to one of the harsher camps where the gas chamber would be their fate. Through Vloka we were even able to send patients requiring special surgery to the well-equipped central HKB at Auschwitz—and get them back again! Although we regularly destroyed potential lives in Vloka's abortion clinic, doing so enabled us to save many of our fellow human beings. We felt that we came out ahead on the trade-off.

I felt powerfully ambivalent about visiting Maria Cielack on any of my trips to the city hospital to get our instruments sterilized. On the one hand, my feelings of longing for a woman had been haunting me ever since I had seen Tina on the operating table. I doubted my self-control the next time I saw Maria. I also didn't want to be tempted with the possibility of trying to escape, contradictory as that may sound. I was so sure the Russians would soon reach us. My fantasy was that the SS would abandon all the camps in our region, and in the ensuing chaos I would escape and find Sonja. I decided to avoid both temptations and not meet Maria again.

News of the Russians' continuing advance was spectacular. My hopes for imminent reunion with Sonja soared. We kept up a steady correspondence and gave each other detailed pictures of our respective camps, although I omitted any mention of the angel factory. In one of my letters I asked Sonja whether she had any new information about the military situation on the front. She knew nothing specific. Her camp duties had permitted her to read the *Volkischer Beobachter* (the same newspaper the priest had given to me in Munich), but suddenly the SS had become exceedingly reluctant to let her see the newspaper. Something big must be happening.

I will never forget the earthshaking date of June 6, 1944—D day. The Americans landed in Normandy!

Early that morning I was urgently summoned to the camp factory. The odd conditions I encountered were extremely puzzling. The German workers were tensely clustered in agitated discussion. The German plant manager sat with two SS officers in his glass-walled office, gloomy faces on all three. The French forced laborers, standing in a separate group, were in a jolly mood. A young French Jew lay near them on the floor, apparently unconscious. His face was badly swollen. Blood gushed from a large scalp laceration. I wondered how he had hurt himself, since the machines hadn't yet been turned on. Commandant Quackenacker arrived on my heels and joined the group in the manager's office. No one offered a word of explanation to me.

The young man opened his eyes and tried to speak. I bent down to his lips and heard, "They are coming, they are coming." I thought he was hallucinating that whoever had beaten him up was after him again. I went to the manager's office to get permission to move the injured man to the HKB. They ignored me. They were all glued to an excited voice on the radio announcing that the Allied Forces had landed at three points on the Normandy beach. My heart began to pound.

Quackenacker irritably asked me what I wanted, authorized my request, and dismissed me. With one ear on the radio as I left I caught a few words describing the German mopping-up operation, predicting that the Allies would be repelled within twenty-four hours.

The three prisoners who helped me carry my young patient to the HKB told me what had happened to him. His injuries were certainly no work accident. The young man, who understood German, worked near the manager's office and had overheard—through the open office door—the radio announce the Allied landing. He began to shout the news exultantly. The enraged German factory manager began to beat him with a heavy piece of metal. The young man would have been killed if the guards had not intervened. I realized that the words I had caught on the radio meant that the Allied landing had not been repulsed. The manager would never have become so violent if the Allies were no threat. It was a real invasion after all. The Germans were in serious trouble!

I got my patient to the surgery and repaired the damage—sutured the head wound, set his broken nose with a makeshift splint of cotton and tape, and pulled out the roots of his four broken teeth. I thanked heaven for the angel factory and our sterile materials.

The next day a hurriedly written letter from Maria Cielack told me of the Allied landing, the disintegration of the German military, and her certainty that our liberation was not far off. From her window she planned to signal to me the daily news from the clandestine radio broadcasts. She repeated her offer to help me escape and ended her note with an encouraging "Hold fast!"

The following day a cannon barrel "accidentally" fell on the factory manager, killing him instantly. It was handled as a work accident, and there were no reprisals. A few days later I saw a copy of the *Volkischer Beobachter* that had been smuggled into the camp. The Allies had broken through the German defense lines in a number of places and were now advancing but only, of course, because the Nazi high command was staging a tactical retreat.

Every morning and afternoon Maria was at her window translating the clandestine news into sign language. There was much I couldn't understand, but she succeeded in communicating that the advancing Allies were breaking through on all fronts. It was now just a question of time before the Germans were beaten! But now that it was finally happening, I wondered if Sonja and I would survive and find each other. Himmler was rumored to have ordered wholesale exterminations to cover up the crimes committed in the camps. Our liberation was coming closer and closer, but we had to make sure that we were still around when it finally arrived. We all felt that we were racing against the clock.

The summer peaked and began to wane. Our liberation had not yet materialized, although some of the camp personnel seemed to feel that they should begin their private preparations for Germany's imminent defeat. One day in September a very anxious Vloka came to see me. He wanted me to have a statement written for him and signed by the HKB patients to the effect that he had treated them well. I said I would try. I wondered out loud why I hadn't heard from Sonja recently. Vloka clapped his forehead in dismay. He had been carrying her most recent letter around for a week. He had become so preoccupied with his own problems that he had completely forgotten about it. I was very relieved and went off to my room to read the letter.

Sonja was deeply disturbed. She had discovered that the Auschwitz Gestapo were investigating my records. The day that she wrote to me a Gestapo agent had marked my card "Transfer to Bunker." That meant interrogation. We both knew that I could not survive another torture session. Sonja wrote that I was safe for the moment because the Gestapo were fully occupied trying to put their fingers on the instigators of the recent inmate uprising at Auschwitz. The revolt had been unsuccessful in terms of freeing the camp, but it had blown up the gas chamber and crematorium and thoroughly enraged the Gestapo and the SS. The Gestapo would not be sidetracked from me for long, so Sonja had managed to arrange my transfer to another camp, a place in Austria called Mauthausen that she had heard was not too bad. Her disappointment at not being able to see me soon was deep, but at least I would have a chance to survive.

The rug had been pulled out from under me and my hopes. I sent a letter to Maria to tell her about my upcoming transfer. I thanked her for everything that she had done to make my suffering more bearable. I gave her the cover name and post office address of Mme. Billard, who would know whether I had survived and where I could be found after the war. I told Bloch and Bela that I would be transferred very shortly. I wanted them to continue the angel factory for their own protection.

Three days passed without any change in my routine. On the fourth day, Quackenacker called me to his office to tell me that I was being transferred to the Auschwitz HKB that same afternoon prior to being shipped to another camp. He dismissed me with no further word. I returned to my room to destroy all of my correspondence. I shredded Sonja's letters, tearing my heart into little pieces as well. These sheets of paper had become my lifeline to her. I felt as if I were destroying a vital part of myself. Then I tore up my precious letters from Maria, my protective and final act of friendship with her.

Vloka arrived. It was time to leave. He was personally to take me to Auschwitz. I embraced my colleagues and left Laurahutte with only the clothes on my back. I had the unpleasant sensation of exchanging a relatively quiet and secure existence for another unknown and dangerous destination.

Vloka sat by my side in the car to Auschwitz. Before we said good-bye he handed me a full pack of cigarettes. I was unprepared for the emotion I saw reflected in his eyes. The man was capable of some compassion after all.

Before handing me over he said, "I saw your wife, and she is well. She still has her hair, and she is a beautiful girl."

Since women's heads were usually shaved, this was Vloka's way of telling me that Sonja was in a highly favored position in the camp and therefore had good odds for coming out of this alive. Even as a prisoner in Auschwitz Sonja was obviously the same Sonja she had always been, cooly evaluating and manipulating whatever situation she found herself in.

I asked Vloka if I could possibly see her, even for a moment. It was impossible, he told me. The men and women were kept in separate sections of the vast camp, and immediate death was the fate of any caught trying to see each other. What a painful irony it was to have come close enough to touch, yet still be so far apart.

Almost as soon as I arrived at Auschwitz I was informed that I would be in a transport leaving soon for Mauthausen, outside the Austrian city of Linz. Mauthausen was badly in need of another Aryan prisoner-physician. So I was vaccinated against typhus, and was to receive a booster shot the next week. But I never got it, because my number was called the very next day.

19

I was packed into a cattle car again. This time I was with about five hundred Polish inmates. We were distributed among five cattle cars, cars that had been built to hold a maximum of "8 HORSES OR 40 MEN" according to the notice prominently lettered on the outside of each one. There was no food, no water. There was barely room for the garbage can we had to use as a toilet.

There was no room to move, and fights broke out because people who had to use the can couldn't get to it. And inmates suffering from dysentery couldn't control their bowels for long. The odor became nauseating. The car grew intolerably warm. I began to feel thirsty. The odor became worse.

Previous transports had taught me to stand near the can, and since we were so tightly packed on this transport I had no choice but to watch everyone else defecating and urinating. I had to be careful when someone occasionally lost his balance in those unstable conditions and missed the can. My efforts to keep from being soiled were wasted, though. Once the can was full, which didn't take long, I was splashed with its contents whenever the train braked suddenly.

There were frequent moans and cries. I couldn't decipher them because everyone around me spoke only in Polish. I heard gasps of suffocation. There was no room for anyone to sit down. I tried to stand on only one leg at a time. I looked around at my fellow travelers and marveled at man's capacity for endurance despite his seeming frailty. I wondered if Sonja had any idea what was really happening to me. She couldn't have. This transport was beyond all imagination. I kept reminding myself that it was still better than another session of interrogation and torture.

Night came. I tried to sleep standing up. We were too tightly packed in to have any worry about falling down. I slept briefly and dreamed confusing dreams. Something warm running down my leg woke me up. A man next to me was urinating on my leg. I tried to push him away, but there was no room. I wondered if this was how the holds of the African slave ships had looked. But the old slave runners were in it for the money, while the Nazis' goal was to kill as many of us as possible.

Voices all around me were complaining. Several men were calling for *wodih* ("water"). Someone was asking for a doctor. I couldn't move. What could I do anyway except dispense encouraging words in a language that none of them could understand? Anyway, I was just as much in need of encouragement as they were.

Finally the train stopped. I heard the distant opening of a door. This sound repeated itself, coming closer and closer. After a long time, our door opened. An SS guard barked at us to descend and we jumped out, blinded by the sudden light. The can was dumped before we had all gotten to our feet. Those not swift enough were showered with its contents. As I looked around me, I saw that we seemed to be fewer in number than we had been when we embarked. I emptied my bladder, relieved that I hadn't had to do so in my pants.

Someone from inside our cattle car was calling, in German, for a physician. I went to the door. An SS guard ordered me to jump up into the car. Inside was a macabre still life of bodies. I was ordered to examine them. I counted seventeen bodies, all dead, most likely asphyxiated or trampled to death.

The SS guards commanded me to pile them in the corner. I stacked two corpses and then, with difficulty, put two more on top of them. With the fifth corpse I felt almost out of breath and had to use my back to push it up, feeling very pleased that I had found the strength to do it. The guards watched me, probably betting on whether I could do it.

The sixth corpse was very rigid, which meant that it had probably been one of the first to die, and was consequently very difficult to move. I was too tired to bend down and pick it up. I struggled to pull it into a standing position and slowly pushed it up onto the pile. As I went to get the next corpse the pile of cadavers came tumbling down, to the delight and amusement of the SS guards.

When they finished laughing they called a few inmates in to help me. This time we stacked the corpses alternately head to foot, with outspread legs, so the pile would hold.

We were herded back into our cattle car. This time I wisely placed myself a little farther away from the can to avoid being showered when the train braked or went around a sharp curve. One of the Poles began to speak to us, and my neighbor explained he was suggesting that we all try to sit down. The seventeen deaths gave us just enough room to sit with our knees pulled up tightly to our chins.

With the door closed, the oppressive heat increased again and the stench worsened. Our hunger was still tolerable, but thirst had become unbearable in this airless furnace. I began to hear many cries for "*wodih, wodih, wodih.*" My own tongue was swollen.

The night passed with nightmarish dreams interspersed with waking periods filled with voices screaming for help. I opened my eyes, I tried to swallow, but I couldn't. My lips were becoming dry and sore. I saw light outside between the slats. An inmate standing at the can urinated into his hands and drank it. I closed my eyes. When I looked up again, the man was no longer there. Had it been a hallucination?

Much later the brakes squealed and the train halted. Flickering lights from outside indicated that we were in a station. As we stood there for what seemed an interminable time I became aware of a sickly sweet, strangely familiar smell. It was the odor of decomposing flesh. I heard the doors opening, one by one. Commands of "Get out! *Schnell, schnell!*" Our door opened and we jumped out.

We looked around us at the remains of a bombed-out station as we were lined up. Four inmates were assigned to pull over a little wagon with two big barrels on it. Each of us in turn walked to the wagon and picked up a bowl into which some ersatz coffee was ladled. We had to gulp it on the spot and replace the bowl for the next person. We also got a piece of bread each, about one-quarter of a pound. I drained my coffee in one continuous greedy swallow without stopping for air.

I had barely begun to eat my bread when an SS guard yelled for the physician from my car. When I entered the car I saw bodies lying everywhere. One lifeless body suddenly jumped up and, laughing hysterically, ran to the door and leapt from the train. He staggered like a zombie across the tracks to the ruins of the station platform. An SS guard yelled *"Halt."* The crazed inmate kept walking. A shot rang out and he fell forward. An SS guard behind me kicked me to punctuate his comment, saying, "He was already dead anyway." Two inmates carried the fresh corpse back to our car. The twenty-seven bodies still on the floor were all dead. Six men were ordered to help me stack them. Before I could go out for a moment's fresh air, everyone was ordered back on the train. I was surrounded by people, and the door was closed.

My nose filled with the nauseating stench of the decomposing bodies. We stood about aimlessly until someone organized our sitting-down procedure. Silence reigned for quite a long time, occasionally interrupted by a heavy, heartrending sigh. The odor continued to torment me, and I desperately wanted to wash the smell of rotting flesh off my hands. My bladder finally filled up. I made my way to the can and urinated on my hands, but the smell remained just as strong.

A voice broke the silence with a few Polish words. No one responded. The voice continued anyway, seeming content in talking to itself, sometimes interrupting itself with laughter. The voice eventually grew ragged, then pleading. Someone called in German for a doctor. I tried to make my way toward the sufferer, but after a few steps someone else cried out that I was walking on his hand. I returned to my place. The Polish voice began again, shrilly alternating between laughing and singing. I heard the sounds of someone being beaten. The voice was crying now. In German I asked my neighbor what was happening. In broken German, he replied, "He lost his mind." The sounds of beating continued. The voice began to beg, then was silent.

My inability to talk with any of the other men increased my sense of loneliness and despair. In a desperate attempt to break out of my isolation I began to whistle Maurice Chevalier's song "Titine." To my joy, two of my nearest neighbors joined me in the French refrain. Then I began talking in a rush of words, in French, German, Hungarian. If I hadn't, I would have lost my mind.

Although no one spoke any of these languages, my Polish neighbor seemed to understand my need. He answered me in his language. We spoke back and forth, his smile telling me that he was listening to me even though neither of us understood the other.

Eventually I recited some of my favorite poems by Verlaine, and in Polish he seemed to answer me in kind. In these conditions, with the stench of decaying bodies permeating the stagnant air of our cage, another human being understood and responded to my desperate need to be listened to. His inability to understand my words was unimportant, for his solace warmed my soul.

When the train came to a stop it was still light. The door opened and we were shouted at to get out. Two inmates were laughing, grimacing and screeching, oblivious of their situation, no longer in possession of their sanity. I saw many more corpses. I asked the SS guards if we could collect the cadavers from all the cars and put them into one, and distribute all the living occupants among the other four cars. My answer was a relatively gentle kick and a sarcastic smile. I tried to figure out how long we had been traveling and concluded that it had been about three full days.

I was ordered to count the most recent corpses. One of them was covered with blood. I wondered if it was the man who wouldn't stop talking to himself the night before. I wondered how many of us were still sane. Some vague snatches of conversation that I overheard among the SS guards indicated that we were nearing our destination, an impression confirmed when we were ordered to lay all our dead out on the embankment by the tracks. Evidently it was time for bookkeeping. The SS guards began to check off the dead men's numbers against the transport list.

The man in my car who had been shot presented a problem for them. He had not yet been tattooed. Blood had seeped through the number on his clothes, making it impossible for them to read it. I watched the guards scratch their heads, discussing what to do, and I felt glad that at least one of us had given these animals some trouble, even though he had had to die to do it. Along the length of the train this same macabre scene was being repeated in front of each car—the SS identifying the numbered corpses in striped uniforms frozen in their glimpse of eternity.

For the first time I noticed two civilian cars attached to our train. Their curious occupants, mostly women and children (the families of the SS men accompanying us), were looking out at us. I wondered what the mothers were telling their children, and what the children thought at the sight of their fathers participating in these horrible scenes. I never saw anything more than simple curiosity on any of those faces, nor did anyone cross himself out of respect for the dead. You would have thought they were watching some primitive rite.

It began to grow dark. We still had to transfer our cadavers back to the car. There were now forty-three dead and fifty-seven still alive. To give ourselves sufficient room to lie down inside the car we made one towering pile of cadavers, placing the bodies in each layer at right angles to those below. By the time we made the fifth layer the pile was very solid and held well. Many of the living inmates were exhausted and sat down to rest, their backs leaning against this wall built of the corpses of their comrades.

When I went out to tell the SS guards that our work was done, they were still standing over the inmate who had been shot. I thought they would leave the corpse where it was, but one of them finally came up with an idea. He ran to one of the civilian cars and returned in a few minutes with scissors, needle, and thread. He summoned a tailor from my car and instructed him to sew his own number onto the dead man's jacket and put the blood-stained number onto his own uniform. What an ironic joke it would be if the tailor died before we reached our destination! Once the exchange was completed and the tailor and I had placed the cadaver on top of our wall of corpses, I returned to my place.

The Pole with whom I had "talked" greeted me with the delighted smile of a close friend who hadn't seen me in a long time. A bond had grown between us that had no need of words. Now that we had more room, my friend was able to turn so that his back was against mine. Supporting each other, we slept.

I soon awoke. Voices coming from near the wall of corpses intermingled with the thud of falling bodies. The stench of flesh rotting in the unbearable heat had worsened. My throat was horribly parched.

Daylight was flickering in between the slats of our cattle car when I woke up to go to the can. As I stood up my Polish friend slid onto the floor, still asleep. When I returned he was snoring peacefully. He woke up when I gently lifted him into a sitting position and was already fast asleep again when I sat down. We slept until the engine's whistle woke us. I felt the train switch tracks and pick up speed.

It wasn't until late in the day that the train began to slow down. The brakes screeched and the train stopped. After a few minutes of silence we heard steps approaching and then the door opened, but we were not allowed to get out.

I noticed the civilians carrying their luggage from the train. The children were running energetically around their parents, not once glancing at us. We weren't people.

At last the discharge of the human cargo began. First the living, then the dead. I looked to our carefully constructed wall of corpses. It was no longer there. Bodies were strewn all over the floor. When we were ordered to get out, several of these bodies suddenly stood up. Some of my companions had dismantled the wall during the night and used the decaying corpses as mattresses.

Our next order was to remove the dead from the freight car and load them onto two-wheeled wagons that stood on the platform. We were filthy and stinking. About a dozen of my group walked like sightless zombies as we carried the rotting corpses to the wagon. The wagons couldn't hold them all. The Austrian station-master and his men looked on as though we were nothing out of the ordinary. They must have been used to such scenes. I located the station sign: "MAUTHAUSEN."

20

We were lined up by fours, two to pull and two to push the wagons on which we had loaded our corpses. The wooden wagons were not covered. The arms and legs of dead men dangled over the sides as we pushed them through the seemingly deserted Austrian village of Mauthausen. I saw people peeking out at us from behind their spotless lace curtains. We climbed a winding road for about an hour, finally arriving at a broad open space that was dominated by a fortresslike building. The fortress gates opened. We marched in under the insignia "Konzentrationslager Mauthausen." The gates closed behind us.

We were lined up, with our wagonloads of corpses, to be counted. The SS soldier in charge of our transport reported that many corpses still remained on the station platform because there weren't enough wagons to hold them all. He was angrily bawled out. Eight of us were called out of line. Everyone else, the corpses included, were taken away.

Two trucks appeared to take the eight of us and four SS guards back to the railroad station, where the work of loading the remaining corpses went quite quickly despite our exhaustion. We had to ride back to the concentration camp seated directly on top of the decomposing cadavers. Many of the little town's inhabitants were now outdoors. How unperturbed they were at the sight of us, filthy and emaciated human beings sitting in these two trucks piled with emaciated and rotting corpses. When the war ended, these were the people who cried in horror that they had no idea of the monstrous crimes being committed under their very noses.

Inside Mauthausen's gates the trucks stopped in a small courtyard, where we were ordered to unload our rotting cargo. I hesitated for a moment, trying to figure out how I could best conserve my dwindling energy. An SS guard hit me painfully hard and screamed, "*Schnell, schnell!*" in my ear. I jumped to the ground too fast to prepare my landing. I heard a bone crack in my left ankle. The pain took my breath away. Damn, I had broken it again! Another inmate helped me get to my feet, and somehow I managed to help unload the bodies.

We were driven to the Brausebad for showers and shaving and assigned to the quarantine barrack to be numbered and processed. The alternately steaming and freezing shower relieved some of my pain, but my foot and ankle were swelling considerably. Luckily, a broken-down and laceless pair of shoes came along with my "clean" striped uniform. I couldn't possibly have squeezed my swollen foot into a properly fitting shoe. Each step on the way to our temporary barrack was an enormously painful effort. My pain became excruciating. The next morning I requested permission to go to the HKB. My comrades were unanimously taken aback. "Are you crazy?" they asked. "Don't you know the risk of admitting that you're unable to work?" Of course I knew the risk. I preferred that to the pain now tormenting me.

I sat in the infirmary and waited. Suddenly I heard voices behind me conversing in French. What music it was to my ears! I called out, "Who here is French?" A man ran over to me and introduced himself as Frédo Ricoll. We spoke for a few minutes. He pointed out a young man in a white coat, a French medical student named Louis Fichez, who would take care of me. (Frédo Ricoll would remain a close friend of mine after the war. We would also work together in organizing the first rehabilitation center for stabilized tuberculosis patients. Dr. Louis Fichez would be instrumental in legislating postwar medical benefits for French camp survivors and would write the classic work on the "kazet syndrome," the premature senility that characterizes many concentration camp survivors.)

Louis came over to examine my ankle and took me inside beyond the waiting room. With the experiences of the journey from Auschwitz still fresh in my mind, I was unprepared for the clean room I saw before me. There were real beds and sheets. Louis left me with no illusions. This infirmary was restricted to the care of Prominents. Only rarely was he able to admit a French national. No Jewish prisoners were ever admitted.

Louis X-rayed my left ankle. While we waited for the film to be developed, he filled me in on Mauthausen. The Allied and Russian successes in the war were having a devastating effect on life in the German and Austrian camps. An already hellish situation was becoming increasingly worse. Droves of inmates who had been hastily evacuated from the eastern camps had swelled these camps to overflowing. Camps built to hold ten thousand prisoners were now housing three and four times that number. Serious disruption of the German supply lines was drastically reducing the normally minimal food supply in these camps. Hygienic conditions, which had never been good, were now abominable. Epidemics of all kinds were breaking out on top of the ever present dysentery. The Hungarian-Jewish transports had brought in scarlet fever. Lice were spreading the dreaded typhus, and the SS were reacting with such terror that entire transports were sent directly to the gas chambers just because a few inmates had been seen scratching themselves from lice bites. At Mauthausen and its many satellite camps, innumerable inmates were dying from illness and starvation as well as the more direct forms of brutality. The Mauthausen constellation, which covered a one-hundred-mile radius, was a vast graveyard.

I learned from Louis that I was scheduled for an immediate transfer to Gusen I, a satellite camp a few miles away. Gusen I and Gusen II were evil sister camps. At least Gusen I was not quite as bad as Gusen II.

My X-ray showed a fractured fibula. Since I was going to be leaving for Gusen I the next day, Louis was unable to do any more for me than help me make a temporary splint out of a few pieces of wood, some bandages, and some string. He promised to contact the medical chief in Gusen I as soon as he could. Dr. Kaminsky had a lot of influence. Louis told me to wait a week before going to see him.

Louis and Frédo wished me luck when I left. I was going to need it. Whoever had given Sonja the impression that Mauthausen was a fairly good camp must have enjoyed a real laugh at her expense. I found myself again the unwitting victim of a Nazi's practical joke.

Several other inmates and I walked the several miles from Mauthausen to Gusen I. Our uniforms, including my bandages, were taken away from us for disinfection before we were sent to the showers. I hid the string and pieces of wood for my splint under a bench in hopes of finding them after my shower. We were not given new clothes. We all had to run naked to the quarantine barrack about five hundred feet away. I had retrieved the materials for my splint, but the run was still extremely painful for me and the hot shower had sensitized me to the cold air.

In the quarantine barrack we were registered and given new numbers on a piece of cloth to sew onto our uniforms, which were to be returned to us before dinner. Dinner was the standard miserable camp soup I had known in Dachau and Flossenburg.

After we ate, the Blockältester called us to attention. Very ceremoniously, he asked all the professionals among the new arrivals to step forward. Thinking that I would certainly become the barrack physician, I stepped forward with the three other inmates who were still as naive as I was. The Blockältester cheerfully announced that the highly skilled job he was about to give us required an advanced graduate education, and a bachelor's degree at the very least. He congratulated us. "You are now the barrack's new *Scheisse Kommando!*" Scheisse Kommando—the "shit unit"—I had volunteered to clean excrement out of the latrines!

Although I had grown familiar with excrement in Dachau, cleaning out these latrines was far more monotonous and foul work. Dysentery was rife, and the lack of septic holding tanks forced us to empty the overflowing latrine trenches constantly. All day long we filled barrels with the liquid excrement and loaded them onto a cart that we pulled to a campsite beyond the camp, from which local farmers carried it away for use as fertilizer. It was impossible to avoid getting splashed as we hoisted the barrels onto the cart. We had no facilities for washing our clothes or our hands, and the odor quickly penetrated to our bones. We all stank so badly when we returned to our barrack that the Blockältester ordered us to sleep under our beds. If we didn't disappear under them quickly enough, we were helped by his boot. But the work had definite survival value. If an SS soldier or Prominent threateningly called us over as we walked down the camp streets, the moment we were close enough for him to get the barest whiff of our nauseating stench he invariably ran away as fast as his legs would carry him. I dubbed my three colleagues and myself "the engineers of the Gusen Express."

A few days into our new job, we were dumping our barrels when the farmers arrived. The large metal tanks in which they collected and carried away their fertilizer gave me an idea. I spoke to the Scheisse Kapo, who spoke to the farmers. I wanted to save us a lot of work by loading the human excrement directly from the latrines into these tanks. The farmers were enthusiastic about this system because it meant that most of their work would be done by the inmates. The Kapo sold my idea to the Lagerältester, and the farmers brought us their tanks. And that is how we modernized the fertilizer delivery system at Mauthausen!

We not only kept pace now with the rate at which the latrines filled up, but the fact that the farmers' mules carried the containers back and forth gave us time for some much-needed rest and a chance to participate in the goods-exchange market that always went on in the latrines. Anyone who had something to trade came to the latrines. Cigarettes were exchanged here for bread, or bread for cigarettes or paper or any other commodity that was important, gold teeth and hoarded coins included.

My work occupied me physically but left my mind free to wonder what was going to happen to me here. Would I be able to work as a physician, or would I die of exhaustion in one of the labor Kommandos? Would Louis Fichez be able to arrange anything for me with the medical chief here at Gusen I?

When I presented myself at the HKB at the end of my first week at Gusen I, the inmate medical chief was expecting me. Dr. Kaminsky, about fifty years old, had been an eminent professor of pathology in Poland until the intelligentsia and military officers were rounded up and arrested by the invading Germans. Now he was a high-ranking Prominent who had managed to accumulate considerable knowledge and influence in the camp. Dr. Kaminsky was very thin, as was everyone imprisoned in the camps, but full of energy. His hair was not in the style dictated for the inmates of Gusen I and II, which was cropped almost to the skull with a strip shaved down the center. Dr. Kaminsky's unusually shiny skull was clean shaven and as neat-looking as his inmate's uniform. He had carried several things with him from his previous life, his professorial manner, a number of pathology texts, and his wristwatch. He was the only inmate I had ever met who had managed to keep his watch. It was a distinction that was not lost on Dr. Kaminsky. He was like a little child in his pride, forever finding an excuse to raise his sleeve and check the time.

Dr. Kaminsky X-rayed my ankle again and put me in a light cast. He told me about the typhus epidemic that was now raging in

Gusen II and promised that he would use this to initiate my transfer there as a camp physician.

I had mixed feelings about Dr. Kaminsky's plan. Yes, I wanted very badly to be able to work as a physician. No, I didn't want to be transferred to Gusen II. I had heard too many terrible things about it.

Although Gusen I and II were sister camps, Gusen I, the older of the two, had all the facilities: It possessed the pharmacy, the well-equipped operating room and pathology laboratory, the gas chamber and crematorium. It distributed all of the materials and services used by the other. At Gusen I people were killed "scientifically" by gas or in medical experiments. At Gusen II they were killed with whatever primitive weapons were at hand.

I did not want to go to Gusen II.

About two weeks after my visit with Dr. Kaminsky I was ordered to appear before SS Hauptsturmführer Fetter, the medical chief of Mauthausen and its neighboring satellite camps. I had already been sent to shower and given a clean uniform, the standard preliminaries that indicated transfer to another camp. Fetter told me that the prisoner-physician who headed the HKB at Gusen II had just died of typhus. The camp was in desperate need of a replacement. Since I was "Aryan" and spoke fluent German, I had been chosen for the job. I was to leave immediately.

On the same day toward the late fall of 1944, I marched the half mile or so from Gusen I to Gusen II. Wintry weather arrived very early in that part of Austria. My thin uniform offered no protection from the icy rain and biting cold.

THE HELL OF GUSEN II

IV

21

It was raining when I arrived at Gusen II. Patches of snow already dotted the ground. Whenever it rained or snowed, the crude, unpaved streets in this primitive camp of rough wooden buildings turned to ankle-deep mud. Gusen II was incredibly filthy, and this extraordinary accumulation of filth had taken only six months to reach its current level.

Gusen II had been hastily built to accommodate some ten thousand prisoners from all over German-occupied Europe. While I was there the number often soared to thirty thousand. To make room for new arrivals whenever another transport was due, the frailest inmates were routinely *selected* and dispatched to the gas chamber. Even without the gas chamber's assistance, conditions in this camp were responsible for a staggering mortality rate. Of the ten thousand inmates brought in when Gusen II first opened, only five hundred walking skeletons remained two months later. In the approximate year of Gusen II's existence before the Liberation reached us, more than fifty thousand men died there from disease, starvation, physical violence, and deliberate murder.

On the cold and rainy morning when I arrived at Gusen II a mountain of cadavers had been piled two stories high outside the morgue shack, which itself was filled to overflowing. This was to be my new home. If Gusen I was purgatory, Gusen II was the ninth circle of hell.

I was immediately sent to the *Schonungblock* (the "convalescence barrack"), which was later to become the infamous barrack number twenty. I soon discovered that the convalescence barrack was really a holding pen for the victims of impending exterminations. Since Gusen I's gas chambers could not keep pace with the growing number of men who arrived to fill these barracks, the green triangle Prominents cooperated in getting rid of the weak and the sick.

I was left in the convalescence barrack with no instructions, no explanation, and no idea of what I was supposed to do as the camp's new medical chief. I was told only that my quarters were temporary. I would soon be transferred to my own room adjacent to the HKB. I didn't even know where the infirmary was. For the first few days I spent my time taking stock of my new barrack. It was an immense room with the Blockältester's private quarters at one end and the latrines at the other. The Blockältester seemed to have come out of the same mold as the ones I had known in Dachau and Flossenburg. He looked like a gorilla and behaved like a sadistic brute.

The food was by far the worst I had encountered in the camps. Breakfast was a small piece of leaden black bread that seemed to be made primarily from sawdust and a nauseating slice of meat they called liverwurst. A rancid dark liquid masqueraded as coffee. Dinner was no better. My one consolation was that I had been given my own bunk, and did not have to sleep two and four in the same bunk space as all the others did.

I was in a room that was filled to the rafters with sick men. There was no need to sit around idly waiting for instructions. I found some blank sheets of paper and made my rounds, methodically visiting and recording the medical history of each of my patients. Whenever any of the barrack's Prominents walked by me, they looked at my notes and laughed contemptuously.

It quickly became apparent that a large number of men in the convalescence barrack were recovering from accidents sustained during their work. I learned that these slave workers at Gusen II were setting up and manning factory workshops in tunnels that had been dug in the nearby mountains by the camp's original inhabitants. Slave labor in these tunnel factories, for the Sauer Erde Industries, went on practically around the clock. The day shift was from 7:00 A.M. to 6:00 P.M., the night shift from 6:00 P.M. until 5:00 A.M. The men were starved and exhausted. It was no wonder that so many of them were hurt during these inhumanly demanding work schedules.

I became puzzled by the regular disappearance of my patients. I would examine a man and return the next day for a checkup or more information, but he would no longer be there. The bed would be empty or hold an unfamiliar face. I soon learned why these patients seemed to disappear from the face of the earth.

Among the camp population, and particularly concentrated in the convalescence barrack, were a great number of so-called "*Musulmen*" ("Muslims"). The term was coined by Heinrich Himmler, the founder and original leader of the SS, when he first visited the gas chamber at Auschwitz. A group of skeletal human beings were sitting on the ground waiting to be taken inside. Each emaciated, exhausted victim was naked except for a blanket wrapped around his head and shoulders. It reminded Himmler of an Arab's cloak. "Who are these Musulmen?" he asked. Himmler's term was now commonly used to denote all inmates who had lost at least half their original body weight and descended to the final stages of exhaustion. They would be in the next *selection* for the gas chamber, if they survived that long.

After I had been in Gusen II for two days, an inmate from the convalescence barrack was discovered to be missing at the evening roll call. Everyone assumed that he had tried to escape. The SS closed the camp, and all of the inmates stood outdoors for hours without food or permission to go to the latrines. As the night went on more and more inmates soiled their pants, and more and more died in the mud. By the time Commandant Pausch arrived, scores of dead bodies were lying in the mud. He ordered them to be collected and stacked. An arm or leg occasionally stirred in this pile of corpses.

Commandant Pausch was in a vile rage. This was the first time I had gotten an extended glimpse of my new commandant. What initially struck me about Pausch was his age. Graying, somewhere in his early to mid-sixties, he was much older than any of the camp commandants I had seen or heard about. He was thin, Prussian, and appeared to take great pride in his elegant and tight-fitting riding boots. He carried a riding crop that he kept tapping against his boot. Later, as I grew to know Commandant Pausch during my eight or so months in Gusen II, what most struck me was his complete inability to look at whomever he was talking to, whether prisoner or SS. He would either watch his riding crop tapping out a rhythm on his boot, or stand with his arms across his chest and stare intently at his boot tips.

Now Pausch was violently angry. The disappearance of an inmate could not be tolerated. He finally dismissed everyone except the inmates from the convalescence barrack. As the barrack physician, I was included. He gave us a deadline of four hours. Either the missing inmate appeared by then, or every tenth man would be shot. I thought this was only a threat to frighten us into divulging whatever information we might have. Several hours passed.

We were ordered to change our lineup from a number of front-to-back rows into one long horizontal line. Cold, stiff, and exhausted, we arranged ourselves side by side. A Musulman from my barrack, a breathing skeleton scarcely able to stand on his feet, was on my left.

An SS guard began to count in a monotone: "*Eins, zwei, drei, vier. . . .*" The count reached ten. A gunshot broke the early morning silence. I was too terrified at what I would see to turn my head in the direction of that sound. I stared at the Musulman.

The SS guard counted to ten again and I heard another shot. For a third time the count was repeated, and the gun was fired. I was still too afraid to look. The count began for the fourth time.

Finally I looked to my right. An SS guard was writing down the number of the tenth man in line. He was pulled out, forced to his knees, and shot in the back of the neck. I quickly counted along the line—the next number ten was me! I moved to my left and pushed the Musulman into my place. My action was so immediate that I didn't have time to think it through until after it was done. The doomed man didn't even have time to realize what was happening to him before the SS guard took his number and pulled him out of line. He was facing me as he fell to his knees and he cried out to me imploringly in a language I didn't know. His eyes fastened onto mine. The gun fired. He was dead. Mud splashed him as he fell sideways, his open eyes still staring at me.

When our punishment was over, thirty more corpses had been added to the pile before we were returned to our barrack. The missing inmate was found on the worksite later that day. He lay behind his machine where he had silently collapsed and died the day before.

I tried to forget the man I had caused to die in my place. What I had done had happened so fast that it must have sprung directly from a primitive impetus to survive at any cost. I consoled myself with the thought that the wretched creature had been only a walking corpse. All I had done was to hasten the Musulman's inevitable death. Today I am deeply disturbed by doubts. Was my evaluation of his moribund condition accurate or merely a rationalization to justify what I had done to save myself? Although it has been thirty-nine years now, I have not succeeded in erasing the sight of his eyes or the sound of his voice, or my own sense of guilt. The memory still haunts me.

Back in the barrack an inmate introduced himself to me as Dr. Henri Desoille. The name belonged to a man I had met in Paris on a number of occasions. The last time had been at his inauguration as chairman of forensic medicine at the medical school in Paris.

Henri was older than I, a highly cultured and widely read shoemaker's son. His slightly stooped shoulders reflected his lifetime of avid reading. What had previously impressed me about Henri was a spontaneous and childlike sense of wonder and discovery that had never been dulled or displaced by the grisly and depressing nature of his constant involvement with death, both natural and unnatural. I soon learned that not even the nightmare of life in Gusen II had succeeded in robbing him of this quality. But the sad child's eyes I knew from Paris had become so much sadder. When I apologized for not recognizing him, Henri laughed sardonically.

"Don't feel badly, Albert! We have lost our more human characteristics here as well as our weight and our formal clothes. I've found a few compensations of a sort. I'm a forensic specialist who was really very limited in my education. I used to witness only the results of fatal crimes. Here I see the actual crimes committed, and my understanding of the techniques of murder has improved substantially."

He looked at me and went on in the same sardonic tone. "And you, Albert, as I understand, are our new surgeon."

I looked at him wide-eyed. "Henri, you must be mistaken! You know I'm no surgeon!"

"That's not important. At Gusen II the trained surgeons don't do much more than the manual labor of carrying out the cadavers and examining the HKB inmates for lice. The green triangle Kapos reserve the privilege of surgery for themselves. But no matter what, don't refuse this position! The last physician given this job wasn't a surgeon either, and he refused it. He was found drowned in a bucket of water. Officially it was said that he died of typhus. I can only advise you to stay put and don't provoke anyone. We Frenchmen who still have our strength are few in number, and we don't want to lose you. I'll be introducing you to the others soon."

Then Henri continued, in a somewhat lighter vein, to describe a further advantage for Gusen II's surgeon. As a camp Prominent I would receive coupons—exchangeable for cigarettes and soda at the Häftling canteen—in "payment" for my work. The exchange rate at Mauthausen and its satellites was much better than at some of the other camps. One coupon here was worth six bottles of soda or fifty cigarettes, or a combination of the two.

And a new coupon treat had become available several months before my arrival. Gusen I now had a brothel servicing all interested Aryan Prominents. The Germans had established irrefutable evidence for the future that their concentration camps were highly civilized. Brothels, after all, are found only in civilized societies. And many of the camp's more brutal Prominents were proudly showing everyone how civilized—and virile—they were by using this brothel regularly. Henri's tongue-in-cheek description was reassuring. The camp had not destroyed his Gallic sense of humor!

I learned that one coupon bought fifteen minutes of intercourse between four-thirty and eight-thirty each evening except Sundays, our day of leisure, when the brothel's doors opened from one until eight. The brothel's two floors—each with about thirty unlocked cubicles along both sides of a central corridor—alternated in use. The girls who had just completed their duty would have the next quarter-hour to wash, disinfect themselves with a dilute potassium permanganate douche, and dress for their next performance.

The *Kapo Poufmutter* (German slang for a madam) and her SS woman supervisor ran this sex barrack on a rigid timetable. As the quarter-hour approached, the waiting Prominents lined up in two's outside. When the SS woman opened the door and ordered them to "March!" they started forward, handed their coupons to the Poufmutter, and paired off with the first available girl on their side. The thirty couples stood at attention in front of the open cubicle doors until the Poufmutter clanged her cowbell and bellowed: "Begin!" Then each couple disappeared inside. Permission for intercourse had been granted. When only five minutes remained the cowbell clanged again to accompany the harsh command: "Dress and leave!" I smiled—momentarily forgetting my anxieties—to think of the cowbell's inevitable impact on all of the erections that had not yet ended in orgasm, and now never could!

I wondered where these poor women had come from, since neither Mauthausen nor its satellites held women prisoners. Henri told me that Aryan peasant girls from the conquered countries of Middle Europe had first been forcibly transported to the rear of the Russian battlefront as a sexual morale booster for the Reich's disheartened frontline soldiers. Now that Germany had been routed from Russia, these women were forced to staff the brothels recently set up at the older concentration camps, among them Dachau, Auschwitz, Buchenwald, and Mauthausen. The prostitutes were elegantly dressed by the Germans in fine clothes taken from Jewish women on their way to the gas chamber. And they were well-fed to give them allure and energy for their job.

But turnover was rapid despite their nourishing diet and relatively short workday. The women were periodically given hasty medical checkups supervised by the SS physician at Gusen I. And anyone with a vaginal discharge (usually from urethritis, a harmless bacterial infection common among male and female inmates) or an obvious pregnancy was immediately gassed to death. The few "old-timers" still in the brothel did not yet have urethritis and had also avoided pregnancy during their vulnerable initial months in the camp when they were still menstruating. (Women inmates lost their period after the first several months of imprisonment.) But it was rumored that they were automatically to be sacrificed to the gas chamber after one full year of service to the principles of civilization.

The Germans refused to let Jewish women be used as prostitutes because of their inflexible "*Rassenschande*"—dishonoring the race—prohibitions. Polluting German offspring with Jewish blood was a capital crime. Yet every pregnant prostitute was killed because she wouldn't be able to work. You couldn't beat German logic!

Brothel clients were encouraged to bring the women gifts of food despite their substantial diet. These gifts, as the Germans foresaw, were stolen from the starvation rations of the nonprivileged inmates, creating another technique for hastening the pace of death. But they hadn't realized that the food stolen from one starving prisoner would, in the end, simply be given to another. Even though prisoners typically hoarded extra food, the prostitutes dropped each quarter-hour's gifts down to the starving Musulmen suddenly appearing precisely at the session's end to fight like animals for these precious morsels. I was to witness many times myself the Musulmen's uncanny sense of timing. These half-dead prisoners had no access to clocks or watches, and would have been shot instantly if caught waiting in the area. But their survival instinct had become so strong that some unerringly accurate biological clock would set off an internal alarm to measure the passage of time.

Henri spoke with warmth and sadness of the enforced prostitutes' unfailing generosity to inmates struggling to stay alive. They were (and still are) among the camps' unsung heroes.

Shortly after this conversation our breakfast was distributed. The dinner we had missed the night before because of our extended roll call was, of course, lost. Henri didn't stay for the meal. He ate at the HKB. Before he left I expressed my wish to see the HKB since I had been sent to Gusen II in the capacity of medical chief.

"Don't be in such a hurry," Henri answered me. "You'll see it soon enough."

All through that day I tried to decide what to do when I would officially be given the responsibilities of camp surgeon. I didn't want to sacrifice my own life by refusing. Nor did I want to run the risk of butchering and slaughtering my comrades by accepting. And I knew I could not stand by and watch the Kapos take pleasure in doing just that. As terrible as Gusen II seemed, I had been somewhat confident that I would be able to find a way to cope with what I would encounter. Dachau and Flossenburg had taught me that I could live through a nightmare, and at Laurahutte I had learned the ropes of daily survival. Now, however, I felt backed into a corner with no way out.

That night, around midnight, a persistent tapping on my arm woke me. I followed the shadowy hand to a man in the bunk next to mine. He was whispering to me in a German so broken and heavily accented that it was incomprehensible. The hand on my arm was hot and wet. I was afraid that he wanted to kill me and pushed the hand away.

He continued to nudge my arm. I began to think he was a homosexual. He refused to leave me alone. Finally I realized that the man was barely able to talk and was having tremendous difficulty swallowing and breathing. I palpated his neck and discovered the badly swollen maxillary glands indicative of a throat infection, most likely a strep throat. I told the man to go to the HKB about his throat first thing in the morning. He made me understand that he had already been there and had been sent to the convalescence barrack, but his throat was worse than ever. I explained that it was too dark to examine him properly. In the morning I would see him in the latrine.

Persistent pressure on my arm woke me up again. I opened my eyes to discover that dawn had begun, though barely, creeping into the shadows. I went to the latrine with my "patient." His face was flushed, he had a severe headache, and his eyes looked as if they were about to pop out of his head. The man obviously had a very high fever. Using my index finger as a tongue depressor I saw a thick, grayish-white mucus clogging his throat so badly that the air whistled through it when he exhaled. If I correctly remembered the one case I had ever seen, the inmate in front of me had diphtheria!

I was appalled by the dilemma that was suddenly confronting me. This man had a highly contagious and very serious disease. If I treated him he might recover. But if the SS learned there was an inmate with such a dread disease in the camp, my patient and everyone who had come into contact with him would certainly be exterminated to prevent the infection from spreading. It was the SS approach to preventive medicine. But if I ignored the man's condition and gave the rest of us a chance to live, he was sure to die. Where did my obligation lie—to the life of the individual before me, or to the lives of the rest of my comrades?

I was overwhelmed. I decided to go to the HKB and talk it over with Henri Desoille since he was already an old hand in the camp. Henri had pointed out to me where the HKB was located, just across from the convalescence barrack. The two buildings were separated by a fence that ran the entire length of the camp. When I finally reached the infirmary I ran quickly through the darkened building. I found Henri bending over a fellow Frenchman who had cut himself on a piece of rusty metal several days before. The man was in the agonizing rictus of the final stage of tetanus.

Henri murmured to me, "I can do nothing for him, absolutely nothing. I just hope he'll die quickly. What a waste, what a tragic waste. . . ."

I told him I was in trouble and desperately needed his advice.

Henri took me into an unoccupied bedroom that we entered from outside the HKB. We could speak in privacy here. I explained how I had tentatively diagnosed a case of diphtheria in the convalescence barrack. Henri was aghast.

"*Merde!* If it's discovered by the SS the entire barrack, including you and the Blockältester, goes to the gas chamber! But before we decide what to do, we first have to be sure it's really diphtheria."

Henri advised me to take the Blockältester aside and explain that one of the men in the barrack *might* have diphtheria, but that I had to get a laboratory analysis of a smear from his throat in order to know for sure. I should emphasize that it was in the Blockältester's own interests to find a reason for me to visit the laboratory at Gusen I right away and to keep our conversation a dark secret. I should be graphic about what would happen if the SS ever got wind of it. Henri gave me a long splinter of wood around which I could wrap a piece of cloth to make a swab for taking mucus from the man's throat, and a small bottle in which to carry the sample.

I returned to the convalescence barrack. My little speech to the Blockältester worked like a charm. When I returned to my patient, he was on the verge of asphyxiation. I tore a piece of fabric from his shirt and dipped it into some kerosene from the Blockältester's lamp. I wrapped this around my index finger and used it to clear away enough mucus to ease his breathing. Then I sealed my sample in the bottle and ran over to Gusen I, leaving it with Dr. Kaminsky.

I returned to my patient as fast as I could. He was gone! No sooner had I left with my sample than the Blockältester and his cronies had taken the matter into their own hands. They had killed him in the latrine to stop an epidemic before it could start! And this tragic affair was not yet closed. For weeks afterward, anyone in the convalescence barrack who complained of a bad sore throat was murdered without delay. I was heartsick.

My last night in the convalescence barrack finally arrived. In the morning I would be moving to my own room in the HKB across the fence. During the night, crying, begging, and screaming voices disturbed my sleep. I got up to see what was going on. I saw a struggling inmate being pulled from his bunk by one of the barrack Prominents and dragged across the floor to the latrine. The door opened to let them in. In the moment before it closed I glimpsed four struggling bodies inside. I went over to the closed door and called out that I was the doctor and wanted to see my patients. I heard the Blockältester's voice say, "Let him in!"

The sudden light inside the latrine blinded me for a moment, but my eyes soon adjusted to a grisly scene of wholesale murder. I saw why my patients had been disappearing during the night. The four inmates on the latrine floor were each to be killed by a different method while an SS officer in the corner timed each death and took notes. I saw terror and agony in the victims' eyes. I turned to leave.

"*Halt!* Stay right where you are!" shouted the SS officer. "Now we have a medical expert to keep our death records." Insubordination in the camps was automatically punishable by death. I did not have the courage to refuse.

I spent my last night as a resident of the convalescence barrack watching and timing the four techniques used to murder four of my patients. One victim's head was held in a bucket of water until he drowned. For the second, two men stood on a flat piece of wood placed across the supine victim's throat until he choked to death. The third victim squatted as if he were sitting in a chair. A rope tied around his neck was hung from the ceiling, then his feet were kicked out from under him. The fourth technique, meant to teach the SS officer to break a man's neck quickly and cleanly, was demonstrated with considerable skill by a gypsy inmate. Then the SS officer practiced the gypsy's method on the other three corpses.

It was dawn before I was allowed to leave the cadavers and return to my bunk. I lay there, sleepless and shaking. I cried and cursed myself for not having had the courage to stand up to those animals and die for what I believed in. I was still awake when "*Aufstehen*" rang through the barrack.

My number was called after breakfast and I was told to report to SS Dr. Fetter at the HKB. Fetter and three inmates, including Henri, were waiting for me in the room where ambulatory patients were treated. SS Dr. Fetter's plump and jovial appearance unsuccessfully masked his supremely arrogant nature. He looked down on everyone. Fetter took great pride in his wit and what he termed his "accomplishments" in science. It didn't take me long to learn that his pride in both these attributes was highly unwarranted.

Fetter informed me that, as of that morning, I would assume my responsibilities in the HKB. They included doing whatever surgery was needed. I would also have to give him a daily list, through Sanitätsdienst Schultze, of which inmates I was transferring to the convalescence barrack.

I asked Fetter for permission to speak. He nodded his head. I collected my courage and tried to make up for my silence of the previous night. I had given up many values in order to survive in the camps, but I would not give up the basic sense of morality that separated me from an animal. I could not participate in wanton barbarism.

"The convalescence barrack is a travesty," I said. "It is no more than an extermination center. I demand a transfer to work in the factory."

The three other inmates looked at me with stunned expressions. Fetter scrutinized my face. A lazy smile appeared on his lips as he said, "You are merely imagining things, Häftling-Arzt. Barrack twenty is well known for its successful rehabilitations. If you don't want to choose the inmates for transfer there, I am sure the HKB Blockschreiber will gladly take over your responsibility."

And with that, he left the room.

"What an astoundingly lucky guy you are!" were Henri's first words. He had expected Fetter to send me directly to the gas chamber. "Why did you do it?" I described my night's experience to them. Henri had known of these torture sessions since he had been in Gusen II. His response to the situation was to warn everyone, and particularly political prisoners, to keep away from barrack twenty no matter how sick or exhausted they were. Henri also told me that these killings went on, albeit on a reduced scale, in all of the other barracks in Gusen II.

It was no better at our HKB. An infirmary was supposed to be for healing, but our infirmary's Austrian Blockältester was a psychopathic murderer and his Blockschreiber was a former pimp. Often all the physicians ended up doing was keeping patients alive long enough to be killed by one of the HKB Prominents or their cronies.

Henri then gave me a tour of the HKB. First he showed me my room, which was attached to the infirmary but entered from the outside. I was given the one bed. The three bunks there were reserved for three other physicians who often slept in the barracks where they had to make up the two daily lists of sick inmates.

Nightshift injuries and dayshift workers who were too ill to go to work were sent to the infirmary after breakfast. Dayshift injuries and sick nightshift workers were seen after 8:00 P.M. The morning treatment session finished by 10:00 A.M., and then inpatients were attended to and surgery performed. Henri told me that evening hours often ran late into the night.

Our resources for treatment were ludicrously meager. For a camp that already held twenty thousand inmates, all potentially vulnerable to serious illness and injury, our HKB was stocked with aspirin, ichthyol (a fish extract used to ripen abscesses for lancing), iodoform (a disinfectant), and iodine. That was our entire stock of pharmaceuticals. Cellulose was substituted for cotton, and crepe paper for bandages. Sterilization was not a concept that entered this infirmary.

I asked Henri where the operating room and the surgical instruments were. His answering laughter held overtones of hopelessness. "What do you mean, surgical instruments! And as for the operating room, come, I'll show you."

On the way from my room to the ward entrance, I asked Henri who actually performed the surgery. He explained that it was done by two self-appointed inmates. One was Nikolai, nicknamed the butcher, a vicious Ukrainian who was the homosexual mistress of the convalescence barrack Blockältester. The other "surgeon" was a sadistic, dark-skinned young Spaniard called Pedro. He was the mistress of the HKB Blockältester. He was the one rotten apple in the camp's population of Spanish Republicans, who had been among the very first inmates in Gusen II. Henri described Pedro as a bloodthirsty murderer who had metamorphosed, self-taught, from unskilled laborer to camp surgeon. He entertained the SS officers with demonstrations of surgical techniques performed on unanesthetized inmate "volunteers." Pedro's greatest performance was a graphic presentation of the heart's pumping action. He had inserted individual wires, each with a colored flag, into the heart's four chambers so that a chamber pumping out blood made its flag wave. He had neglected to clamp the "volunteer's" opened blood vessels, but his subject's death did not lessen his audience's enthusiasm.

The unlimited power that came from having the right friends gave Nikolai and Pedro the freedom to operate on infirmary patients whether they needed it or not. It really made little difference in the end. If the patients didn't die under the butcher's knife, they would be killed during the night or as soon as they arrived in the convalescence barrack.

As we arrived at the entrance to the HKB Henri pointed to the barrack number over the doorway and said, "Instead of that number Dante's words should be written over this entry—'Abandon hope all ye who enter here'."

Inside we passed by the doors to two opposing *Stuben,* or "small rooms." The one on the left was the Blockältester's room. The one on the right belonged to the Blockschreiber.

I asked Henri to help me understand what my role as physician here was really supposed to be. He sighed deeply.

"How can I explain the futility of our profession in this place? I'm sure you were also an idealist as a student, with dreams of conquering the fatal illnesses. You worked hard, you struggled to understand the human body, and you finally felt ready to accept the perpetual fight against death. Your successes in this fight gave you a tremendous sense of fulfillment and the lofty moral satisfaction of living up to the Hippocratic oath. When you see the patients inside this HKB, you'll ask yourself if the Germans haven't made this oath into the cruelest of mockeries!"

As we entered the ward a pungent, sickening odor assailed me. The beds were filled with tangles of semiconscious bodies, fleshless arms and legs moving slowly and aimlessly, bodies defecating and urinating on one another. Each bed held a pile of six to eight of these living skeletons writhing in slow motion, skin stretched over bone, burning eyes monstrously huge in proportion to the hideously emaciated head and body. It was not possible to believe that these had once been healthy, happy human beings.

Henri told me that the SS who ran the camp called this ward the *Bahnhof* ("railroad station waiting room"). The inmates with uncontrollable dysentery waited here for death. They received no medical treatment and only half of the paltry food rations given to the other infirmary patients. They usually died from starvation or suffocation, and sometimes actually from the disease itself. The *Pfleger* ("male nurse") and *Leichentragerie* (a common camp expression corrupting the German *Leichentrager,* meaning "pallbearer") made their rounds of the Bahnhof early every morning. The Pfleger dragged each corpse to a loading platform and recorded its number, and the Leichentragerie carted them out to be trucked to the ovens in Gusen I.

I heard a thumping sound that was to become very familiar to my ears, the noise of a moribund body hitting the floor, pushed off a bed in the final efforts of a dying man still clinging to the illusion of his own survival.

As we passed into the regular infirmary ward the atmosphere and conditions "improved" somewhat. There were several tiers of bunk beds, each narrow bed with a straw mattress, pillow and blanket, all soiled with blood and puss. There were three men to a bed.

"These patients aren't completely exhausted yet," Henri explained. "Once they recover from their illnesses—and avoid being murdered by the Blockältester and his cronies—they'll return to their barracks and continue with their forced labor until their strength is gone. They'll either die on the job or be sent to the convalescence barrack and killed."

He took me into the surgical area, a tiny rectangular space marked out by four beds. The term "operating theater" was never more appropriate, since Nikolai's and Pedro's surgical performances were carried on here in full view of the sick inmates who occupied the bunks surrounding the operating table.

Henri explained that the surgery consisted primarily of amputations. The two self-appointed surgeons had done so many that by now they weren't really as bad as one might expect. There was no anesthesia. Lucky patients lost consciousness. Those who didn't had cellulose stuffed in their mouths.

My voice rose in angry disbelief. "But they could choke to death!"

"That," said Henri, "is their only relief. . . ." He lowered his voice and changed the subject. "Four other physicians and I have organized a resistance group of sorts to fight the green triangles and the SS. The political prisoners in our group have already wrested some positions of authority from the green triangles. Now that you'll be with us, since you're an HKB Prominent and you speak both French and German, we will be in a better position to develop effective resistance against those who make our lives intolerable. Since you're the only one of us to have a private room, we can hold our meetings there."

Henri's group included a French cartoonist (who later became famous under the name of Aldobert), two French career soldiers from the St. Cyr Military Academy, several Spaniards, a French-speaking German anatomist, two Rumanian physicians, a Hungarian pathologist, and several German political prisoners.

I was skeptical. How could we protect ourselves against the green triangles when they, as the executioners for the SS, were so firmly protected by them? How could we possibly hope to counter the SS? We had to make these assassins need us somehow, but it seemed hopeless to find an indispensable service that we alone could provide. Despite my doubts it felt good to be part of the Resistance again, looking for ways to fight the people who had enslaved us.

Thinking of some of my more immediate problems, I asked Henri if there might by chance be a qualified surgeon among his group. There was none. We left for the treatment room. There I first met Luis Bournasell, our nurse, an intelligent, resourceful and sensitive former Spanish medical student. He had remarkably retained his idealistic sense of humanity despite his long incarceration in France after the Spanish Civil War and then his internment by the Germans. Bournasell was all arms and legs, yet he possessed a considerable amount of Latin charm that was accentuated by his dark Latin features.

I also met Sanitätsdienst Schultze, the SS administrative overseer of the HKB. Schultze's thick Bavarian accent, combined with the pipe that was always in his mouth, often made him impossible for his fellow Germans—let alone me—to understand. He was a corpulent and heavy-drinking brute. He beat everyone within reach and was always eager to send patients to the gas chamber. Schultze was the most viciously sadistic SS in Gusen II. He always walked with an attack dog at his side to make sure no inmate could get close enough to touch him accidentally. Whenever the mood seized him, he indulged his pleasure in watching the dog's lethal jaws rip an innocent inmate to pieces.

Most of the problems I noted during my first morning hours were representative of the medical problems prevalent in the camp. The most common problem I saw that morning was gangrene, a consequence of malnutrition. Most of the inmates were undernourished to the point where they developed hunger edema. Then a beating of any kind would rapidly result in necrosis. I would debride and bandage the area, but the lack of any antiseptic meant it was only a matter of time before the patient developed general septicemia and died in the HKB, either of his infection or by the Blockältester's hand.

Many inmates came in with phlegmon, a severe infection involving the lymph nodes that can develop from an untreated wound. We would open the infected wound to drain the pus and treat the affected lymph glands with wet packs. Lancing phlegmons was one of Nikolai's adopted specialties. Without anesthesia the intense pain caused inmates to jerk involuntarily, sometimes accidentally slapping Nikolai. He would continue, unruffled, until the wound was drained and neatly bandaged. Then he would call a crony over to kill the inmate in return for the slap.

We could do nothing effective for inmates who had dysentery except to give them water mixed with aspirin and some words of encouragement so they would not feel totally abandoned. For typhus victims all we could do was prevent severe dehydration during the fever stage.

All of the prisoners were in various stages of exhaustion and debilitation. Our daily camp diet provided us with six hundred to eight hundred calories, even for those who did heavy physical labor. With the two separate work shifts, the number of work accidents brought in to the HKB were staggering.

When I had finished with my ambulatory patients I returned to the ward and saw the green triangle Blockältester. I told him that I was to be given Pedro's surgical instruments immediately, and that SS Dr. Fetter had forbade Pedro and Nikolai to perform further surgery. The Blockältester believed me. Pedro appeared, and I asked for his instruments. The Blockältester told him that the SS physician had ordered it. Pedro went to his room and returned with a dirty and bloody cloth-wrapped bundle that I sorted out in my room: a butcher's knife, several scalpels, some hemostatic clamps covered with dried blood, a long forceps, two surgical scissors, and spreaders. None of them had recently been disinfected. There were no surgical needles and nothing to use for sutures.

And so I began my professional career as doctor of the damned in Gusen II.

22

The Resistance group met in my room. Henri Desoille was elected as our group liaison and would be responsible for recruiting additional political prisoners. We also elected our action bloc, old-timers in the camps whose successful existence attested to their survival ability. They had all gained positions that permitted access to information and influence. There were two Germans, one who was now secretary in Commandant Pausch's office and one who kept the books for the SS' food supplies. There were also two Spaniards, one who had become an engineer in the construction office and another who headed the camp's fire department.

We had three basic goals. We would try to replace the green triangles with ourselves in positions of authority. We were determined to find medication and extra food for inmates who were strong enough to survive with this added support. We also decided to spread our own manufactured good news about Allied progress in the war. We had to keep alive our comrades' hope, and thereby their motivation to survive.

Before we dispersed I asked the French-speaking anatomist for his help. I knew that I must review anatomy in detail if I was to have any chance of performing successful major surgery. We began my lessons that night. Our classroom was the cadaver shack. It provided an endless source of subjects. After "operating" on my hundredth cadaver, I finally stopped counting. I learned to do appendectomies, hernia repairs, amputations, gastrectomies. The only occasions on which I had to go beyond the cadaver shack were in learning the vascular and neural networks. In order for these structures to stand out clearly we needed fresh corpses, and it was loathsome drudgery carrying fresh excrement-covered cadavers from the Bahnhof on our backs. I worked far into the night. Then when I slept I repeated endlessly in my dreams the names of the nerves and blood vessels and muscles. I was driven by the urgency of mastering surgical technique as quickly as I possibly could.

We urgently needed supplies that were not on our authorized list. Such supplies as we had were picked up every two weeks or so from the SS pharmacist in Gusen I. I discovered that the pharmacist came from Cluj, which had been part of Hungary until after World War I. I also found out that he was addicted to ether and was usually under its influence. When the time came for my first visit there, I took along three assistants—Henri, Bournasell, and the Hungarian pathologist.

The SS pharmacist, sitting in an ether fog in the midst of his impressively well-stocked pharmacy, smiled stupidly at me as I approached him. I explained in Hungarian that I was picking up the medication supplies for Gusen II. Surreptitiously I fingered several small colored boxes lying close to me and hid them under my loose shirt. I had no idea what they contained. Whatever it was I was sure we would find a use for it. While I kept the pharmacist occupied my companions stole everything they could lay their hands on. A metal can in a back corner caught my eye, and I motioned Bournasell to steal it.

We returned to my room in Gusen II to survey the fortune of pharmaceuticals we had stolen: Evipan, Dagenon (a French sulfa drug), Rubiasol (a German sulfa drug), papaverine (an opium derivative), medical charcoal for diarrhea, a half-gallon of elixir of paregoric, vitamins, hormones, cough syrup, and several powerful disinfectants commonly used in urinary infections (potassium permanganate, urotropine, methylene blue, and salol).

As great a windfall as this was, it could not begin to meet the overall needs of the camp. Our committee met a few days later to discuss how we should distribute the additional medications we had acquired. It was decided that since I had been responsible for stealing it, I should shoulder the responsibility of deciding who would and would not receive it. I refused. It was too great a burden for me to cope with. But the others also shrank from the responsibility. No one wished to be accused of favoritism, or risk the accusation that he had allowed others to die by not selecting them for help.

After a long and heated discussion we agreed that each case would be decided by a full committee vote. We also agreed that none of the green triangles should receive any medication at all unless it would advance our Resistance efforts.

For some reason, the metal can Bournasell had stolen from the pharmacy was not opened right away. Almost a week passed before Henri and I got around to it. The can contained Peruvian balsam, a sticky brown substance with an unusually pungent odor. I suggested to my friend that we improvise and use it to solve a very bad problem in the HKB.

The problem was scabies, a skin-loving parasite that thrived in the filthy conditions under which we were forced to live. It could be picked up directly or simply by sleeping in the unsanitized bed of someone who had it. The parasite caused the skin to itch unbearably, particularly at night. Scabies had become a life-threatening condition in Gusen II because of the HKB Blockältester's murderous reaction to any noise that disturbed his sleep. The wooden bunks squeaked noisily with even the least movement, and so did the floors. If the Blockältester was awakened by the noises from inmates scratching in bed or walking around when the itching became intolerable, he would come out of his room and kill them. It was as simple and terrible as that.

I planned to smear our scabies-infected patients with the Peruvian balsam, hoping that the Blockältester and his sadistic friends would find the odor so unpleasant that they would leave the recuperating inmates alone. Henri agreed it was certainly worth a try.

The next day we applied our Blockältester-repellent to each patient who came to us for his scabies. The following morning each one returned to tell us that the itching had disappeared and he had slept through the night. To my amazement and delight, I felt I had actually discovered a cure! We used our new treatment for scabies until our supply of Peruvian balsam was exhausted. Most of those we cured of scabies went on to recuperate fully from their more serious medical problems.

I thought I had discovered a miraculous treatment, but was crestfallen when I eventually learned that I had only rediscovered an ancient remedy used several centuries ago!

Necessity often dictated other very unorthodox medical practices. In treating diseases in the camp, even in merely attempting to decrease pain or comfort a sick patient, we had to be ceaselessly inventive. One of our more innovative treatments was the one we soon developed for gangrene. If the affected area was not too large when the patient presented himself, the wound would be debrided (removal of the dead tissue), then treated with powdered charcoal to absorb the pus and help purify the area, followed by permanganate (an antibacterial agent), and a few drops of peroxide (also a disinfectant). Then the wound was bandaged and the patient sent back to his barrack. Despite our consistent success, though, we were always surprised to find a clean and uninflamed wound when the patient returned in two days for rebandaging. But the rate of mortality in the camp was so great that we rarely had the opportunity to observe the long-term effects of our new concoction.

SS Hauptsturmführer Fetter was so impressed with the short-term success of our new treatment that he gave our little group one hundred cigarettes and "allowed" us to write an article for a German scientific journal under his name. When it was published, the article of course contained no acknowledgment of any assistance whatsoever.

The green triangle Prominents were a scourge. They had been criminal misfits in the outside world, motivated only to abuse the rules that permitted normal society to function. In the concentration camps they were kings with nothing to limit their power. Their greed and sadism tyrannized our lives and constantly threatened our very existence. Our Resistance committee held several planning meetings to decide how we could begin eliminating our killers. Our basic credo had to be "kill or be killed."

Some of the criminal Kapos in charge of tunnel digging could easily fall victim to work accidents. If they did not die immediately, we could finish them off in the HKB. The green triangle Prominents whose responsibilities remained within the camp were too well protected for such obvious tactics. We would have to find other means of achieving our goal to replace them with men of our own kind.

We also needed some means of bargaining directly with the SS. They were constantly in short supply of drinking alcohol. Our Spanish engineer suggested that we work out a way to distill alcohol from potatoes, sugar, or yeast. Alcohol would give us a powerful bargaining weapon. He could design and build the equipment, but first we needed someone who could help us put this plan into operation.

Luck played into our hands, and I found the ally we needed. The Blockältester of the convalescence barrack asked me privately for a medical favor. He had a very embarrassing personal problem. When he urinated he had a terrible burning sensation, and afterward a thick secretion oozed out. He would be deeply grateful for whatever I could do—in secret—to help him.

I examined the man. All he had was a general urinary tract infection, but his ignorance presented a golden opportunity to win his confidence. I regarded the Blockältester with as much gravity as I could muster and told him, "Brother, you have a *Triper!*" (*Triper* is German slang for "gonorrhea.")

He was shocked. How could he have gotten it with no women in the camp? I pointed out that he must have contracted the disease from his boyfriend, Nikolai. The Blockältester paled, seized by a jealous rage at the thought of his Nikolai having slept with someone else. But he calmed himself and asked me if I could cure his gonorrhea. I told him that he must refrain from any sexual activity for a while, and I would start urethral washings if he could get me a supply of syringes from his friends at Gusen I.

Two days later my anatomy professor and I found Nikolai's fresh corpse in the cadaver shack. His neck had been broken. The professor suggested that we work on him since it was a rare treat for us to have a healthy, nonatrophied body to study. The next day when the Blockältester brought me the syringes, I mentioned that I hadn't seen Nikolai around. The Blockältester answered casually that he had last seen him with the gypsy (whose murderous specialty I had seen demonstrated in the latrines when I first arrived). I made a weak solution of permanganate to wash his ureters and combined methylene blue, urotropine, and Salol for him to take orally three times daily.

While I was debating how to enlist the Blockältester's help for our alcohol project, I had to visit the Gusen I HKB again for an X-ray to see if my ankle cast was ready to come off. I got permission to go. At the HKB I was sent to see a Dr. Anton Goshchinsky, a Polish physician of whom Dr. Kaminsky had spoken well. I glanced longingly at his equipment. Dr. Goshchinsky left me alone in his treatment room while he went to see if the X-ray machine was free. By the time he returned, all of the important instruments, the surgical needles, catgut, and silk that had been set out on his table were safely inside my shirt. He looked at his empty table but didn't say a word. Much later Dr. Goshchinsky told me that he had deliberately arranged the situation so that I could steal the surgical supplies he knew we desperately needed in Gusen II.

The X-ray showed sufficient healing, and my cast came off. I spied a brown glass jar on my way out of the infirmary. Back in my room at Gusen II I was terribly disappointed to find that it contained only arsenic rat poison. I felt I had wasted my energy in stealing it and carrying it back, but my committee comrades were sure they could put it to good use against our assassins. The deadly cocktail of alcohol and arsenic that they envisioned was only one of the ways in which we eventually used it.

The Blockältester reported that his problem was improving after only a few days of treatment, and thanked me for what I had done for him. I realized that the iron was hot. I told him there was also something else I might be able to do for him.

"Could you use some homemade alcohol?"

"And how!" was his enthusiastic response.

I told him I could set up a distillery in the HKB if he supplied the raw materials I needed, and if he granted me a favor. Would he authorize my colleagues and me to use the latrines for making charcoal? We had found charcoal to be extremely effective in treating diarrhea. Inmates who needed this treatment brought back small pieces of wood from their work sites for us to turn into powdered charcoal, but we had no place in which to do this.

The Blockältester gave us free use of the latrines. As soon as our distillery was ready to go, he set up the supply lines for our potatoes, yeast, and sugar. Several sacks of potatoes could be turned into potato schnapps, a very rough vodka. By aromatizing some of it with cough syrup, we also had after-dinner cordials. The Prominents and the SS quickly came to depend on our alcohol to liven up their parties, and our daily lives improved considerably. The SS even continued our regular supply of raw materials when the entire German nation was forced to decrease their sugar ration. I was an established moonshiner.

Our distillery provided other important benefits as well. I distilled water and used it as a placebo, telling patients in pain that I was injecting them with morphine. We used the alcohol to sterilize our equipment in the infirmary. We also drank it when we felt in need of consolation ourselves. We mixed it with the anise-flavored elixir of paregoric to provide a crude Pernod. Since nothing even remotely edible was ever thrown away, we took the foul-smelling mess remaining from the fermented potatoes and made a sort of mashed potatoes for our dinner. It was a horrible concoction that seemed delicious in the midst of our incredible deprivation. The fermented potatoes would have given us diarrhea had they not been counteracted by the constipating effect of the paregoric in our homemade Pernod. One balanced the other, and we maintained our biological equilibrium.

We built our newly won privileges in the latrines to the point where they became a kind of HKB outpost. After we began the secret manufacture of charcoal, we also used the latrines as our unofficial information center. News was disseminated and encouragement handed out, including the morale-boosting rumors of Allied victories spread by our committee members. Most importantly, the latrines came to serve as a temporary hideout where debilitated inmates could find a brief but life-sustaining respite from their heavy labor.

We had gained some victories, but in the essential aspects of concentration camp life we remained helpless. One of these was *selection.*

When a *selection* day was scheduled I would have to be at the convalescence barrack before breakfast, because on those days the doors would always be locked before the inmates' meager morning meal was distributed. At 9:00 A.M. the presiding officials arrived: SS Dr. Fetter; Commandant Pausch; Commandant Ziereis, who was the overall SS chief of Mauthausen and its satellites; Commandant Bachmayer, SS chief of Mauthausen; and Sanitätsdienst Schultze.

I had to give SS Dr. Fetter a list of the barrack's inmates and their numbers. The Blockältester would call out the numbers and each inmate in turn would stand with his back to Fetter. If the inmate's thighs were so emaciated that the Blockältester's meaty fist could pass between them, he was ordered to the left—*selected for death.* If he still had some muscle on his thighs he was sent to the right—*delayed sentence.* Fetter's job was to assign the official "diagnosis" to each selected inmate from the convalescence barrack. "Häftling number so-and-so died of 'cardiac arrest;' Häftling number such-and-such died of 'respiratory failure. . . . ' " The "Cause of Death" column in the camp records could be properly filled out. The selected inmates would be taken away for *Sonderbehandlung* ("special treatment"). We were never told the nature of these "special treatments." Those inmates never returned. We all knew what that meant. The only thing we did not know was how they were killed.

On one of the periodic visits Henri and I made to the pharmacy in Gusen I, we found the SS pharmacist too busy to fill even our small order. He mumbled that he was fully occupied with preparations for the inmates who had just been taken upstairs to the special treatment area. Henri and I exchanged glances. We left the pharmacy and headed straight for Dr. Kaminsky's room.

Kaminsky's insistence that we forget the pharmacist's remarks served only to whet our curiosity. We stubbornly persisted. Finally Kaminsky relented. We could hide under the roof and watch what was going to happen to the selected inmates, but only if we promised to conceal his complicity if we were caught. Henri and I reinforced our word with ten cigarettes apiece.

The three of us went over to the pharmacy, and while Kaminsky engaged the pharmacist in conversation we stole up the stairs and climbed a ladder to the crawl space under the roof. The three special treatment rooms below us, separated by drapes, were clearly visible in the chinks between the rough planks under our bellies. The "treatment" had not yet begun.

Naked victims sat huddled in the antechamber. It opened into a cubicle furnished with two chairs and a table. Two syringes and two jars—phenol and benzine, we soon discovered—were on the table. The large bare room beyond the cubicle was broken at the far end by a windowlike structure that opened onto a chute. Two Polish inmates sitting on the floor near the chute joked while they waited for their work to begin.

Two naked inmates conversing in the antechamber caught my attention. One was a prematurely aged man who looked at least one hundred years old. The other was a hideously emaciated adolescent. I couldn't hear their words, but the gestures were those of a father comforting his son (which I later learned was accurate).

I thought involuntarily of François, and for an aching moment felt myself in this helpless old man's place.

"Thank goodness my son is safe with my mother and sister in Villon," I confided to Henri.

He cut me off brutally with his whispered reprimand.

"Stop it, Albert! Stop thinking of your family while you are in this place. If you don't, you'll never get out of here alive!"

Suddenly the main door opened. A Polish inmate-physician passed through to the cubicle to finish the preparations. He filled the larger-needled syringe with phenol and the other with benzine. Then his inmate assistant entered, followed by SS Sanitätsdienst Schultze and his entourage.

Two naked patients were summoned into the cubicle. One was ordered to place his hands on his head so that the Polish physician could locate the interspace (over the heart, between two ribs) and push in the large needle of the phenol syringe. The contents were emptied into the man's heart. The needle was withdrawn. The man staggered into the end room, collapsing onto the floor.

The second victim had to offer his arm to the assistant for an intravenous benzine injection. He, too, collapsed in the end room. The other assistants interrupted their jokes to dump the two bodies into the shaft, which we later discovered led directly to the crematorium.

A few turns later an inmate went into terrible convulsions and then a coma after a poorly placed phenol injection, and may have been cremated alive.

Henri wanted to compare the intervals between injection and death for the two types of injections. He suggested that we measure the time by counting our pulse rate. Henri counted 60 seconds for the intravenous benzine. My observation indicated 120 seconds to die from phenol injected directly into the heart. We found it puzzling that an injection directly into the heart should take so much longer. After I counted 120 seconds and Henri only 80 seconds for the same interval between a poorly placed phenol shot and the onset of convulsions, we realized that my pulse was racing. Henri's work in forensics had accustomed him to violent deaths. My pulse was reflecting my raw emotional response to the "special treatment" below.

As the father and son neared the cubicle I admired the old man's apparent courage. I wondered if his placid face reflected strong discipline or merely exhausted resignation. The father entered first and gave his arm for the benzine injection. In the outer room he collapsed and lay motionless. The boy entered the cubicle looking for his father.

We heard the Polish physician remarking, "This one will be easy —he's so thin you can't miss it!"

Then to the boy, "*Hands über Stirn*" ("Hands over the forehead").

Only then did the boy realize that he was about to die. He begged desperately to return to his barrack just long enough to get his only fortune, the spoon and fork by his bed. This pitiful ruse to gain a few minutes of life was so painfully absurd that even his executioners were momentarily silent.

While the needle was placed and inserted the boy continued to plead, "*Ich brauch mein Löffel, ich brauch mein Löffel*" ("I need my spoon"). As the syringe emptied, the bone-chilling shriek of a brutally wounded animal tore out of his emaciated throat. I froze. Tears stung my eyes. I marveled that his beaten and exhausted body had found the strength to produce such a sound. In the few seconds of terrifying silence that followed, his life ended. The old man and his son were dumped into the shaft together.

The special treatment continued until all of the naked inmates huddling in the antechamber had been disposed of. That day's treatment was only one of many. Special treatment was a euphemistic catchall: death by gas, death by hanging, death by shooting, death by clubbing, death by torture, death by medical experimentation, death by design, death by caprice. The official record usually indicated "S.B."—Sonderbehandlung.

When SS Dr. Fetter had initially outlined my responsibilities as medical chief, I refused to play a role in the *selection* of convalescence barrack inmates. Soon I realized that I could be somewhat more active in my fight to save lives, or at least delay death. When I learned that a *selection* was scheduled for the convalescence barrack I discharged those inmates I felt were strong enough to survive the next few days out in the camp proper. I would readmit them to the barrack as soon as the *selection* was over.

The problem was that many of these inmates refused to be discharged. They were convinced that I was sending them out to certain death from starvation or physical assault. They would not believe me. Many *selections* took inmates who had refused my help.

23

Our distillery soon became an indispensable asset to the social life that the green triangle Prominents enjoyed in Gusen II. Orgies were their favorite entertainment after dark. Everything was available—food, music, sex (homosexual), and now, plenty of alcohol. In my unending effort to improve my leverage over these murderers, I invented a homemade lipstick for their male mistresses—mercurochrome mixed with the cocoa butter from suppositories.

The Prominents dangled privileges for those who helped make their parties a success. They would hand out an extra piece of bread or a bowl of soup and a reduced work assignment to musicians and singers who agreed to entertain.

These privileges were just as often the special favor preceding the prisoner's execution. I learned of two incidents in particular that I can never forget.

Eisendradt was an inmate in Gusen II who had been a leading member of the Jewish opera company in Warsaw. His voice was so outstanding that he sang occasionally with the Warsaw Opera despite his being a Jew. The miseries of camp existence had not ruined his rich baritone, and he was frequently "asked" to sing at parties. During one of these orgies Eisendradt impressed a young male prostitute who desperately longed to be able to sing with such a voice. The prostitute and his lover were the Stubendienst and Blockältester of Eisendradt's barrack. The Blockältester asked the singer to teach his "mistress" to sing a few popular German songs. The fee would be extra food and two weeks without outside work assignments. Eisendradt agreed without a second thought.

For the next two weeks he spent his days working with his new pupil and he received two life-sustaining ladles of soup in addition to his daily ration. But the pupil could not sing on pitch. The lessons were an abysmal failure and the Stubendienst's recital was a humiliating disaster. Eisendradt was held responsible. The Blockältester, his "mistress," and several assistants threw him on his back in the latrine and placed a board across his throat. The assistants stood on the board while the Stubendienst repeatedly stuck a knifepoint into the singer's flesh and commanded him, each time, to sing like a bird. Eisendradt finally died of asphyxiation from a crushed windpipe.

Another Jewish inmate had been first violinist of the Berlin Symphony. He had survived in Gusen II because the SS sadists treasured his music at their parties. The priceless violin carried with him when he was deported had been carefully preserved by the SS officers and taken out whenever his professional services were needed. The violinist was eventually asked to play at a Prominents' orgy in return for extra soup. He could not refuse. It turned out that these partygoers expected him to play popular and bawdy German songs. The violinist knew very few. As he continued to repeat his limited repertoire, dissatisfaction grew until his audience dragged him outside and hanged him.

We actually had enough musicians at Gusen II to form an orchestra that was grotesquely integrated into the camp's daily routine. Some of the most famous musicians in Europe were stationed by the camp gates every morning, playing as the inmates despondently filed out to their slave labor. These living skeletons walking morosely to the aristocratic music of a ghostly orchestra formed a macabre vision in the half-light of dawn.

Rampant homosexuality was the primary hallmark of the Prominents' orgies. Their fiercely passionate competition for the adolescent prostitutes at times approached the tragicomic. One day the Blockältester of the convalescence barrack decided to throw a big party and asked the tailor-shop Kapo to contribute women's dresses for all the "ladies." The barrack Blockschreiber asked me for schnapps, cordials, and lipstick. When the party was over, he confided the evening's dramatic events to me.

The party had begun with an unusually good dinner provided by the SS kitchen Kapo, a green triangle psychopath who had been chef in a first-class Berlin hotel until he was arrested for murder. After the meal everyone began to drink heavily and the evening's mood shifted into high gear. The Kapo chef began flirting with the

Blockältester's mistress, an adolescent Ukrainian of corrupt sensibilities. After intercourse with the Blockältester, these two lovers would often leave their room in hopes of finding an inmate out of bed to torture to death. It was their stimulant to further lovemaking. They would leave their door open to broadcast their groans of pleasure.

The Kapo chef was tall, handsome, and held the keys to an unlimited food supply. He was a superb catch. The young Ukrainian came over to sit in the chef's lap and happily responded to his advances. The Blockältester calmly followed this seduction scene through a haze of alcohol until the Kapo chef opened his fly and began lifting the boy's skirt in order to have intercourse. The Blockältester jumped up in a sudden rage and tried to drag his mistress away from this handsome seducer.

The Kapo chef was not to be thwarted. He grabbed a carving knife from the dinner table and stood there defiantly, the boy gripped firmly in one hand, the knife brandished in the other, and his erect penis cleaving the air below. The Blockältester also picked up a knife and lunged. He was too drunk. His opponent wounded him immediately, bringing the party to an abrupt halt. The Kapo chef left with the prostitute.

I was called in to treat the Blockältester's superficial but bloody wound, which was the first I learned of the destructive turn the party had taken. As soon as his arm was bandaged he ran after his lost love, promising him the moon if only he would come back. The Ukrainian boy was too well aware of the unbeatable advantages of his new lover to return to his old bed, and the Blockältester suffered a genuine *chagrin d'amour* as he vainly tried to recapture the prostitute's affections. Two weeks later he suffered a nervous breakdown and ran into the electrified fence surrounding the camp, ending his misery. Even murderers want to be loved.

I found the Kapo chef a fascinating study in contradictions. He was a gifted cook, a tender homosexual lover to his mistresses, and a ruthless killer. He once told me about the first murder he had committed, the one that had sent him to jail. He had become disgusted with married life and decided to end it by putting an end to his wife. It seemed to him the simplest thing to do. He planned to cover up his crime by dismembering and cooking her body to make it easier to bury. He actually chopped up his wife's corpse and put the pieces to stew in a huge kettle on a stove in his cellar. It wasn't long before his neighbors called the local police to investigate the atrocious smell emanating from his house. He was caught and sent to prison. If he had it to do over again, he said, he would have masked the smell by cooking her with lots of garlic and good vegetables!

In the world of the concentration camp this murderer had become the all-powerful Kapo of the SS kitchen in Gusen II, keeper of the keys to all the food stores, and courted by the SS for his culinary abilities. He used his power as he pleased. He would bring me soup left over from the SS meals to be distributed to the dying patients in the HKB. Whenever one of his former homosexual mistresses became badly run-down, he ordered the fellow's Blockältester to let him rest in the barrack for a few days and give him extra food. If the Kapo chef caught an inmate stealing potato peelings, particularly if the culprit was still in fairly good physical shape, he would let him go after forcing him into sexual submission. If he was especially satisfied with the inmate's performance he would send his victim off with two slices of bread, a large serving of artificial margarine and a bowl of soup. It was a payment comparable to a dinner at Maxime's in the outside world.

The Kapo chef never killed for amusement as did the other green triangle Prominents. When he wasn't angry, he was, in his own fashion, a decent and concerned human being and a startling exception to the rule. But if he was in a bad mood, any inmate who didn't answer his summons immediately was killed on the spot. Yet the Kapo chef considered himself a good Catholic and kept an inmate priest alive with extra food rations in return for the guarantee of a spiritual confessor and absolution every time he killed. The priest eagerly accepted, exchanging absolution in the hereafter for bread and sausage here on earth.

I was not only a reliable moonshiner and lipstick maker to these green triangle Prominents. They had also come to regard me highly as a doctor. My reputation with them had begun with my successful treatment of the convalescence barrack Blockältester's "gonorrhea" and grew each time it was to our advantage to care for their medical problems. My medical reputation among the green triangle Prominents eventually reached the ears of Commandant Pausch, who was also aware of the respect with which

Fetter regarded me since my refusal to participate in the *selection* from the convalescence barrack.

Fetter even addressed me now as *Herr Kollege.* And Commandant Pausch, slow-witted as he was, must have suspected that SS Dr. Fetter was an incompetent physician. After I had been in Gusen II for a while, Pausch informed me that I would be his personal doctor. Fetter was more than willing for me to take over his responsibility.

One day Pausch called me in complaining of severe headaches and morning dizziness. The commandant's symptoms, in light of his reputation for consuming great quantities of hard liquor, pointed to cirrhosis of the liver. During the physical examination I asked if he had recently experienced any nausea and perhaps some difficulty in walking. His affirmative answer confirmed my diagnosis. I realized that I had another golden opportunity to broaden our edge on survival.

I knew Pausch was a passionate horseback rider and favored the classic German riding boots, which were so tight they often produced subcutaneous bleeding. I crossed my fingers and asked him if he had ever noticed blood under his toenails. "Yes!" he exclaimed, regarding me in amazement. I breathed a sigh of relief. My plan might work. I assured Pausch that I was quite confident I understood his problem. I wanted to do a little checking and would see him the following day with explanatory materials and his treatment program.

What I hoped to achieve was a means of stopping one of the major forms of punishment practiced in the camp, twenty-five lashes on the buttocks. The prisoners were too debilitated to sustain such physical abuse. Their severe emaciation inevitably led to the development of gangrene, and ultimately the gas chamber. Pausch took a great deal of pleasure in ordering and watching these whippings. I hoped to scare him into stopping the beatings by presenting him with a persuasive relationship between his tendencies to violence and his physical problems.

I had set the stage for this hoax at his physical examination. The next step was a visit to my friend Dr. Kaminsky in Gusen I. I knew he had been able to keep some of his books with him in the camp. I wanted some pathology textbooks with grisly colored illustrations of diseased brains. I returned to my room with two fat books under my arm.

In the morning I went to Pausch with one of my books and announced that I had definitely identified the source of his troubles. He was desperate for my answer. As was typical of these sadistic masters of life and death, Pausch was terrified by his own pain and mortality. I told him that he was suffering from subconscious conflict and guilt feelings.

"Hauptsturmführer Pausch, everyone knows that you are basically a moral person and a civilized German citizen. But you also have a streak of violence. The violent part of you enjoys ordering and watching the inmates' whippings, but then your moral side reacts with terrible guilt. This subconscious struggle and consequent guilt reaction produce exactly the physical problems that you described to me. This happens because of a medical condition you have. It was obvious to me after I examined you yesterday that you have an angioma."

I opened the pathology book to the most frightening pictures I had located and accompanied them with the following medical explanation: "When you witness a prisoner being beaten you experience a strong surge of sadistic pleasure and become highly excited. Because of your angioma the flow of blood in your brain is increased, which then increases your intracranial pressure. The pressure that you have subjected your brain to has made it look just like this picture."

Pausch, turning pale, was hanging on my every word. I continued: "That's why you have headaches and dizziness and nausea. The extra pressure begins to affect your whole body. You can see that has already started to happen to you because of the blood under your toenails and the difficulties you complained of in walking. Your severe guilt feelings make all of this even worse.

"Nothing can be done about the angioma. There is no surgery or medicine to cure such a brain condition. But thank goodness we can keep your angioma from getting worse and finally killing you, and we can also treat your symptoms. Take the medication that I have prescribed for you, try your best not to get excited, and stop drinking for the time being."

Pausch gratefully accepted my explanation and treatment program. He never thought of consulting another physician. Again the typical SS, he was so confident in his total power over us that it never occurred to him one of us might have the audacity to try to trick him. He dutifully abstained from alcohol and put a stop to the whippings. He faithfully took the medication (merely aspirin mixed with a mild laxative) I had prescribed for him. His headaches, dizziness, and nausea gradually disappeared, and his walking improved.

My success so impressed Pausch that when anyone from his family or coterie of boyfriends complained of a medical problem, I was immediately summoned. He developed a thriving private practice for me, and I made sure to receive my fee. I always found a reason for prescribing medication, always had the patient bring me three times the ingredients I planned to use, and immediately skimmed my two-thirds off the top for our own uses in the camp.

Commandant Pausch even brought his pet German shepherd to me for treatment when the dog became ill. Once again he played into my hands. Benno was vicious, one of the most dangerous of the SS dogs unleashed to add their brand of menace to our environment of constant peril. These dogs were trained to rip an inmate to pieces. They had been taught to recognize our smell, which sent them frothing.

When I was called in to see Benno I didn't need training in veterinary medicine to diagnose him. It was obvious that he was suffering severe indigestion from something he had eaten. I made a big show of examining the dog while several SS privates struggled to hold him down. I told Pausch that his pet had a digestive tract malady that would require a special diet and constant supervision by one of his men. He answered as I had anticipated.

"I can't possibly spare any of my guards to watch over Benno, but I will accept anyone you recommend for the job."

My recommendation had been prepared in advance. George would take care of Benno. George was a slender, gentle and scholarly young Greek who had been castrated after the Greek Jews had been delivered to Gusen I. He was one of the few to survive forced participation in mutilating medical experiments. He was becoming dangerously weakened, and I feared for his life. I introduced George to Commandant Pausch.

The special diet I prescribed for the dog was just what George needed to regain his strength, a healthy mixture of chopped meat and noodles. George followed my instructions and regularly ate most of Benno's meals, giving the remains to the dog. He was careful never to let the dog catch him in the act of stealing his food, and the two gradually became good friends. Benno responded to his pathetically inadequate diet in the same way the inmates responded to theirs. He became weak, despondent, and frighteningly emaciated.

The day arrived when Pausch summoned me again about Benno's worsening health. I conducted a very thorough examination of the starving dog. Calling a concerned and somber expression to my face, I sadly informed Commandant Pausch that his dog was dying of cancer. The kindest thing he could do for his pet would be to put him out of his misery immediately. Pausch was distraught at this news and adamantly refused to take his dog's life. I spoke of the cruelty in forcing the poor animal to suffer. He wasn't swayed and decided to call in another physician. This unexpected turn of events terrified me. Only something drastic could save us.

As I talked the problem over with our committee, I remembered the brown jar of rat poison I had stolen from Gusen I when I had gone to have my cast removed. We mixed a small amount with Benno's next meal. When the dog had convulsions and vomited and even bit Pausch's hand, the Commandant sadly agreed that Benno's irreversible deterioration was so obvious it would be best to put him away without wasting any more time. I asked him for enough morphine to kill the dog painlessly. Although there was absolutely none available for inmates and very little even for German soldiers, I knew Pausch would find some. I was not prepared for the full box of twenty-four ampules he brought. I needed only six ampules to put bloody Benno to rest, and I kept the rest for our own needs.

Another victim of our underground war on our killers was Otto, a psychotic sadist who was our green triangle camp Kapo when I came to Gusen II. Otto was short and muscular, insignificant in appearance except for his piercing blue eyes and the thumb that was missing on his left hand. He had also developed a prominent facial tic every time he killed. Otto kept a precise running tally of his murders and bragged endlessly to his cronies about the ever growing number of inmates he had eliminated. His supreme pleasure in life was killing. He took his greatest joy from murdering Jewish and French inmates. Jews he considered subhuman. The French were dirty. Otto prided himself on his own meticulous habits and always washed his hands after he murdered.

He came to see me one day complaining of a minor infection in his right index finger. His ignorance made it easy to convince him that it was severe enough to require lancing to drain the pus. My extensive experience with cold-blooded killers had made me well aware that Otto would have such terror at the prospect of his own pain that he would beg me to anesthetize him during the procedure. He pleaded with me to knock him out. I promised that he would feel absolutely nothing. He promised me an entire pack of cigarettes in return. I scheduled his surgery for that afternoon, in the privacy of his empty barrack. Bournasell would assist me.

I anesthetized Otto with some of the precious Evipan from our secret stock. Bournasell was puzzled to see me giving this sadistic butcher a pain killer.

"Wait, and you will understand," was all I said.

I surgically draped Otto's right arm, tied a tourniquet, and prepared my scalpel and hemostatic clamp. Bournasell's bewilderment grew. Why was I using such elaborate equipment for such a simple procedure? When I picked up my scalpel and began to make an incision in Otto's elbow, sudden realization dawned in Bournasell's eyes.

He whispered to me in French, "You're crazy! You can't be serious!"

"Of course I am," I replied, and proceeded to amputate Otto's right arm.

We carried Otto to his bed. When he came out of the anesthetic, Bournasell sent for me. As soon as I appeared Otto said, "Doctor, I am so grateful. I didn't feel anything during the operation!"

"Just as I promised you," I answered.

Otto tried to raise his right arm to shake my hand, and froze. He stared in horror at his bandaged stump. I explained that his infection turned out to be so deep that I had been forced to amputate his arm in order to save his life. Otto accepted this without argument, and we put him in the HKB for an uneventful recovery.

When Otto returned to his barrack he was no longer the same man. He lost interest in his homosexual lovers and spent more and more time brooding. How could a right-handed murderer practice his chosen profession without his right arm? One afternoon when the barrack was empty, he hanged himself. It was understandable, even predictable.

I had done a terrible thing to Otto, but life in the concentration camps had necessitated a cruel reinterpretation of the values that had inspired me to become a physician. In order to save lives, I now had to take them. Survival in these death camps was extraordinarily precarious at best. We were forced to take advantage of every edge. Being forced to play God, whether an avenging God who destroyed evil or a supportive God who chose the inmate to receive medication and extra food (which also meant choosing those who would not get them), created a terrifying uncertainty that I lived with day after day after day. When I felt particularly overwhelmed, it was of some help to be able to remind myself that I was, in the end, saving many lives. But this solace was small, and very temporary.

At one of our committee meetings to work out additional safeguards for protecting our chosen patients, we decided to enlist the unwitting assistance of SS Dr. Fetter. The next afternoon I spoke with him frankly about some of the unusual and fascinating cases that often passed through the HKB.

"If you would authorize me to keep them alive for the sake of science," I offered, "I will gladly write up the cases for you. After the war you can publish them in scientific journals and greatly enhance your standing in the professional community."

As we had hoped, this so appealed to Fetter's vanity that he gave me the free hand we sought. Dr. Kaminsky picked out a number of exotic diagnoses, because we knew we could rely on Fetter's basic ignorance. He would never suspect that there was something fishy about so many rare diseases cropping up in the same place at the same time. It was a fitting put-on for the man who had dreamed up the fake diagnoses for so many of the inmates selected for the gas chamber.

By January, Allied air raids on the surrounding countryside had begun to increase in frequency. I got permission from Commandant Pausch for several assistants and me to carry a supply of first aid materials to the tunnel workshops. Pausch did not stop to think that first aid supplies were unnecessary because the tunnels were adequate shelters. My real purpose was authorization to visit the work area with a large—and empty—box on my shoulder in which I could carry away everything I could lay my hands on.

When we arrived at the central work site the foreman's office was empty. Other than a piece of bread and some dried sausage, there was absolutely nothing worth stealing. I could not take the bread knife because an innocent inmate would certainly be killed for the theft. I cut off half the bread and sausage and took one last look around. Several unmarked blue metal containers stood in the shadows on a high shelf, and one of them found its way into my box as I left to rejoin the others. Bournasell whispered to me that he was sitting on some goodies. When no one was looking I took from under him a jar and an oilstone for sharpening tools and added them to the contents of my box. Suddenly the air raid alarm sounded. We had to take shelter in the tunnel and delay our return to the HKB.

Once back at the HKB we discovered that the Blockältester and his cohorts had, again, found it necessary to distract themselves during an air raid. They had killed bedridden patients to take their minds off their own anxieties. Ten inmates, all with their necks broken, had been sacrificed to tranquilize these green triangles. Two of them had been chosen by our committee as having a better than average chance of survival, and their "exotic diseases" had been entered in Fetter's file. We had been giving them much extra care, food, and precious medication. With the air raids rapidly increasing in frequency now, it was imperative to stop these "tranquilizer" murders.

I went to Commandant Pausch's office and told him of my arrangement with SS Dr. Fetter "for the sake of science." Then I informed him that the Blockältester and his friends had just murdered two of Fetter's special patients. I asked Pausch to take up this dreadful situation directly with Fetter and work out a way to stop the wanton killings of his important cases. Fetter was outraged. He ordered the cadavers sent immediately over to Dr. Kaminsky in Gusen I. Dr. Kaminsky's autopsy confirmed that each one had died of a broken neck.

As for the HKB Blockältester and his cronies, Fetter consigned them to the lowest caste of inmates. Now that they were no longer protected, it was child's play for our committee to take care of them. To prevent their replacements from daring to repeat the crime Fetter issued an order, countersigned by Pausch, that every single infirmary death would now have to be accompanied by a death certificate, signed by me, indicating the cause of death. I was also empowered to request an immediate autopsy whenever I felt it necessary. Even though the routine *selections* and nightly murders in the barracks continued as before, this was still an immensely exciting victory for us.

This very victory was almost my undoing. I did not anticipate the vengeful retaliation I would provoke among the Blockältester's coterie. The night after he was removed from the HKB I was awakened from my sleep by pounding on my door and voices calling me to attend a work accident. I jumped out of bed and grabbed some of my instruments, including a scalpel, that lay soaking in their alcohol bath. I stepped outside.

Before I could get my bearings in the dark, I was jumped and viciously assaulted from all sides. My right hand shot out without any direction from my bewildered mind. I felt my scalpel penetrate flesh. The wounded man screamed in agony and began shrieking uncontrollably. The camp firemen—Spanish Republicans—came on the run to investigate the disturbance. They separated us and grabbed my attackers. The thugs were the HKB Kapo and his partners.

Back in my room I pondered my suddenly vulnerable situation. I contacted the head fireman, who was part of our committee's action bloc, and discussed the spot I was in. He promised me the full support of his men. One would always be at my disposal as an unofficial bodyguard, and if these thugs dared try the same trick again, his men would accompany me en masse to and from the HKB. In the morning I called on the Kapo and bluntly warned him that if anything at all happened to me in the future my friends would know exactly who was responsible for it. He would never live to enjoy his revenge. From that point on I had no more "accidents." (This brutal Kapo did not have long to live in any case. When the Liberation finally reached us, he was the first Prominent in Gusen II to lose his life at the hands of the newly freed inmates.)

In the excitement of all these events I completely forgot about the blue metal container I had stolen from the foreman's office in the tunnel factory. Once I had the Kapo in hand, Henri and I sat down to open it. Can you imagine the utter joy of discovering that we had stolen a can of real coffee! We convened the entire committee as soon as we could and shared our dividend as we began to discuss further plans for resistance.

But our discussion strayed in no time at all to an elaborate verbal exploration of the universe of food. It inevitably happened whenever something occurred that would carry us back to our previous lives. Henri had organized weekly gatherings to keep our minds alive by discussing books, theater, music, politics, etc. Yet it was never long before one of us took center stage with an obsessively detailed dissertation on the piece of potato or meat he was lucky enough to have found in his latest bowl of soup.

My old friend from Laurahutte, Robert Bloch, had by then joined our committee and was able to share in the pleasure of our stolen coffee. Laurahutte had been evacuated about two months after I left, and many of its inmates were transported to Mauthausen and sent to Gusen II. I had taken Bloch into my private room, which I was already sharing with Henri, and introduced him to the rest of the committee.

24

By then I had become aware of the true nature of the so-called medical experimental work that was being carried on in the camps at the request of German scientific institutions. Respectable German pharmaceutical firms and long-standing medical institutions looked upon the camp inmates as a gold mine of human subjects for a variety of otherwise impermissible experiments. The healthy inmates selected for these experiments usually ended up as more fuel for the camp ovens. If they survived and were permitted to live, they were severely and permanently crippled.

Much of this work was done in the name of advancing technical frontiers to better the condition of the Aryan race. In one typical instance a famed Viennese orthopedic clinic was interested in improving the existing technique for repair of a fractured hip. They conducted a critical research project at Gusen II. The clinic's medical researchers smashed prisoners' hip joints in order to try out a variety of surgical techniques and artificial joints. Others, such as the famous I. G. Farben chemical company, used inmates' bodies as test tubes to gain information on toxic reactions to new and untested drugs.

I became involved in one such experiment, an investigation of TB1, one of Farben's crude early sulfa drugs used in the treatment of tuberculosis. My colleagues and I strongly suspected something nasty was in the works when we were notified by Fetter to assemble all of the inmates throughout the camp who had symptoms of tuberculosis. We immediately spread the warning through our communication channels that, at the coming assembly, prisoners should remain silent even if they were coughing up blood.

When the convalescence barrack was assembled the next morning and Fetter asked if anyone in the group was coughing a lot and perhaps bringing up blood, about thirty-five men stepped forward despite our warnings. Before they were taken away Fetter informed them that they had made a wise decision. The SS were putting them in a specially constructed hospital barrack for high-quality treatment.

I had no idea what was really going to happen to these men and the others who had disregarded our warnings. The fact that I had heard nothing about the new hospital barrack, or anything else about this project, was a warning to me that something awful had been planned. If I could get official sanction to become involved with it, I might have a chance to influence the course of whatever the events were to be. The following day Fetter agreed to my request—ostensibly motivated by my interest in chest diseases—to be put in charge of the new hospital barrack.

The building was relatively small and had obviously just been built. I could not believe the sight that greeted my eyes when I first walked inside. The barrack was quite clean, and instead of the standard triple-tiered planks partitioned into primitive bunks I saw genuine hospital beds with good mattresses and pillows, bed linen, clean blankets, and even individual night tables! Three male nurses had been assigned to assist me, and an SS Sanitätsdienst was attached as a kind of administrative supervisor. The Blockältester and Stubendienst were almost mild-mannered.

The humane environment was not all that left me puzzled. The traditional list of patients (120 altogether) lacked the standard order for blood and urine analyses, specified an unnamed medication, and noted dosage only as a quantity to be calculated by individual body weight once the patient had regained adequate weight and strength.

My bewilderment mushroomed when I saw the prescribed diet. Breakfast: one egg, one-half liter of milk, bread, real butter (the Germans themselves were already rationed to margarine substitutes). Lunch: soup, meat and vegetables, bread, an occasional artificial sweet. Snack: milk in the late afternoon. Supper: noodles with meat or fish, an artificial sweet. I estimated 2,500 to 3,000 calories daily, comparable to what the civilian population was allowed.

Considering the prolonged inhuman treatment to which these prisoner-patients had been subjected, they achieved a satisfactory condition surprisingly quickly. With the enforced regime of rest and good nutrition, even those patients whose tuberculosis was hopelessly advanced improved somewhat.

Just prior to beginning medication, we took the human guinea pigs to Gusen I to use their fluoroscopy equipment for recording the status of each inmate's disease and the specific pulmonary regions affected. Sputum analyses were done as well. When medication was started the next day, we received orders to start daily blood and urine analyses. The still unnamed medication was delivered in two plain boxes, with half of the patients' names on one and half on the other. Obviously, one box held an experimental medication and the other held a placebo. Dr. Kaminsky found out for me which was which.

Several days after medication had started, Dr. Kaminsky reported to me that the blood and urine tests indicated a number of my patients were beginning to show toxic symptoms. I checked their names against the boxes. As I expected, they were all receiving the higher dosages of the unidentified drug. We drastically decreased their medication levels and planned to return them gradually to the prescribed level. We kept our decision secret. Soon the laboratory reported evidence of liver damage in these same patients. Five of them died several weeks after medication began. The autopsies revealed kidney failure and severe jaundice, the classic picture of a toxic death. Patients on the lower dosages were now showing initial toxic symptoms. With so many mortalities resulting directly from this drug, I questioned the wisdom of continuing to administer it.

We decided to stop. Since we could not actively disobey our orders we secretly substituted our own form of treatment, meanwhile maintaining our records as if we were still administering the drug. The SS weren't the only masters of illusory bookkeeping!

We were exceedingly fortunate in locating and successfully stealing an old pneumothorax machine from Gusen I. Judging from the imprint of the Rothschild Hospital on the machine, the Germans had stolen it from France. It seemed fitting for it to end up in my hands. We were going to use therapeutic pneumothorax for those patients whose tuberculosis was still amenable to treatment. For those whose disease was incurably advanced there was nothing to be done, but at least they were more comfortable because of the rest and nourishment the experiment was providing for them. Now they would not have to suffer the toxic effects of the drug.

Pneumothorax is a procedure that introduces air into the pulmonary chest cavity to collapse a lung. The collapsed lung is able to rest, permitting the diseased area to heal more quickly. A possible complication from this procedure is empyema, an infection of the fluid in the chest cavity that is life-threatening even under normal conditions. In Gusen II it would be fatal. We introduced the pneumothorax to the first of our experimental patients and crossed our fingers against complications.

All went well. Soon about half of the patients had tolerated the pneumothorax and could be maintained without the drug. In the event that our periodic fluoroscopy inspections were ever changed to X-ray evaluations, we protected ourselves in advance via a deal with the X-ray workers not to divulge any evidence of pneumothorax. I wrote up all of the patients' records and entered the observations that would be appropriate in terms of our orders.

The time finally came when our worst fear was realized. A patient developed empyema. The inmate was convinced he had become ill specifically because we had withdrawn his medication, and threatened to tell Fetter that we had stopped giving the drug to the entire group of patients. But if this became known, every one of us faced extermination.

We convened the full committee. After a very lengthy and soul-searching discussion we finally concluded that our only reasonable perspective lay in balancing the life of this one man—who was going to die of his empyema in any case—against the 123 lives of the remaining patients and staff. We saw no alternative but to sacrifice him.

We had to act immediately. Fetter was due for an inspection visit the very next day. We couldn't steal enough Evipan in time and we did not want to deplete our precious reserves. We pooled our valuables—about twenty cigarettes, two loaves of bread, and a piece of artificial honey—to barter for additional Evipan and morphine. That night we injected the poor fellow with five times the normal dose of the drug. He was pronounced dead early the next morning.

When Fetter visited later that morning he was extremely surprised to see how well our patients were looking. I informed him that, appearances to the contrary, we had noted some toxicity and that one of our patients had died during the night of toxic complications. I asked if he needed an autopsy report for documentation. He did. We sent the corpse to Gusen I after explaining to Dr. Kaminsky that the patient had died from empyema, and why we would appreciate his neglecting to indicate it as the cause of death in his autopsy report. Dr. Kaminsky was more than accommodating.

As early February approached, the surviving patients were looking almost like normal men. They had all gained weight, seeming to indicate that their lesions were responding favorably to the drug. Fetter visited us once again to glance over the patients and review their records and histories. He was delighted with the drug's success and our excellent job, and promised us a good reward. The next morning I received five hundred cigarettes, and each of my three associates received three hundred. Each reward was a veritable fortune.

But the fact that these patients had recuperated sufficiently to be productive workers again was, to the German mentality, totally irrelevant. They had been selected specifically to test the effectiveness of the new drug, and now the test was finished. These inmates were of no further use, as we learned three days later.

At 5:00 A.M. on an unusually bitter February morning, the Blockältester roughly woke all of my patients and tore off their nightgowns. Easily reverting to the brutality that had been so blissfully absent, he kicked the terrified herd of naked inmates out into the subzero cold. I watched the Blockältester and the Stubendienst bring out some garden hoses, as innocent-looking as the ones you would use for watering your flowers. They opened the nozzles and began to spray the naked men, continuing until the last of my special patients had frozen to death.

German industry, as we now know, worked hand in hand with the Nazis to help them run their concentration camps. German firms vied for contracts to manufacture the gas pellets that fed the extermination chambers, the gas chambers themselves, the furnaces specifically designed to handle human corpses, and the machines created intentionally to grind human bones and teeth to dust. And they enthusiastically developed new products to meet new needs. One of these innovations, which became an all too familiar sight in the camps, was still in an experimental stage the first time it was used in Gusen II.

A *selection* had been announced for the convalescence barrack.

That in itself was nothing out of the ordinary, but the grapevine rumored that the Germans might be using this *selection* to try out a new method for these group disposals of the weak and sick. As always when a *selection* was scheduled, I tried to discharge those inmates I thought were strong enough to survive outside the barrack for the few days until the *selection* was safely over. And, as always, quite a few of them refused to leave what they saw as their only refuge.

By the time I arrived at the convalescence barrack that morning, the Blockältester had already awakened the inmates. Breakfast was brought in and the doors were locked. The SS men who always attended on these occasions arrived promptly at 9:00 A.M.: SS Dr. Fetter, Commandant Pausch, Commandants Ziereis and Bachmayer, and Sanitätsdienst Schultze. There was an additional spectator on this particular morning, a man in civilian dress.

As each inmate's number was called out by the Blockältester he stood with his back to Fetter. Forty of the most emaciated inmates were ordered over to the left. Among them were several of the men I had tried to discharge a few days earlier.

A black van, soon to be dubbed the *Schwartze Mina* (Black Mina) by camp inmates, pulled up in front of the barrack for the first time. The rear doors opened. Fetter told me this van was taking the forty selected inmates to Mauthausen itself "for the sophisticated medical treatments they needed." I knew what that meant. But as they were gently helped in and seated on two roomy benches rather than beaten and herded in like so many naked sardines, I was dumbfounded at the absence of the usual brutality.

An inmate reproached me as he passed on his way to enter the van. "You wanted to discharge me into the camp to die! Look how they're treating us. We're finally going to get proper medical attention now!"

As the doors of the black van closed on the smiling faces, I noticed heavy rubber insulation around the openings and a closed ventilation hatch on the roof.

The empty black van returned about an hour later. The SS Sanitätsdienst accompanying it remarked cheerfully to Fetter that the new invention "worked wonderfully."

The SS physician smiled at me. "You see, your patients are already in good hands and getting the best of treatment!"

The *selection* was not over. The black van took a second load of forty inmates from the convalescence barrack, making thirteen round trips in all before that day was over. The exhaust pipe of the van opened into the comfortable interior, filling it with carbon monoxide during each ride to Mauthausen's crematorium. Before feeding the ovens, each load of corpses was carefully examined by the German medical and industrial personnel who had developed the special van. The civilian I had seen at the morning's first *selection* was an important participant in these examinations. He was the engineer who had designed this death machine. His goal had been to develop an effective gas chamber that presented an irreproachably innocent face to the outside world, rendering its victims docile by deluding them into thinking they were being saved from extermination.

That day 520 optimistic human beings were helped into that specially constructed van, and 520 corpses arrived at their destination. The convalescence barrack was empty, ready now for a new batch of victims.

25

By February 1945 the war was going so badly for Germany that her own newspapers could no longer deny it. Air raids were occurring with ever greater frequency. Although the Allies were careful not to bomb the camps themselves, the decimated ranks of German manpower were so dangerously low that the SS desperately took every precaution not to sacrifice factory workers needlessly. Commandant Pausch ordered the barrack Prominents to take their inmates to hide in the tunnels whenever an air raid was signaled. Only those in the HKB and the convalescence barrack could be left behind.

As the Allies and the Russians closed in, more camps were evacuated. The Germans sacrificed their passion for order to their need for haste. Camps receiving the evacuated inmates became badly overpopulated. And male camps, such as Mauthausen, suddenly had to make provisions for female prisoners. Gusen II was now overflowing with inmates of many nationalities who had survived their evacuation. They told stories of forced marches during which the SS routinely shot every inmate who was slow in keeping up with the column. Rumors ran through the camp. According to some, Buchenwald, Dachau, and Gross Rosen had already been evacuated. According to another, American soldiers had already captured at least one concentration camp and were so horrified by what they found that they lined up all the SS they could find and shot them.

As February continued, an enormous influx of Jewish prisoners evacuated from Auschwitz in the face of the advancing Russians temporarily swelled our camp population to its all-time high of forty thousand. These Jewish prisoners had all been designated for the gas chambers. The sudden urgent need for evacuation of the camp had only given them a brief reprieve. The SS accelerated the pace of the *selections,* reducing our population to twenty-five thousand in just several weeks. It was during this time that the *Schwartze Mina* was brought into regular use.

Typhus began to rage, adding significantly to the death toll, but our camp population remained stable due to the constant stream of incoming prisoners who replaced the dead. The SS became hysterical when this typhus epidemic began. A single louse found in a barrack sent the entire barrack population for permanent disinfection—in the gas chamber. Our committee had to use considerable ingenuity to extricate some of our comrades from certain death.

Our food rations were cut in half, but the labor schedule at Gusen II continued at its abusive pace. Work accidents became so numerous that my arrival to attend the incoming day or night patients was invariably greeted by fifteen or twenty seriously injured inmates laid out on the bare and bloodied floor.

During this time a Dutch inmate lied his way into my morning treatment schedule at the HKB. He and his brother had been extremely wealthy businessmen before the war. His brother had already died in Gusen II.

The man now in front of me had been my patient in the HKB once before, in November, shortly after his brother's death. At that time he had come in with a self-mutilated pinky finger. He knew I would have to amputate it and put him in the HKB for two life-giving weeks of respite from heavy labor. Now he was back, emaciated and exhausted, asking me to amputate the pinky on his other hand.

"I can't possibly do it," I told him. "Nothing is wrong with your finger. I am a physician here to help the sick and wounded, not to mutilate. I cannot do what you ask."

He pleaded with me. "It's my only chance to survive until we are liberated from this nightmare. Without two weeks of rest, I'll never make it."

"I can't," I said, and turned to move on to another patient.

"Look," he continued. "In Amsterdam I am a businessman, not a laborer. I won't need my finger to survive once the war is over. But I have to lose it now to survive this camp. I beg you—help me live!"

I did it. I diagnosed a severe infection in his remaining pinky that necessitated amputating the entire finger. He spent the next two weeks in the HKB. (I saw him return to Holland after the Liberation.)

During one of the regular bombing attacks on nearby Linz I was standing watch outside the HKB with Bournasell and another committee member. A sudden patch of white in the sky above us grew into a falling parachute. To our astonishment it landed right in front of us. An English pilot was strapped into the harness. We ran over to him, quickly got him out of his harness, and dragged him out of sight. I told two of our Spanish firemen to fetch a fresh cadaver about the pilot's size from the morgue shack. There was such an abundant supply that I knew their task would be easy. We took off the pilot's uniform, dressed a Ukrainian cadaver in it, and strapped the corpse into the parachute harness. We dressed the Englishman in the dead Ukrainian's striped camp uniform.

The pilot had been watching all of these rapid proceedings with a dazed and uncomprehending expression. He finally found his voice. In broken German, he asked where he had landed. "Gusen II," we answered. I explained that we were saving him from immediate execution by the SS, and that we would do our best to help him stay alive. The pilot expressed deep gratitude to us for saving his life. (As the Liberation approached and the prisoner exchange began, we turned our English pilot from a Ukrainian into a citizen of Luxembourg to make sure that he would be returned to an Allied country. It was Luxembourg purely by chance. We took the registration number off the one Allied inmate who had died after that morning's roll call and therefore was not yet recorded as dead. The pilot survived Gusen II and finally returned home to England.)

We knew by now that the war must end in the next several months with Germany's defeat. Medications had become so scarce that even the SS pharmacy had none left for us to steal. We had to reserve the few ampules of morphine left in our private supply for patients who seemed to have the strongest probability of surviving until then. We prayed frantically for some miracle to end the war tomorrow. Selecting candidates for survival was impossibly painful. The silent faces of the rejected inmates are still in front of me, their eyes begging, pleading, beseeching me for something—anything—to give them relief, to let them live.

The daily weight of these constant decisions began to take its toll. I became tense and irritable. Sometimes, even during surgery, I would break down in an uncontrollable fit of sobbing. I lost all patience for inventing encouraging war news. My emotional suffering had become too great to absorb.

Yet I had developed a reputation among the inmates as a medical miracle worker, a reputation I hadn't sought and one that was certainly not based on any real achievement. I am sure it would never have happened if all of the witnesses with relevant testimony could have been called: those who had died because they were not among the "chosen"; those who had died because they had had to wait for my attentions; those who had died because I had nothing with which to treat them; those who had died because the Blockältester had murdered them during the night.

Food had become such a dire problem that we had to be constantly on the lookout for an opportunity to augment the continuously reduced rations we were grudgingly given. One day I was summoned to treat an SS soldier who worked in the SS kitchen. I forgot the medical aspects of his case long ago, but I will never forget what was on the kitchen stove!

A tantalizing aroma beckoned seductively from a huge simmering kettle. My empty stomach ached. While I tended my patient my mind was riveted on the kettle outside. On my way back out of the kitchen I hurriedly detoured by the kettle and saw the leg of some animal stewing in it. No one was watching me. I grabbed the steaming leg of meat, shoved it under my shirt, shot out the door and ran back to my room. I ran as fast as I could, oblivious for the moment of the heat searing the flesh on my stomach.

I called our committee together. When we sat down to contemplate our bounty, I was given the honor of the first taste. It turned out to be a delicious and nourishing leg of mutton that provided enough meat for all of us for two days. It took an entire week of treatment, though, to alleviate the excruciating pain from the burns on my stomach!

Our severe deprivation sometimes drove us to less traditional food sources. One day I was walking along by the electrified camp fence just as a beautiful young SS attack dog urinated on the fence and was electrocuted. Suddenly in front of me was a potential meal. I quickly found Sanitätsdienst Schultze and told him that we were very anxious to autopsy the dog to study the systemic effects of electrocution. The results would be given to Dr. Fetter. As soon as the dog was brought to my room we cooked it and stuffed ourselves. There was still enough left over to stew for a second meal. Dog meat is fairly good, particularly if you are starving.

Different inmates found different ways to get enough food to stay alive. The Bahnhof was producing even greater numbers of corpses because the HKB Blockältester was now regularly stealing the paltry rations of these terminal dysentery patients to feed himself and his friends. And Gusen II had become a world in which even the eating of human flesh lost its horror.

The anatomy professor and I were still working on improving my surgical technique. We began to notice something very strange going on. Hearts and livers were being secretly removed from corpses. I alerted our committee that someone in the camp was practicing cannibalism.

Two days later one of our German members reported seeing a Ukrainian inmate in his barrack eating something during the night. When the Ukrainian left his bunk to go to the latrine, our man looked under his straw mattress. He found a piece of human liver.

I informed Commandant Pausch of our macabre discovery. He did not seem very perturbed by my revelation, but he sent a man to the barrack to investigate. The SS man discovered the fresh remains of a human liver under the Ukrianian's mattress and questioned him about why he was eating human meat. The Ukrainian replied in his broken German, "You do same in my place." He paused and smiled. "And does not taste bad if bake it. You want try, let me know!"

The SS man smiled back at him and left the barrack. I was incredulous that the matter was closed. Everything in the camps was designed to turn us into the basest of animals. We decided to put our own guard at the cadaver shack starting that same night. But in the morning we found our watchman dead, minus his heart and liver. We gave up.

When I looked into the darker recesses of my own character I realized that I would probably do what the Ukrainian had done if that became my only alternative, and I could no longer blame him. My one final regret was that I had denounced a fellow human being for secretly eating human flesh in the concentration camp in order to stay alive.

As we progressed into March my surgery teacher was delighted with the continuous improvement in my skills, though I'd had little opportunity to gauge my own progress. Patients recovering from surgery couldn't work to earn their food, so their lives held no value to the SS and their survival was very unimportant.

Inmates wounded by the Americans during a close-by bombing raid were carried to the infirmary on improvised stretchers and laid out around my operating table. I began surgery on the inmate who was most seriously wounded. Fetter entered the room as I reached a crucial part in the procedure. He called all of us to attention as I was in the midst of clamping a bleeding vessel. I continued what I was doing. Fetter removed his gun from its holster and shot my patient on the operating table. He didn't blink an eyelash. Still cool as a cucumber, he slapped my face with such force that I was thrown down onto a bed. He pressed the smoking barrel of his gun against the side of my head and snarled, "And now—you!"

I felt suddenly at the end of caring. My endurance crumpled. I answered Fetter in a monotone of despair.

"Do it. Death is exactly what I am waiting for. Now my torture will finally come to an end."

"It would be a waste of a good bullet," he said.

Fetter put his gun back in its holster, turned sharply on his heel and left the infirmary.

As the end of March approached, rumors started flying that Germany had agreed with the Allies to exchange captured German civilians for Allied nationals in the concentration camps. I shrugged it off as another false rumor being spread by the morale boosters, yet I too repeated it to prisoners whose morale was badly sagging. With our comrades rapidly disappearing from typhus and the gas chamber, we were in dire need of optimistic news.

Our only consolation in the steadily worsening typhus epidemic was the apolitical nature of the lice that spread the disease. Most of the SS who developed typhus were sufficiently well nourished to recover, but some of them did succumb. We were especially pleased when one of the sharpshooters stationed in the watch-tower died of the disease. The SS became so alarmed when typhus deaths began to appear among their own ranks that they ordered all of the green triangle Prominents to have a series of typhus inoculations and boosters. The SS felt that their immunity to typhus would be ensured if the inmates with whom they had their greatest contact were made immune. Since the lice that spread the disease didn't discriminate, it was like building a seawall for avalanche protection.

It finally became very obvious that the Nazis had exhausted their resources. They were scraping the bottom of the barrel. They had even begun to ask the inmates for volunteers to don German army uniforms and go out to fight the Russians. Their first pick were German and other Axis political prisoners. When only a few volunteered they turned to their allies, the green triangles. A significant number of these criminal Prominents volunteered, but they were only seizing the opportunity to get out of the camp and save their skins. The moment they were out, they deserted from the German army and surrendered to the advancing Allies. In this way many of our murderers found an anonymous shelter for quite a while after the war. Some were never discovered.

One day during the first week in April, Henri expressed concern that I wasn't looking well. He was upset that I had begun to lose weight again despite our fairly adequate diet. His comments suddenly reminded me that I had never gotten my booster shot for typhus after leaving Auschwitz.

From the steady stream of good news that began pouring in, we now knew that the end of the war was about one month away, two at the very most. Holding on became more critical each day. Keeping our patients alive was a frantic race against time.

April 10, 1945 was a red-letter day. Henri came running into the infirmary. Large tears of joy glistened in his eyes. He cried out that the rumor of prisoner exchange was true! Our committee member who worked in Pausch's office had just confirmed it. A preliminary group of Red Cross officials was already in Mauthausen setting up the repatriation machinery. They had food packages to distribute among each incoming group of inmates designated for exchange.

Despite Henri's extraordinary news I didn't feel much excitement. For the past several days I had been headachy, nauseous, and depressed. Henri remarked on my pallor. He put his hand under my armpit and discovered that I had a fever. He ordered me to bed. My fever soared as the day wore on. I was on fire. My head was splitting open. Henri brought me water and some milk and eggs stolen from the SS kitchen. I couldn't eat anything. He made me mashed potatoes from our distillery supplies and some kind of tea that we had stolen from the pharmacy. I could barely eat. The next day there was no change in my condition. On my third day in bed I woke up with typical typhus eruptions on my skin. I couldn't go to register myself as a French citizen for the prisoner exchange because my illness would have been obvious. Henri took care of it for me. If I was to survive, it was crucial that my disease be kept a deep secret.

My thoughts wandered as I lay in bed. A sudden commotion outside caught my attention, and I ran to my window. In front of the convalescence barrack on the other side of the fence the inmates were all being lined up. They were stark naked. Then they stood and waited. As I continued to watch, the *Schwartze Mina* arrived. The first naked prisoners in line were herded inside the gassing van with kicks and blows to hurry them. The *Schwartze Mina* returned periodically for new passengers. By the end of the day, the famous barrack twenty was no more.

Despite all of our precautions, the HKB Blockschreiber eventually got wind of the fact that I was sick. He lost no time in reporting it directly to Sanitätsdienst Schultze. Schultze came to my room and ordered me to undress. As soon as he saw the typhus eruptions on my skin, he smiled. His smile made my skin crawl. He wrote down my number and ordered me to go to the infirmary in Gusen I. If I went there, I knew I would never leave it alive.

Henri came in the moment Schultze left. I told him that I had reached the end of my rope. This time there was no way out for me. But Henri absolutely refused to accept the inevitability of my situation. He didn't lose a moment in contacting our committee's action bloc. Someone had to be in a position to do something about the dead-end turn my fate had suddenly taken. It was Pausch's secretary who found a way. Even though inmates were being processed for exchange by alphabetical order, he managed to include Henri and me in the first transport of exchange prisoners leaving Gusen II for Mauthausen. We were scheduled to leave the next morning.

I was very weakened by my high fever. Henri insisted—much as Père Bonaventure had at Royallieu—on carrying me on his back.

THE ROAD BACK TO LIFE

V

26

The war was ending as Henri carried me back over my route of last fall. Now we marched from Gusen II to Gusen I for the last time. The SS sent us to the showers—the standard precaution against typhus-carrying lice—and gave us clean uniforms. The Swiss Red Cross sent us food packages that we sat down to eat in the dusty Assembly Place.

Then Henri supported me on the march from Gusen I to Mauthausen. We were showered again, then sent to Mauthausen's quarantine barrack where we would await repatriation by the Red Cross. I remembered when I had arrived here in terrible pain from the leg I had broken when an SS soldier had pushed me off the truckload of corpses. Now I was put in the section of the barrack that was a makeshift hospital area for repatriation-bound inmates who still required serious medical attention. The Red Cross would give us priority for the optional two-week rest and recuperation stay in Switzerland that was allowed the repatriated inmates.

This barrack and the camp commandant's office were all that the Red Cross officials were permitted to see of the death camp and its satellites. It had been furnished to create the proper impression. My new bunk was luxurious. The straw mattress was made up with clean sheets, a clean blanket, a pillow, and a clean pillowcase. My fever heightened the already dreamlike quality of so sudden and drastic a change. I caressed my sheets with my fingertips again and again, vacillating between joy and disbelief in my surroundings. But it was only a hollow image, as I would realize very soon.

In the bunk to my left was Armand Petchot-Bacqué, a French military physician who had been responsible for the health and medical care of the famous General de Lattre de Tassigny, a kind of MacArthur figure. Armand was a Basque, typically brown-eyed and dark-haired, with a compact body that radiated energy and cheer. The more we talked, the more I liked him. He was a wonderful neighbor and became a close friend.

That first night as we lay in bed, thunder began to rumble in the distance. It was still storming when we fell asleep and had not yet stopped when we woke up the next morning. The thunder grew louder and louder as the day went on. We finally realized that we were listening to fighting at the front, which was moving closer and closer to us. The German army was disintegrating. The Americans were breaking through from the west for their eventual meeting with the Russians advancing from the east. We were listening to the final battle between American and German guns. It was the first of May, 1945.

That afternoon a Red Cross ambulance and trucks returned from having transported one of the first groups of repatriated inmates to Switzerland. Mr. Heflinger, the Red Cross transport chief, sadly informed everyone that the next transport scheduled would not be able to leave. It had become impossible to avoid the battle area. Traveling had become such a dangerous proposition that he could no longer take us out. At least we would be comfortable until the situation improved.

Mr. Heflinger did not know what we had already learned, that the highly restricted entry of the Red Cross onto the scene had really changed very little in Mauthausen and its satellite camps. The Red Cross hadn't the faintest notion of what conditions in the camps were like. And the SS, cruel and bloodthirsty as ever, were desperately afraid of these horrors reaching the outside world. Except for the brief visits of the Red Cross ambulance, even we in the quarantine barrack were as much prisoners as ever. Liberation had not really reached the gates of Mauthausen, and we were in grave danger as long as we remained behind them.

On May 2, the German inmate-secretary to Commandant Bachmayer of Mauthausen contacted the camp's well-established Resistance committee. He had just overheard an order that jeopardized everyone's survival unless something could be done. Speed was crucial. Commandant Bachmayer had been told to exterminate the entire camp population—put all of the inmates in the factory tunnels and blow them to smithereens with explosives, burying their remains and their testimony under an impenetrable mountain of rocky debris.

Emile Vallé of the Mauthausen Resistance committee met with Mr. Heflinger immediately and told him of the twenty to thirty thousand men who were in peril. He had to get word out to the American forces before it was too late. Mr. Heflinger promised to find the Americans and bring them to save us. He jumped into his white jeep painted with the Red Cross insignia and tore out of the camp like a modern-day knight on his white charger.

Even before Mr. Heflinger left, the Resistance committee had surreptitiously begun arming inmates. Committee members had taken advantage of the presence of the Red Cross and temporarily decreased SS surveillance to steal weapons. They planned an uprising against the SS for the following day.

Before the sun rose on May 3, the SS officers stormed in from their barracks and attacked us. We were debilitated and exhausted, but we far outnumbered the SS men and our motivation to survive nourished and sustained us. The battle raged.

I was now only semi-lucid from my still rising fever, yet when someone put a gun in my hands somehow I found the strength to use it. I had to participate in this last battle for our freedom. I ran toward the camp gates and suddenly came face to face with an SS man. He threw up his hands in terrified surrender and begged me, "Please don't shoot! I didn't do anything!"

His blanket denial of any guilt violently liberated all of the anger I had been accumulating and storing for so long. I emptied my gun into him.

Other than that, I remember only foggy patches. My feverish state left me with almost no recollection of these events. I know there was much shooting and our uprising continued into the morning of May 5. Well before dawn, we attacked the SS and the green triangles who had joined forces with them. In the chaos someone opened the camp gates and those of us who had guns swarmed out through them. The SS massed inside the gates. We were afraid to go far. If the human chain of SS that always encircled the camp was still in place, we would be massacred. We spent the rest of the night in the bushes and kept watch.

Somewhere around ten o'clock that morning, our attention was riveted by the whining and clanking of a huge machine. From out of nowhere, an American tank drove into view. Mr. Heflinger had kept his promise!

The tank stopped. The hatch opened and an American officer in clean battle fatigues jumped to the ground. He called out, "Does anyone here speak English?" Belivier, a young French artist, volunteered his services and hurriedly brought the officer up to date. The American officer addressed three words to us—"Take it easy!"—and threw a hand grenade into the area where the SS men were concentrated. The Germans must have thought the entire American army was right behind that one tank and grenade, for in the next five minutes they had all surrendered.

To this day I find "Take it easy!" one of the most comforting phrases in the American language. Yet the memory holds a pathetic irony as well. My comrades and I were a group of horrifyingly debilitated men, skeletons in torn and filthy clothes who were obviously just emerging from a life of deprivation and abuse that was beyond all imagination. How ridiculously inappropriate, however well-meaning it was, for this well-fed and -dressed American casually to tell us, "Take it easy!"

The American tank stood guard at Mauthausen's gates. American soldiers disarmed the SS and tried to protect them and the green triangles from the inmates, but the Americans were too few in number to save them all. The green triangles were now completely defenseless with the loss of their SS protectors. Their sudden vulnerability unleashed their victims' long pent-up rage. Little mercy was shown as the tables violently turned, with those who had barely escaped slaughter turning into a mob of madmen greedily butchering their tormentors. This bloody scene was repeated at every single camp on its Liberation day.

I did not want to participate in this bloodbath, and I was too exhausted to get back to the hospital barrack. I was still outside the camp gates and completely drained of strength. A group of buildings to my right offered temporary shelter and a place to rest. I went into one of the buildings and found myself in the SS enlisted men's quarters. I was in a large dining room. A long wooden table held the undisturbed remains of a hurriedly interrupted breakfast. I ate some bread and margarine and drank from a half-empty pitcher of cold coffee. My stomach rebelled. I could not keep any of it down.

I heard a noise in the doorway and turned. A wary-eyed American soldier was pointing his rifle directly at me. I froze. He called out something that I was not able to understand. Finally he shouted, "Up, up!" I painfully lifted my arms, terrified that this very young and very jittery soldier was going to pull his trigger at any instant. How could I make him understand who I was before he shot me?

I called out in German: "I am a *Gefangener* ("prisoner")." No response. I repeated it in French. Still no response. I wanted to show him my tattooed inmate registration number. I was trembling so badly that I could not pull my sleeve down my upraised arm. The American soldier cocked his gun.

If I had not been so horrified I would have laughed. Here I stood in my striped uniform, pale and thin as a sheet of paper, finally witnessing the Liberation after all I had endured, and on this very morning I was going to be shot dead by a very young, very nervous soldier from a faraway country!

My loose sleeves began to slip down my thin, upraised arms. My number was finally exposed to view, but it seemed only to increase the young soldier's angry suspicion. He pushed me roughly with his rifle butt and drove me out to the tank commander. His hostility bewildered me.

Belivier's face took on an incredulous expression and he came running over to me. The commanding officer listened attentively to the young soldier's report, then Belivier spoke with him. Finally Belivier explained to me that the young soldier was sure that he had caught an SS officer camouflaged in an inmate's uniform. The soldier, in his inexperience, had mistaken the number tattooed on my forearm for the blood type that all the SS and Gestapo officers had tattooed on the inside of their upper arms. The tank commander apologized to me for the unfortunate incident, and the young soldier returned to the SS barracks to continue his hunt.

Belivier realized that I was on the verge of collapse and had two comrades carry me inside the camp to the hospital barrack. Armand Petchot-Bacqué welcomed me with a glowing smile.

"How does it feel, at last, to be free again?"

I related the experience that had almost turned me into a liberated corpse! Armand told me that a Frenchman had come to see me during the uprising. He was a tall, thin inmate who said he had been in Gusen II, and he seemed in urgent need of talking with me. But everyone in the camp was thin. I was at a loss as to who this man could be.

I was not feeling at all well. My fever was climbing again and I began to shiver. I was too tired to think.

My tall, thin visitor from Gusen II returned later that day. I had known him as Paul, one of those chosen by our committee to receive the supplementary food that had enabled him to survive. Although he was unusually tall for a Frenchman, he had the classic elongated hooked beak of the French noblemen, the *nez Bourbon* as it was called. His steely eyes belied a basically warm nature. Some of Paul's talk of his earlier life reflected a highly educated man whose two primary nonprofessional pursuits were women and art. He spoke in a voice of authority, as if he were very used to giving orders. Paul would always tell us that he was a high-ranking French officer and would eventually reward us for our help, but in those times practically everyone was a millionaire or bigshot promising fabulous rewards after the war for an extra bowl of soup in Gusen II. I ignored Paul's claims as I had ignored all the others. I told him the only promise I wanted from him was that he would survive.

Now Paul had brought another Frenchman to the hospital barrack with him, another one of our chosen ones. I said that I was deeply gratified to see that both of them had kept their promise to survive.

"Here is my gratitude," Paul said, and he stooped down and kissed my forehead tenderly. "I owe my survival entirely to you. And now, finally, I can be honest with you. I recognized you from the very first moment I laid eyes on you in Gusen II. We had met a long time before, but of course you had no way of knowing that. I was the voice that talked to you and Sonja from the darkened room in London. I am General Paul Guivante de Saint-Gast, chief of your *réseau.* And here at my side is Captain Jean Veit, my aide-de-camp. We had both returned to France on an assignment shortly after you and Sonja were arrested. Someone denounced us to the Germans almost immediately and we were arrested too."

My illness and extreme exhaustion had rendered me incapable of functioning lucidly. This seemed too crazy to be real. My head was so hazy I decided that I must have dreamed the Liberation. We were still in the camp and Paul was still telling his stories and making his false promises! I just wanted them to leave me alone so I could sleep.

"I'm in no mood to listen to fairy tales," I told them both. "I'm happy that you were both gentlemen enough to keep your promise to survive. Beyond that, you owe me nothing."

My eyelids were too heavy to hold open. As they closed I heard Armand telling my visitors, "He's almost unconscious from his fever. He has temporarily lost his hold on reality."

"Can't the Americans do something for him?" a voice asked.

So the Americans actually had taken over Mauthausen. We really were free of the Nazi nightmare after all!

My realization was the last thing I remember.

27

When I opened my eyes again, the first thing I saw was my left arm stretched out at my side and a length of tubing running from the needle in my vein to a bottle hanging by my bed. I looked around me. Armand Petchot-Bacqué was still in the bed to my left. He was also attached to a hanging bottle. We had been moved close to a window.

"*Bonjour*, Armand," I said. My voice was weaker than I had expected.

Armand greeted me with one of his warm smiles and explained what had been going on while I was in the throes of my illness. The Americans had arrived with medical personnel. In the past twenty-four hours they had fully supplied and organized this hospital. Armand was awestruck at the feat. I had spent that time drifting in and out of consciousness, semi-lucid at times and incoherent at others, sometimes babbling in French, German, and Hungarian all at once. General Paul Guivante de Saint-Gast and his aide-de-camp had visited me again and left some fresh eggs and sweet apple cider under my night table. The cider was in a gallon jug, but I was far too weak to lift it by myself. I also learned that Henri Desoille had been visiting me regularly, and would be back as soon as he could.

Armand also had deeply satisfying news about Commandant Bachmayer, the SS head of Mauthausen. Bill Aurix, an outstanding organizer in the Belgian Underground who had been caught and imprisoned in Mauthausen, had cunningly worked his way up to become a highly privileged Prominent who gained Commandant Bachmayer's confidence. When Bachmayer's secretary had overheard him, on the eve of the camp uprising, discussing the order for immediate extermination of Mauthausen's inmates, Bill

Aurix was quickly warned by the Resistance Committee. Then Aurix went to see the commandant with a bargain. If Bachmayer delayed execution of these orders for three days, Aurix would save him from the Americans when they arrived. Aurix had accurately gauged the German's fear of being captured. Bachmayer postponed the extermination.

When the Americans arrived at Mauthausen, Aurix dressed the commandant in an inmate's dirty striped uniform, shaved his hair, and smuggled him out of the camp to the nearest wooded area. As Bachmayer was changing into civilian clothes, Aurix shot him and led the Americans to the corpse he had "accidentally" stumbled upon. Later he told his friends that he had kept his promise to Bachmayer not to deliver him to the Americans!

Then Armand told me that the repatriation process was starting up again. A lot of inmates were going to Switzerland first: inmates who needed to convalesce, those who wanted to rest before going home, and those who did not want to return home at all. Frenchmen who were strong enough and wanted to return home right away were being taken to the Orching Airfield by the Swiss and flown home by the French Red Cross. Paul had been the first Frenchman to be repatriated directly to France. Armand had heard that he left Orching with the fanfare accorded a returning general.

"What are you still doing here, Armand?" I wanted to know. "Don't you want to leave the umbilical cord that's feeding you?", I teased him.

Armand told me that he had been passing blood in his urine. Although he felt okay enough to leave, the American physician wanted him to stay a little longer.

Our American saviors were compassionate and generous down to the lowest private. An army private helping out in the hospital noticed that I had nothing to wear except a ridiculously large and very old shirt. He ran out and returned with a military shirt and briefs, put them on me, and stood back to survey my new outfit. He didn't look entirely satisfied. He ran out again, this time to return with a pair of brand-new pajamas for me!

That afternoon, while Armand and I were talking, his eyes suddenly moved beyond me and he interrupted himself, "Ah! There he is! The man who has been looking for you!"

I saw a tall, lean man in his mid-thirties, smiling, wearing eyeglasses. He looked like the stereotypical New Englander, though he was from the Midwest. He was accompanied by a beautiful blond girl and an interpreter. They were all in American battle dress. They stopped at each patient's bed and said a few words to him. When they reached mine, the man sat down and introduced himself through his French-speaking interpreter.

"I am Major Donald Vollmer, chief of the 130th Evacuation Hospital. This is Miss Helen Baker, my head nurse. I'm delighted to see that you're finally on the road to recovery, Dr. Haas. In addition to your typhus you had typhoid, an intestinal infection, and severe bronchitis. I don't have to tell you how lucky you are! At least your bronchitis was easy enough to clear up with penicillin."

"What is penicillin?", I wanted to know. "I've never heard of it!"

Major Vollmer laughed. "It's the greatest miracle drug you ever heard of, and it costs a fortune just to look at. You don't have to worry. We have plenty on hand!"

We chatted for a few more minutes. As soon as the Americans left, Henri Desoille walked in and came over to Armand and me. He removed our plasma needles and stuck thermometers in our mouths while he told me about his work as a physician in the female hospital barrack. The Americans had created it from one of the SS barracks outside the camp gates. It was rapidly becoming a very difficult situation because most of the physicians who worked there longed to return home to their families. Henri himself couldn't wait to get back to his wife and children. He had promised himself that he would stay at Mauthausen until he got word from them.

Our thermometers came out. My temperature was almost normal. Tomorrow Henri would have an ambulance take me over to the field hospital for a thorough checkup.

"Henri," I asked, "do you think I can go outside to sit in the sun for a few minutes?"

"Sure," he said, "why not. It's nice and warm outside. Let's go!"

This was easier said than done! I stood up on legs so unsteady that I thought they could never become strong enough to support me again. Henri helped me outside to a bench in the sunshine and sat me down next to three other convalescent inmates. He left to get fresh plasma bottles for Armand and me. The three other men were strangers. I was not in any mood to start up a conversation.

I basked in the gentle sunshine, slowly filling my lungs with the fresh springtime air. My thoughts began to turn homeward—to Sonja, François, my mother, and my sisters. For the very first time it dawned on me that I might not have any family left to return to! What would I do if I returned to an empty house?

These sudden thoughts were terrifying. I began to fall apart inside. I sat there, lost, my elbows on my thin knees, my head buried in my hands. I was oblivious to the three men beside me. Some photographers from *L'Agence France-Presse* arrived and took photographs of the four anonymous and emaciated inmates sitting on the bench in the spring sunshine after their Liberation.

Henri returned and helped me back to bed. While he was starting the fresh plasma I.V. he noted the change on my face. "Are you angry that the photographers took your picture?" he wanted to know. "No, no," I protested. "I'm just very tired." I didn't want to tell him that I was afraid to go home.

I had no appetite for lunch—stew, pineapple, and poundcake. Dorothy, our nurse, sat on my bed and coaxed me to eat. I said I couldn't swallow. She fetched me a meal of applesauce, soft boiled eggs and milk, and started to feed me as if I were a baby. I tried to fight her. Finally, with her motherly encouragement, I cleaned my plate. She pepped up my milk with a little sugar and a few drops of brandy from a small bottle in the pocket of her army uniform. I began to feel a little less morose after finishing my milk.

Armand jokingly observed that "pissing blood obviously isn't serious enough to warrant being fed by such a good-looking nurse!"

"Armand," I answered him, "you eat like a horse. Stop eating and she'll feed you, too."

"That much she ain't worth!" was his reply.

Suddenly I blurted out, "Armand, I'm scared to go home! I'm scared of what I'll find!"

He turned to me, and said very softly, "So am I, friend, so am I."

I had to interrupt our conversation. It was imperative that I leave to find the toilet. My bowels were about to betray me again.

The bathroom was clean. I sat down on the toilet. My eyes traveled over all of the clean surfaces and came to rest on something that I had forgotten ever existed—toilet paper. I was so moved and excited by my discovery that I longed for a poet's lyrical facility so I could write a tender poem to this long lost friend I had finally found again.

The next day the ambulance took me to the field hospital. The very first thing the Americans did was weigh me. When the scale pointed to "84" I said, half aloud, "That's not so bad." The nurse who was weighing me immediately saw my mistake.

"This scale is in pounds. You weight eighty-four pounds, only thirty-eight kilos!"

I passed a mirror and stopped to look at myself. The face that looked back at me was a total stranger. An eternity stood between me and myself. I saw an exhausted scarecrow, an emaciated face of skin on bone with large, haggard, dead eyes sitting deep in their sockets. I hadn't realized that I looked just like all the others!

After my checkup Major Vollmer sat down and talked with me. He said that I could go home with the next transport. I admitted to him that I was afraid to go home. He seemed to understand my feelings. I was very grateful and relieved. The ambulance returned me to the hospital barrack.

A group of inmates I knew from Gusen II were sitting on my bed waiting for me. One of the younger men gave me some very painful news about a comrade who had been one of our committee's chosen ones.

"Père Jacques is in the hospital in Linz dying of bronchopneumonia. Nothing can be done for him. He received the last rites while I was there. He asked me to tell everyone not to be sad when he dies because he is dying in happiness knowing that God let him live to see evil overcome."

If there are such creatures as saints, Père Jacques was certainly one of them. Père Jacques was our affectionate name for Jacques Brunel, a barefooted Carmelite who had headed a monastery in Avon, near Fontainebleau. He hid the Jewish children whose parents had been deported, gradually finding safe homes for them with neighboring farmers. Someone eventually denounced him to the Germans. His monastery was going to be raided. Instead of saving his own skin, Père Jacques had worked throughout the night to remove all of the children then in hiding with him to safety. When the Germans arrived and found no one, they arrested him. The Gestapo beat him savagely and deported him first to Mauthausen and then to Gusen II.

He was the most charming and devotedly unselfish man I have ever known. Whenever our committee voted to give him surplus soup, Père Jacques always gave it to a growing young boy. One day during the bitterest part of the winter I found a sweater, which I gave to this gentle priest to wear. The next day I saw him minus the pullover. I should have known better than to ask him where it was!

"I gave it to a young fellow who works outdoors so that he doesn't catch cold," he said.

It was this same young man who now gave me the bitter news. Père Jacques died that next morning. I don't cry easily, even though I am always deeply affected by sadness and tragedy, but this time tears streamed down my cheeks. The young man accompanied Père Jacque's coffin back to Avon.

Armand's kidney infection showed no sign of change, and he became increasingly restless. He didn't want to wait any longer to be repatriated. Then he found out that General de Lattre de Tassigny had been searching for him for weeks. The general hadn't found him because Armand had been registered under his Resistance cover name. The very next day the general's private plane arrived to take him home. Before Armand left, I made him repeat an earlier promise to telephone my mother in Villon as soon as he landed in France.

As my strength slowly returned I began to assist Major Vollmer. Henri was now in full charge of the women's infirmary. One day toward the end of May he finally received word from his wife and children. It was time for him to go home. Yet close to one hundred French-speaking women inmates still remained in the infirmary. Since I would be the only French-speaking physician left, Henri thought I might want to take his place. I agreed, since I had not yet gotten any news from Villon.

Henri left for France. I was moved outside the camp gates and given a private room in the old SS quarters. I found the women's infirmary only half full. To avoid constant traveling back and forth between the men's infirmary inside the camp and the women's infirmary outside, I got Major Vollmer's approval to transfer the French-speaking men to this external hospital barrack. I was the inmates' chief physician again.

An American nurse worked with me in the infirmary. Her name was Rosalie, and she was in her late twenties or early thirties, blond, with large, expressively intelligent blue eyes. Rosalie was warm and motherly, and always in good spirits. We did not speak each other's language, yet we understood each other somehow and worked together very successfully.

Occasionally Rosalie invited me to join her on a ride into Linz, and we would have coffee together at the PX snack bar. Frequent socializing with no common language for conversation can become awkward and unappealing, but it wasn't so with Rosalie. Although we each picked up only a smattering of the other's language, she had an unusual gift for getting across what she wanted to say and sensing what I was attempting to communicate. Words seemed only an accessory for her. She had more basic resources to draw on. That quality, I think, is what made her the gifted nurse she was.

One day over coffee, about three weeks after I had assumed control of the infirmary, Rosalie handed me a small box wrapped in paper. She made me understand that I was not to open it until I returned to my room. On our way back to Mauthausen Rosalie pulled the jeep off the road and stopped at a quiet spot overlooking the Danube. She began speaking to me. I heard loneliness in her voice. Without conscious decision I pulled her close to me and tenderly kissed her eyes. As she straightened up her lips reached mine and we kissed each other. We sat for a long time, secure in each other's arms, watching the flowing rhythm of the river. I felt a peace of mind that I had not experienced in a very, very long time.

When we returned to Mauthausen I went to my room and listened to her jeep drive away before sitting down to open the little package. Inside I found a watch that she had bought for me. Two notes, which she had had translated into French, lay under the watch. The first note read: "To remind you that there is a future." The second: "What kind of doctor are you, with no watch to take a pulse!"

It was an emotional day for me, the first time that I felt it might be possible for me to become a whole human being again.

The next day provided me with a very mixed surprise. A French repatriation commission arrived in the late afternoon in an impressive convoy of ten cars. I watched from a distance as a tall man in a general's uniform stepped out of the lead car and straightened up. As he walked toward me I suddenly realized that he was Paul or, rather, General Saint-Gast. I was delighted to see him. We embraced each other.

He looked at me and laughed. "Do you believe me now?"

"I have no choice," I joked. "You are a general, and I am only a lieutenant!"

Saint-Gast told me that Armand Petchot-Bacqué had kept his promise. He had called my mother from the Paris airport and informed her that I was fine and well on the road to recovery. My mother had asked him to let me know that my two and a half-year-old son was a beautiful and healthy boy.

I was afraid to ask Saint-Gast if he had any other news for me, but he realized my unspoken anxieties. He put his arm around my shoulder as he told me, "Sonja's name has not yet appeared on the Red Cross list of survivors. I posted the photo from her military file at the Hôtel Lutetia in Paris, which is our official receiving center for all returning French survivors. Everyone returning from the camps has to stop there. I included instructions for anyone who has seen her to report directly to me. So far, only two people who knew Sonja at Auschwitz have contacted me. The last time either of them saw her was in January, when the camp was being evacuated. One of them believed she had been shot for falling behind the marching column."

The sun went out for me. I repeated over and over, "It can't be, it can't be, it can't be!"

"The report could be wrong," Saint-Gast said. "I'm leaving her picture up at the Lutetia until everyone has returned."

He offered to take me home with him. His plane was waiting at the Orching Airfield. He could easily cut through all the red tape and I could be back in France the next day.

"It's time for you to go back home, Albert. Your son needs you."

I protested that my son was in excellent hands. Here at Mauthausen were still some one hundred French-speaking people who really did need me. Saint-Gast did not press me further. I think he realized that I was not yet able to face the reality of losing Sonja. As long as I remained in this environment I could live in the past and hold on to all of the fantastic hopes and dreams that had sustained me since my arrest.

Back in my room I threw myself on my bed to cry. Tears refused to come. I wanted to sleep in order to forget, but my eyes would not close. I absolutely refused to accept the reality of Sonja's death, never hearing her laugh again, never seeing that playful sparkle in her eyes. Yet I was helpless to erase the bloody picture of her lying in the snow, dead eyes fixed on the sky.

I walked over to the American quarters and called Rosalie. She sensed that something terrible had happened, and she stayed with me. I unburdened my feelings to her. It took me a full hour to talk myself out. I went back to my room and took a sleeping pill and fell asleep at last. I woke up with a start in the middle of the night wondering why Saint-Gast had not said a word about Sonja's sister, Paule, who had been at Auschwitz with her. Sonja was fiercely protective of her. Wherever Paule was, Sonja should be. I couldn't fall asleep again.

In the morning Rosalie drove me to Linz so that I could see Saint-Gast before he left for France. I almost broke down his door in my eagerness to ask, "Is there any news of my sister-in-law, Paule Nadel?" Saint-Gast shook his head but promised to inquire again as soon as he returned to Paris.

My desire to keep alive at all costs, which had been my mainstay for the past year and a half, was deserting me. My face must have shown my devastation. Saint-Gast, obviously very upset, tried to rally me.

"Don't break down now, Albert, after all you have survived. Don't commit some stupidity. You have to promise me that I will be able to pin your decoration on your chest. See you at the ceremony!"

Saint-Gast's words called up painful memories. Sonja and I used to discuss our Resistance activities and kiss each other and bet which of us would receive the most distinguished decorations after the war. I always told her that I hoped the man who pinned a medal on her chest would be a doddering old legionnaire who smelled of chewing tobacco and alcohol. Who would tease me about my decoration now? Who would I show off for? Who could I share my pleasure and excitement with? I was more depressed than ever.

Back at Mauthausen I secluded myself in my room. Someone knocked on my door. I called out, "*Entrez!*" It was Rosalie, wanting to know if I was okay. She came over to the bed and placed her hand on my forehead, ran out of my room, and returned with Major Vollmer. I was running a fever again. My body was shaking from the chills. Major Vollmer ordered Rosalie to undress me and put me to bed. He turned me onto my belly and gave me a shot that put me to sleep immediately.

It was dark when I awoke. My chills were gone, and on the whole I felt much better. I reached out to turn on the light and touched a woman's hand. Rosalie had stayed with me. I asked her what this sudden attack had been. She shrugged her shoulders to signify, "I don't know." She bent over and kissed me before she left.

The next morning I was much improved both in body and spirit. By afternoon my temperature was normal and I was able to resume the rounds of my patients. The women who still remained in the infirmary were primarily eastern Europeans, a sad group of people. Most of them did not want to return to their own countries because their families had been wiped out before their eyes. In the camps they had watched, in anguished helplessness, the deaths of their parents, their husbands, their children, their brothers, and their sisters. Now they were waiting while the Red Cross located relatives who had previously migrated to western countries.

The men's ward today, in contrast, was a place of great joy. Many of my patients would be leaving for home with a Red Cross transport. I had to decide this morning who could return directly home and who had first to spend two recuperative weeks in Switzerland. I took my final lists to Major Vollmer, then went with him for my first official meeting with the American camp commander, Captain Levy. Captain Levy never looked as if he had the ability to take charge, but he well deserved his reputation as a superb organizer.

Captain Levy spoke Yiddish, so we managed without a translator. He had heard about Sonja and offered his help in finding out what had happened to her. A Russian repatriation commission was arriving at Mauthausen the next day, and he would see what he could learn from them. Hopefully they would have a list of the numbers tattooed on the corpses they had found on the road along the path of the Auschwitz evacuation, and he could see if Sonja's and Paule's numbers were among them. I thanked Captain Levy for his help and his concern.

I had recovered from my initial overwhelming shock because I was convinced that it was totally out of character for Sonja to give up and be shot. She had been an unusually resourceful fighter all her life. Whenever I felt that we had come to a dead end during our Resistance work, she had always found a way out. She had never lost her head during our most dangerous missions. Now I was counting on her quick mind and formidable command of languages to have kept her alive.

I spent the afternoon preparing the medical papers for the male inmates who would be leaving that day. I also wrote a letter to my mother. I had to explain to her and François why I had not jumped at my first opportunity to be repatriated. Because my mother was suffering from an incurable cancer, I did not want to burden her with the truth of Sonja's probable fate and how frightened I felt of the future without her. I put together the excuse that I was not yet fully recovered from my illnesses and the Americans here were able to give me much better medical care than I could get at home. I added that I was still needed by the group of French prisoners who could not yet go home.

Major Vollmer visited me the next morning. He hoped I could facilitate the repatriation process for the Russian nationals and Polish Jews awaiting the Russian commission's arrival. I was delighted to help out because I could justify staying on at Mauthausen still longer.

That afternoon I went inside the camp to meet with this group of prisoners. The camp was completely empty. Major Vollmer explained that these inmates were in the "lower camp." This was the first time I had heard of a lower camp. He told me that the SS had put it together at the last minute, shoving all the eastern European Jews and Russians there to make more room in the main camp. The Russian authorities had demanded that the lower camp remain intact, prisoners and all, until their commission arrived at Mauthausen. They were bringing newspaper photographers to document conditions there and show the world in black and white what these people had suffered.

I took two of my patients, Poles who had settled in France, along with me as interpreters. The conditions I found in the two barracks of the lower camp were even worse than Gusen II had been.

We went first to the barrack of Polish Jews. We counted twelve hundred inmates in the one barrack. We began to make up a list of basic information for each inmate. Two things quickly became clear: one, it would take us a full day just to complete our simple list; two, not one of these people wished to return to Poland. Every single one wanted to emigrate to Palestine.

I was deeply moved by their unshakable faith in the basic humanity of mankind. They were convinced that the world would unite in support of the Jewish people because of what had happened to them, and unconditionally sanction the existence of a Jewish state.

It was dark when I returned to my room outside the camp gate. The American quarters were brightly illuminated. A large number of jeeps and trucks were parked outside. I heard music, which grew louder as I approached. I knocked on the door. No one seemed to hear me above the noise, so I entered timidly. Inside I saw Major Vollmer with two strange American officers and a group of men and women in Russian uniforms.

Major Vollmer greeted me and introduced me around. I drew him aside to talk to him about the situation with the Polish Jews in the lower camp. He listened carefully and said that he would have to discuss it with the liaison officer working with the Russian commission. He pointed out a tall, blond, blue-eyed Texan sitting near him.

"My God," I said to myself. "The Aryan prototype from *Mein Kampf.* He probably never saw a Jew in his whole life! How can I expect sympathy from such a man for a group of poor Jews from Poland?"

I excused myself and left the party. I returned to my room steeped in discouragement and frustration. Despair crept into my bones. I felt powerless. God, how I missed Sonja's presence and her shrewd sense of what to do! Before me were the hopeful eyes of these twelve hundred people. Never in my life had I felt so Jewish and so alone.

I finally dozed. A tapping on my window aroused me. It had to be Rosalie, but why the window instead of the door? I jumped up to open it. The blond Texan was standing outside, smiling at me, holding a bottle of whiskey in his hand. In very rusty college French he said, "I am coming in."

Before I could open my mouth he had put one long leg through the window and sat straddling the windowsill like a cowboy on his horse. The only item missing was a cowboy hat. After several attempts he succeeded in pulling his other leg inside. It was very hard to keep from laughing.

The Texan sat down on my bed and asked for two glasses. I gave him the only one I owned. He filled it to the brim before speaking again. "Major Vollmer told me about the Polish Jews. Why are you so concerned about them?"

"I was in the camp too, and I suffered through the same miseries."

As soon as the words were out I was filled with shame at my fear to admit to this man that I was Jewish, and that because of this I cared deeply for those Jewish people from Poland. The Texan threw me for a loop with his next question.

"*Bist du ein Yid?*" ("Are you a Jew?")

I could not believe my ears. I stared at him, speechless. After so many years of having to hide my Jewishness to protect my life, it had become second nature to deny it. I didn't know what to say. I continued to stare at this tall blond man. He seemed to grow irritated with me.

"*Bist du, oder bist du nicht?*" ("Are you, or aren't you?")

"Yes!"

"*Redst der mamalusche?*" ("Do you speak the mother tongue?")

"No!"

"What kind of a Yid are you, not speaking Yiddish?"

I explained that where I had grown up, the assimilated Jews did not speak Yiddish. We learned German. He laughed at the irony.

"And how come a gentile from Texas speaks Yiddish?" I asked him.

He laughed again. "Because my parents emigrated to Texas from a Jewish *shtetl* in Russia!"

So we sat there, we two Jews, a Hungarian Jew from France and a Russian Jew from America, and finally discussed the problems of the twelve hundred Polish Jews who were still imprisoned in the lower camp of Mauthausen.

The situation looked hopelessly complicated. The Russian authorities insisted on the return of all Russians and all people from the newly Russian-occupied territories. The British authorities were so sensitive to hostile Arab reaction that they had refused to open Palestine to all but a tiny handful of Jewish survivors. We had some serious thinking to do.

In the meantime my new friend talked to me about himself, describing his life before the war and his experiences in the army. He had been with the American group that liberated Dachau. Everything he saw had convinced him that Nazi Germany had to be understood as a criminal conspiracy that would not end with this war. I agreed with him. We finally returned to our immediate problem of the Polish Jews in the lower camp. He said he would meet with their leaders in my infirmary in the morning.

It was 3:00 A.M. by now. The prospect of spending the rest of the night alone made me terribly anxious. I had developed a horror of being alone since the frightening news about Sonja. The Texan seemed glad of my company, too. He poured another drink, and we went on talking. The time flew by, and the next time I looked at my watch—the watch Rosalie had given me—it was 7:00 A.M. It was time for me to go down to the lower camp and arrange the meeting.

I spoke with the three young men in the barrack who appeared to be the leaders of their group. I say "young" only because of their voices. Their faces belonged to broken old men. I told them that an American officer wished to meet with them. The meeting would have to take place secretly in my infirmary. They could come back with me, ostensibly for medical treatment.

The meeting took place shortly afterward. I had very little idea of what had occurred until it was all over, because most of the discussion was in Yiddish. Later the Texan explained to me that he had found a way out, after all. If they signed affidavits stating that they did not want to be repatriated to any country under Russian domination, they could be transferred to a U.N. camp to wait until they found a safe way to enter Palestine. But there was still an obstacle. We needed more time. We must find some way to postpone the start of the Russian repatriation.

I returned to the lower camp, anxious to discuss this problem with the Polish spokesmen without delay. First I inquired about their arrangements for medical care for the sick inmates among them. To my knowledge, nothing had ever been done for them. One of the young men pointed to a large section of the huge barrack where bedridden people were being cared for by several Jewish prisoner-physicians. I felt terrible that I had been unaware of their existence all this time. I introduced myself and asked if there was anything I could do for them.

One of the physicians answered with quiet, cold hostility, "We need nothing. We have already been well taken care of by the Americans!"

What I heard was "Where were you when we really needed you, you traitor, you hypocrite?" His misconstrued hostility hit me like a blow. The young man with whom I had been speaking overheard our interchange and saw my reaction. He interrupted to explain at length to this physician the reality of the situation.

The physician joined our discussion as we tackled the problem of inventing a pretext for delaying their repatriation to Russian-occupied Poland. He disclosed that he alone still wanted to return to Poland. He had hidden his wife and son there before he was deported, and he had to find out if they were still alive. He also mentioned a small group of twenty or so people in the Russian barrack—Russian nationals and a few Jews carefully disguised as gentiles—who passionately wanted to remain in the West. One of the young men went to get them. We decided to blend all of them in with the Polish Jews.

We continued to search for our foolproof delaying tactic. The obvious suddenly dawned on me. We could tell the Russian committee that typhus was spreading among these inmates because of the filthy conditions. Many were already bedridden, and the whole lower camp had to be quarantined indefinitely. My Texan friend got Major Vollmer's consent to our plan.

Once the Russians heard "typhus," they quarantined the camp before we even had time to suggest it! Then all that remained to do was disinfect the main camp and move everyone up there out of the incredibly filthy conditions of the lower camp.

I held a strategy meeting with Major Vollmer on the logistics of the transfer. It was essential to keep the two barracks well separated. We wished to prevent anti-Semitic incidents as well as vengeance toward the Russians who had joined the Polish Jews. He suggested that I discuss it directly with Captain Levy, since he was in charge of running the camp. I met with Captain Levy, and once everything was worked out he gave the order to evacuate the lower camp and burn it to the ground. Every last trace of this terrible camp was destroyed, but the fire couldn't touch the bitter memories that the inmates would forever carry within them.

My Texan friend was thrilled with the outcome of our plans. Before he left he made me promise that I would visit him in Dallas if I ever came to the United States, warning me, with a smile, that I had better learn English or Yiddish first. I invited him to look me up in Paris so that I could teach him to speak a decent French. He left Mauthausen with a big smile on his face and a bottle of Russian vodka under his arm.

My days now passed in a quiet routine interrupted only by jeep trips to Linz with Rosalie. Despite my healthy diet I was still painfully thin and terribly depressed. I wasn't gaining weight as I should have been. Rosalie was concerned for my emotional and physical recuperation. She worried about me constantly.

I watched romances blossom between some of our younger female patients and the American soldiers. It was touching to see their tender care for these women, who were still marked by the ravages of their suffering and the ugly short hair that was just beginning to grow in on their camp-shaved heads.

The refreshing normalcy of these relationships was a reassuring contrast to the perverse romances that had flourished between some of the brutal green triangles and enforced prostitutes in the camp brothel. Such ogres had actually found "sweethearts" among these prostitutes. Men who took pleasure in killing their fellow inmates suddenly showed the tenderness and concern of young men in love. Even their liberation fantasies revolved around a respectable future with their sweetheart. They were convinced that a miracle would prevent her otherwise certain death in the camp. But I will never understand how any man involved in the camps' endless struggle against degradation, terror and death could have had any physical or emotional reserves for either sex or romantic love. These callous murderers were obviously not affected by those devastating conditions in the way that the rest of us were.

My nights were now very difficult. Unless I took sleeping pills, the little sleep I did have was filled with fearful dreams. While a prisoner, I had slept like a baby. I could have slept for days if I had the opportunity. Since the Liberation I needed barbiturates to sleep at all.

Finally the U.N. commission arrived to begin the transfer of the Polish Jews and the Russians hidden in their midst. They were evacuated on the same day that Nazi Germany formally surrendered. It was a day of satisfaction for me. I wondered if peace was finally to come to God's chosen people.

Rosalie came to me several days later with exciting news that a French repatriation commission had just arrived and they carried two letters for me. I flew to get them! One was from my mother, the other from Saint-Gast.

My mother's letter detailed the events of the war years, ending with the comforting news that everyone was well. She included a snapshot of my son. François, with his large, clear, intelligent eyes, was beautiful.

Saint-Gast's letter said: "I am sending you a letter I just received about Sonja. Don't build up your hopes too much, but don't forget that miracles have been happening."

Enclosed was a letter from a Mr. Mandel, who had seen the photo of Sonja posted by Saint-Gast. He had known Sonja and Paule back in Antwerp and recognized both of them during the evacuation of Auschwitz. They had disappeared from the column as it passed through Pless near the Czechoslovakian border. He felt sure they had managed to escape.

I was afraid to raise my hopes too high, but they soared anyway. I ran to see Rosalie and jabbered away. Impulsively I hugged her and tried—unsuccessfully, with my weak arms—to lift her off the ground. I began to sing. My exuberance was contagious. Her eyes began to sparkle with her joy for me.

I made such a racket that Major Vollmer came out to see what was going on. He was delighted to hear the good news about Sonja, and invited me to come with Rosalie to the American celebration of the official German surrender. It would be my celebration, too. I eagerly accepted. Ten minutes later I was in a jeep heading toward Linz with Rosalie, Major Vollmer, and a Lieutenant Helen Baker.

General Reynolds was hosting the celebration party on the yacht that had once belonged to Rear Admiral Horthy, the former Hungarian regent. I was very excited about going aboard his former yacht. The contradictions of my country of birth had always intrigued me—a landlocked country that called itself a kingdom, but was ruled by a regent with the rank of rear admiral. The royal yacht had been his entire fleet! One of my most cherished childhood memories was of a victory parade in Budapest with Horthy on a snow-white horse, looking to me, at eight years old, like a knight in shining armor.

General Reynolds had been detained by an urgent matter. His adjutant greeted us and introduced us to the delegations representing the Allied nations. The meal was served from a buffet table by American sailors in spotless white jackets. After my camp deprivation it was still astonishing to see my plate being so amply filled. Two kinds of meat, mashed potatoes, corn, and pineapple were added in turn, carefully separated one from the other. Then the last sailor spooned gravy over the entire plate. It did not go with the pineapple!

When we finished our soggy meal Rosalie asked me to dance. At my best I had never even been mediocre on the dance floor, and Rosalie's attempts to teach me a new dance she called the boogie-woogie were doomed. When the orchestra finally played a tango at least I knew the basic steps. Even at that, I must have murdered Rosalie's feet. Perhaps it was selfish of me not to give her poor feet a rest, but I couldn't deny myself the heavenly pleasures of holding her warm, very real body in my arms. I felt myself so responsive, and finally so alive.

I tired very quickly, though. I located a dark corner where I could sit unnoticed, and made Rosalie understand that I didn't want to interfere with her pleasure. She ran off, apparently offended. I buried my head in my hands and wanted to cry from the tormenting loneliness of Sonja's absence.

The pressure of a hand on my head brought my eyes up. Rosalie began talking to me. I recognized the word "jeep." I hadn't offended her. She realized how down I was, and she had gone to ask Major Vollmer for permission to drive me back to Mauthausen.

We left quietly. In the jeep I rested my head on her shoulder and must have fallen asleep that way, because we were pulling up in front of my room when I opened my eyes. Once inside I undressed, got into bed, turned off the lights and stared at the ceiling, trying to fall asleep for once without taking a pill. I put the little snapshot of François on my pillow, imagining that he was really there in bed with me.

My ear picked up footsteps outside my door. Someone knocked. I called out, "*Entrez.*" Rosalie stepped inside. Her arm moved through the dimness toward the light switch. I said, "*Non!*" I heard her moving around, then her unclothed silhouette approached the bed. She lifted the blanket and lay down beside me. Her body was warm. I didn't dare break the silence and chase this moment of peace. Ever so gently I shifted Rosalie's head to rest on my chest. Then for the first time since my Liberation I slept through the night without pills.

I woke up with the sense that I was being stared at. It was Rosalie, already up and dressed, smiling at me. She kissed me before she left. I felt that she would be offended or feel cheated because we hadn't made love, but once again I had underestimated her. Rosalie slept innocently by my side every night for the several weeks until I left. I had no more need of sleeping pills.

28

The last transport of French-speaking inmates left the camp on June 28. There had been no more news about Sonja, but I had no further pretext for remaining at Mauthausen. Major Vollmer had arranged for me to be on the plane carrying inmates back to France on the last day of each month. I was to leave for the airfield at Orching the next day.

That evening the American medical staff gave me a farewell party. I drank to bolster my courage and quiet my fears of returning to a normal life. I hazily remembered being carried back to my room, and Rosalie undressing me and holding me in her arms. I fell asleep that way and spent a night of pleasant dreams.

She was still sleeping when I woke up the next morning. I tried to fix her features in my mind. I was filled with the sadness of leaving this woman who had so generously helped me adjust to the world of healthy, normal, living people. For an eternity I had been surrounded by the dead, the dying, the emaciated and disease-ridden. That hell had become my norm. Rosalie could not offer me words of solace, but she gave me a primordial emotional and physical closeness without demands or pretensions. She had nurtured me back to life.

Rosalie woke up. Tears filled her eyes as she looked at me. She threw her arms around my neck and called me a "baby," one of the few English words I had finally learned. She helped pack my few belongings in luggage that the Germans had left behind and wrote down her address in the States before we went to breakfast. My emotions canceled my appetite.

Alone, I made a last tour of the deserted infirmary, silent now except for my echoing footsteps. I could almost hear again the last sighs of those who had died here after their Liberation. It was a stark, powerful moment that has lived through time. Whenever my footsteps echo in an empty room, I suddenly feel myself back in that deserted, sighing infirmary.

I wanted to make one final visit to the camp of Mauthausen before I left for home. Arm in arm, Rosalie and I walked to the Assembly Place where I had first arrived, the smaller one where I had unloaded the corpses and broken my ankle, the crematorium, the Bunkers, the wooden platform where inmates knelt to be shot through the neck. I showed Rosalie the meat hooks and other instruments of torture. She became so pale that I grew frightened. We passed through the "shower room," the gas chamber with the fake shower heads. I refrained from explaining what it was. I didn't want to upset her further.

We crossed the deserted camp to the rock quarry, where more than eight thousand inmates had died just to build the 160 steps up from the bottom. Thousands more had died working the quarry, and untold numbers were executed at its rim. The gentle villagers of Mauthausen had always been a very responsive audience, especially when Jews were murdered. Would this spot become a place of pilgrimage, another Via Dolorosa?

When we passed back through the gates my luggage was on the jeep. Major Vollmer and a male nurse were waiting to wish me good-bye. Major Vollmer handed me a formal letter of recommendation that he said would help me if I ever wanted to emigrate to the United States. They presented me with a little package of goodies, and then Rosalie and I left for Orching.

In the hospital tent at the airport, Rosalie made sure that my cot was in a location that would afford me some privacy and instructed the recreational officer to take special care of me. We sat over our last cup of coffee together, drawing it out as if that could postpone our inevitable separation.

A group of men in French military uniforms were passing outside the window. I called out a hurried "*Pardon*" to Rosalie and ran out to see who they were. They had come, under the command of Dr. Falaise, to repatriate the remaining French inmates at Mauthausen. I told them that the last group had already departed, leaving only myself.

My eyes began to play tricks on me. Two of these men reminded me so strongly of Michel Lubchansky and François Hauser, my companions from Sathonay. But they were staring at me as if I were a ghost. My mouth dropped open. I was, indeed, looking at my two friends in the flesh! We embraced as if we had bumped into one another on the Champs-Elysées after dinner together only the evening before.

Michel, auburn-haired and lightly freckled, told me about the crazy French physician all the other repatriation commissions had joked about, the guy who absolutely refused to be repatriated until the very last Frenchman left Mauthausen.

"If I hadn't heard that you were dead, I'd have known immediately it was you!"

I introduced Rosalie, who had joined us by then, to my good friends. François was a tiny fellow with eyes as blue as Rosalie's and warmth and compassion to match. He delightedly shook her hand. Michel was far more subdued, still as painfully shy as ever. Then we all went to meet Dr. Falaise, a lean man in his mid-forties, and enjoyed a good laugh at the ridiculous aspect of an entire commission repatriating one single deportee!

Instead of returning to France alone by plane, my compatriots were determined that I drive back to Paris in their company. I was glad to go with them by car. The trip would take at least a week because of all the slow-moving army traffic on the roads. That meant a week of no opportunities to receive bad news about Sonja, a whole week's more time for good news to arrive in Paris and welcome me home.

Dr. Falaise insisted that I spend the night at their hotel in Linz. We all left Orching. Rosalie and I were silent during the jeep ride. At Linz, Falaise disappeared into the hotel to reserve my room. He returned to the jeep and told me, discreetly, that he had prevailed upon the quartermaster to give me a room to sleep two. I asked Rosalie, through an English-speaking nurse in the French commission, to remain with me that last night. Her voice held tears when she said she couldn't. She asked the nurse to tell me that her duties made it impossible. And she added that I was among my own people now and didn't need her anymore. Rosalie kissed me tenderly on my cheek and held me for one last moment with her eyes. Then she was gone.

In bed that night I could not fall asleep. It was too soft, too empty. I didn't have my sleeping pills. After endless tossing I dragged my sheet and pillow to the floor. In the dark I stared at the ceiling for a long and anxious time, and finally fell asleep.

The next day my friends put together an army uniform for me to wear home. Michel found a spencer, the short American army jacket, and contributed a pair of his olive-green military slacks. He also loaned me his insignia, which showed my lieutenant's rank and the physician's caduceus. My military shirt came from François Hauser and my well-fitting shoes from Falaise.

I had agreed, at Falaise's request, to give a formal talk about my experiences to the full membership of the French repatriation commission, who were all still in Linz. I described in simple words what had happened to me from the time of my arrest in Nice. The faces of some of my listeners began to express obvious disbelief, as if I were inventing incredible horror stories in a pitiful bid for their attention and sympathy. I felt increasingly uncomfortable and angry. I spoke of the Auschwitz procedure of tattooing our registration numbers, branding us like so many cattle. I showed my number to the audience. A woman as much as called me a liar, remarking to my face that I must once have been a sailor and tattooed my girl friend's telephone number on my arm. I was seething with anger by the time I left the meeting.

It had not occurred to me that most people as yet knew nothing of what the Germans had perpetrated in their concentration camps. Many simply found these horrors too incredible to believe on first hearing. Others would steadfastly continue to deny the truth.

29

July 1, 1945. I was going home at last. We would have to stop for the night at several points en route: Munich, Augsburg, and Stuttgart in Germany; Strasbourg, Nancy, Châlons-sur-Marne in France. One moment I felt impatient for the trip to be over, the next I was praying for it never to end. At each stop I was that much closer to home. Every night I woke up to the sound of my own screaming. My nightmares had returned. One in particular recurred every night. (This nightmare disturbs my sleep even now, usually after a particularly stressful day, and always for several nights around the anniversary of my Liberation.)

A faceless SS commandant at Gusen II interrogated me, tortured me and finally hanged me by my fingers. I would wake up screaming from the pain, drenched in perspiration, as if the experience were real. The nightmare always resumed as soon as I fell asleep again. The SS soldiers were chasing me down a camp street. The distance between us shrank. My legs became heavier and heavier. A mountain of corpses appeared in front of me. I began to circle it. The SS wouldn't find me because I could always stay on the side opposite them. Suddenly I saw myself at the top of this mutilated human mountain. Why was my corpse up there while I was still running down here? The SS soldiers were almost on top of me. A gun was sticking out between two dead bodies. I pulled it out and shot the nearest SS soldier shouting, "I killed! I killed!"

Then I would wake up.

No single rooms were available at the hotel the night we arrived in Munich. Michel and I shared one. When I woke up screaming from my nightmare, he was sitting by my bed. He thought I should talk for a while instead of going right back to sleep. We discussed my dream. I feared it meant I had lost my sanity in the camps. Michel was very reassuring. He described my nightmare as a safety valve for discharging my anxieties. I was going to be all right.

Michel prompted me to talk more about what had happened to me, but as I described my escape from Dachau something in his behavior gave me the distinct impression that he too didn't quite believe what I was telling him. He suggested that, since we were in Munich and so near Dachau, it would be good for me to revisit the places I had described to him. It would help me to exorcise my terrors and leave them behind. We could borrow the car and leave right after breakfast.

The next morning we went out for a walk before getting started. The square on which our hotel was situated was bustling with market-day activity. It had a strange hint of the familiar. Michel and I walked around the square. Just past the first corner I suddenly recognized the church where I had hidden during my escape. The Frauenkirche emerged clearly from the anonymous facade of buildings.

We went in. The priest was a stranger to me. I introduced myself to him, and he readily agreed to take us downstairs.

Nothing in the cellar had changed. The table and chairs were in their remembered places. The can that had been our emergency toilet was still in the catacomb. I wanted to know what had happened to the Polish priest who had been in the church during the winter of '43–44. The current priest could tell me nothing. He had only been in the Frauenkirche for several months. So many priests had come and gone in the past year and a half. One of them was rumored to have been arrested for hiding refugees in the church cellar, but his fate was unknown. I inquired after the nun who had helped take care of us. He said no nun had been regularly associated with the church for a long time.

I could find no record of either of my benefactors. Either the clergy had permitted the Underground to take over the church, or the church itself had organized an Underground organization that had become infiltrated by the Gestapo. I never learned the truth.

Michel was going to drive me to Dachau so that I could revisit the camp and the other neighboring places of my nightmare. On our way back to our hotel, I stopped a middle-aged businessman and asked for directions to the Dachau concentration camp.

"Never heard of it!"

I asked how long he had lived in Munich.

"All my life. But I assure you I was always against *Der Führer* . . . er, ah, Hitler!"

It was to become the typical German response to any question related to the Nazi atrocities.

Outside in the sunny marketplace as we drove off, the hum of activity and the lively faces of the buyers and sellers betrayed no sign of the costly war their country had just lost. The innocent and timeless quality that touches all marketplaces belied the Bavaria that gave birth to Hitler and his barbaric designs. How guiltless his "former" supporters looked!

Falaise and François Hauser made the drive to Dachau with us. The widespread devastation from Allied bombs that I saw on all sides gave me a sense of angry satisfaction. The factory in the village of Dachau lay in ruins. I retraced our route, from the bus station to the ditch where we hid during the bombing of Munich, back to the garbage dump. It all seemed so long ago. The American soldier at the gate looked at our papers and called over a French liaison officer. Falaise explained to him that I had been a resident here and wished to see the place once more.

Nothing inside had changed. The gallows stood to the right of the entrance, empty noose swinging lazily in the breeze. The pole from which I had hung on exhibit still stood in the Assembly Place. Our next stop was the quarantine barrack and the courtyard of the rabbit hunts.

The French officer guiding us began to recite the story of a nameless Frenchman and a Polish officer who had been recaptured after hiding in the garbage ditch and escaping to Munich. How bizarre, I thought, to hear a total stranger recount to me the story of my escape with Kalka! Our guide said that the Polish officer had died under torture and that the Frenchman had been transported to another camp. The stunned look on Michel's face pained me. It was very obvious that he hadn't believed a word of my story the night before.

The French officer pointed to the familiar garbage containers standing in a corner of the courtyard. What tender feelings and memories they evoked of my enigmatic Polish companion and our brief interlude as relatively free men! I walked over to the corner and caressed one of them nostalgically, vaguely aware of dialogue in the background. The officer's sudden exclamation—"No! How incredible!"—drew my attention to him. He ran to me and asked to shake my hand. Clearly he thought of me as a hero. What different views we had of what constituted heroism in the camps.

From the courtyard we went to the interrogation Bunker. The bucket was still in its corner, but the door to my cubicle had been ripped out. We passed the official torture chamber, which was identical to those in all the other camps, as if they had been mail-ordered from the same catalog.

The flower gardens that the SS had assiduously cultivated, killing any inmate who accidentally stepped on a flower, lent a morbid note of color to my visit. I left Dachau hoping I would never have to return. (I was back within a year to testify in the secondary war criminal trials. Dachau was filled with people and activity. The flower gardens still flourished, and garlands decorated the crematorium. I was calm when I first arrived, but the remorseless, even bragging, testimony of the war criminals rekindled my rage so unbearably that I left in the middle of the proceedings and returned to Paris.)

We drove back through Munich on our way to Augsburg, where we planned to spend the night. At the outskirts to the city I recognized the roadside tavern, now closed and shuttered. A neighbor told me that the original owner had died fighting in Stalingrad early in 1943, months before my visit, and that the next owner had suddenly disappeared before the Nazi defeat. I wondered if he had been a stool pigeon for the Gestapo. The neighbor knew nothing.

We drove into the city and found hotel rooms. Michel and I walked around until I found the house in which my escape had ended. I knocked. No one answered. I tried several neighboring doors before finding someone home, a woman who appeared somewhat intimidated by my army uniform. All she knew was that the house, now empty, had been taken over by the police after they arrested the owners for black marketeering. A couple who worked with the police had moved in as ostensible janitors. They, too, had disappeared. Their teen-aged son had belonged to the Hitler Jugend. They weren't natives of Augsburg. She couldn't recall their name.

Michel and I set out to find the police station where I had first been taken. From there we got directions to the former Gestapo headquarters, now the American C.I.D. office. At both places I scrutinized every face I saw and obsessively compared it with my memories.

At the C.I.D. office we were taken to a German-speaking American officer. I explained who I was and what I wanted. The officer brought me to his commander. They wanted me to identify some of the torture instruments and inspect some of the self-styled "innocent" Germans they were holding. They were having a difficult time finding witnesses who had survived interrogation in this chamber of horrors.

We entered the gymnasiumlike room. The ropes still hung from the ceiling. The graduated dumbbells were neatly racked on the far wall. The horse and the electric torture instrument my interrogator had called the "singing machine" stood in a corner. I described its efficiency as an instrument of torture. A stenographer recorded my words as I spoke.

The American commander asked me to look over some of their German prisoners. To my disappointment, none of the scared-looking faces were familiar to me. I turned to go. The face of the man cleaning the hallway floor stopped me dead in my tracks. I pointed to him. The commander identified him as a German war prisoner who was doing janitorial work in the building. I asked for him to be stood under a light so that I could see his face clearly.

The man had taken only a few steps in our direction when I exploded. My rage was savage. I pounded my fists into the face that had so calmly orchestrated the atrocities of my interrogation! He fell to the floor, his nose bloodied, before the Americans pulled me off him. I was so overcome that at first I could not speak. I finally managed to stammer, "He is the one! He is the Gestapo sadist who interrogated me and had me tortured in this very room!"

I yanked up the man's sleeve. The fine scar left from the recent removal of his blood type tattoo was clearly visible. I asked for permission to question my interrogator. I sat him down opposite me. I reminded him of how he had burned my hand with his cigarette. He looked at my scar with the same bland expression that he had worn on his face when he interrogated and tortured me.

"Well, Hauptsturmführer," I asked, "and what do we do now?"

Suddenly his assurance collapsed like a deflated balloon. I think he really believed that I was about to use the "singing machine" on him. He confessed everything. The stenographer typed up his statement, and I added my signature to his. (At this Nazi's trial, I was the only living witness. I testified in detail to my torture, Kalka's death, and the numbers he had bragged about killing. He received a sentence of ten years in prison.)

On my way out of the room I looked at the wall where I had been beaten when I could no longer stand on one leg. Some of the brown spots spattered on its surface might have been my own dried blood.

My restlessness grew during the remaining days of our homeward journey. I was irritable, tired, and morose. I avoided my friends and their conversation as much as I could. Something was wrong with me, both physically and mentally. I strived to diagnose it, and at the same time I kept denying the true cause of these symptoms. I felt that I must not brood on the past and what I had endured. I had been tortured, enslaved, and robbed of my human dignity. But the Nazi terror had been defeated and I had survived—and now I was going to resume my interrupted life. So what was the matter with me?

I did not realize then how much I was in need of the special rehabilitation care which, later, I was to develop for others who had survived the camps. I just went on feeling alienated, apprehensive, and full of pent-up anger that sooner or later would have to break out in some violent action. The one thing that kept me going was the certainty that my wife was still alive and I would soon be rejoined with her.

At long last we arrived in Paris. We stopped on the Boulevard Raspail, near the rue de Sèvres, in front of the Hôtel Lutétia. We were at the official *centre d'accueil*—"welcoming center"—for repatriated French deportees.

I walked up the steps rehearsing my fantasy of finding news waiting for me from Sonja. I was sure she was already back in France, waiting for me at my mother's in Villon. Before I could get to the information desk I had to submit to DDT spray against lice as I had done in every hotel between Mauthausen and Paris, pass a brief medical inspection, and get my room assignment. My repatriation card was stamped "apparently free of disease." I would be given a more thorough checkup in a few days.

Finally I was free to get the letter I knew would have arrived from Sonja. I ran to the information desk. There was no letter for me. The anguish of my disappointment was like physical pain. I looked for "Sonja Haas" on the alphabetically organized board of missing deportees. Photographs were pinned above names. I recognized the face of my old Resistance friend from La Camargue —Lieutenant Konrad, who had visited us in La Napoule. The note beneath his picture stated that he had been deported to Gross Rosen and asked for news from anyone who had seen him. It was painful to see his picture, his face smiling from a happier time.

My eyes found Sonja's face, as beautiful as the last time I had seen her. I tried to picture her dead. I couldn't. My chest tightened. I felt chilly. Looking at her picture I found it impossible to accept that she was no more. Yet two months had passed since the Liberation. It was July 6. Sonja would certainly have given some sign of life if she were still alive.

I walked out into the street with Michel and watched the Parisians going about their business. Despite my suffering and loss, nothing had really changed. Soon everything would be forgotten and reduced to a few sentences in the history of mankind. Michel quietly suggested that we go up to my room. I picked up my military blanket, then we went upstairs and found the room assigned to me. Michel was hopping mad.

"Those crummy, insensitive bastards! They've given you no sheets or pillowcases!"

I hadn't noticed. When I did, I didn't much care. Michel refused to let me stay there. He insisted that his mother would be delighted to have me as a guest in their apartment. I stopped at the desk to pick up my repatriation card and the day's pocket money due me. I left Michel's address and telephone number in case anyone tried to reach me.

Once at Michel's, the first thing I did was call the post office at Villon. My mother had no phone, but her house was right across the street from the post office. I asked the Villon operator to let my mother know that she would receive a call from me at the post office at seven that evening. I spent the afternoon with Michel and his mother, relaxing somewhat in the forgotten warmth of family life.

At 5:00 P.M. I got a telephone call. It was Saint-Gast chiding me for not having notified him as soon as I arrived in Paris. He would have picked me up. I reassured him that I was in the good hands of an old friend. He wanted to know if I needed anything, money included, and said good night after promising to have his driver pick me up in the morning. The telephone operator could not get my call through to Villon that evening. Michel gave me a mild tranquilizer to calm me, but I was still fairly agitated when I went to bed.

Mme. Lubchansky woke me the next morning. Michel had already gone down to his unit, and I had a visitor in the living room. I quickly washed and dressed. I found General Saint-Gast himself waiting to welcome me home in person. He asked how I was. I described the frustration and anxiety of being unable to get my telephone call through to Villon. Saint-Gast immediately called his aide-de-camp, ordering him to call the Villon police station and have my mother and son there for an 11:00 A.M. call from his office. He made it all so simple!

Saint-Gast was newly installed in offices on the Boulevard Suchet, near the Bois de Boulogne. He was now second in command of French Military Intelligence. During the drive over I avoided any talk of serious matters. Once he had shown me around the building and we had settled into chairs around his large desk, he flatly insisted on straight talk about my future.

"You must do as I've done, Albert. You must pick up your life from where it stopped. After you've had your medical checkup, you can take two weeks to rest up and decide whether you prefer to return to civilian life or remain as a physician in military service for an additional year."

I wanted to postpone any major decisions until I was absolutely convinced that Sonja would never return to me. I asked Saint-Gast if he would extend my deadline until the end of the month. He willingly did so.

The telephone interrupted our conversation at exactly eleven o'clock. Saint-Gast handed me the receiver. I said, "Hello." At the other end I heard my mother's voice, soft and quivery with emotion. "Just a moment," she said. "Someone here is dying to talk to you." After a moment a little boy's voice said, "Hello, Papa Bélu! We are waiting for you! And please don't forget to bring me a toy car!"

No one had called me Bélu since I had last seen Sonja. It was my Hungarian family nickname. How can I begin to describe the love and joy that burst my floodgates. I was on the verge of breaking down over the telephone. I quickly told my mother that I would take the first train that day, then I put the receiver back in its cradle.

Saint-Gast advanced me money against my captivity back pay and gave me the name of a good toy store and the use of his driver for getting around. I bought a little pedal-car for François and a game set for my eleven-year-old nephew before returning to Mme. Lubchansky's to pack my things and leave word of where I would be. She warmly assured me that the extra room was mine whenever I wanted it.

30

The train from Paris arrived at Tonnerre around two in the afternoon. A bus took me from Tonnerre to Villon. My sister Violette and my little son were waiting for me at the bus stop. There he was, my François, healthy, handsome, dark-haired, smiling green eyes looking me up and down. Sonja's and my flesh and blood! He ran to me. I scooped him up in my arms and held him against my chest, his surprisingly strong little arms wrapping around my neck. I cried—from joy in holding my son again, from having survived, and from the sadness of Sonja's absence. I sobbed, my sister began to cry, and the townspeople, who had all gathered for the long-awaited homecoming of their neighbor's son, were wiping their eyes.

I was introduced to two gentlemen who had helped my family survive the Nazi occupation. The mayor of Villon had defied Pétain and refused to stamp the large "J" on ration cards issued to Jews. The village priest, who had helped out during François' ear infection, had also been constantly ready to hide François and his cousin if the Germans discovered their identity. I thanked them warmly for helping my family endure their hardships. I walked the ten minutes to my mother's house with one arm around my sister's shoulders and the other stretched down to clasp François' hand. My mother was waiting for us by her front door. She had become a broken, cancer-ravaged old woman. I was grateful for the opportunity to be with her again.

I sat down with my family. We talked and talked and talked and talked—before dinner, during dinner, after dinner. Our reunion was a joy marred only by the empty place that Sonja should have occupied.

When we talked about Sonja I began to cry. My son remarked,

"Look, Mama Violette—Papa Bélu's eyes are sweating!"

My sister hadn't yet told him about his real mother. Gaining a new father and losing a new mother all in one day was too much for anyone to handle, let alone a small child. My son fell asleep in the middle of the festivities, his head resting on my shoulder.

He didn't wake up when I carried him to my bedroom. In the dark I gently took the clothes off his small body and covered him with a blanket. I sat by him for a while, staring at the dim white outline of his face. Then I undressed and crept under the blanket as quietly as I could. Holding his vulnerable little body against mine, I fell asleep.

I stayed at Villon for two days, then my restlessness assailed me again. To avoid hurting my mother's feelings I told her that I had to return to Paris for my medical checkup and my military back pay.

Back in Paris my first stop was Michel's apartment to drop off my bags. My second stop was the Hôtel Lutétia. No news. But I learned that a group of convalescent Auschwitz inmates were expected from Switzerland in the afternoon. I stopped by to see Saint-Gast. He reinstated me on the army's payroll for the time being and handed me my food ration card and a military pass for the trains. He wrote down the address of the place where I would collect my military back pay and get a new army uniform. I was still walking around in the improvised outfit my friends had assembled for me in Linz.

I stood at Saint-Gast's door, ready to leave. He asked if he could continue to reach me at Michel Lubchansky's. I admitted I was too restless to stay in one place for long. I had decided to go down to Nice to see my Resistance comrades again and spend some time with cousins of Sonja's who had settled there before the war. Saint-Gast asked me for a favor. The military decorations for slain Resistance fighters had to be delivered to Nice in time for the parade and memorial ceremonies on July 14. Would I carry them down for him? He went to get the package for me. One of the medals in that box was for Sonja.

My final stop before returning to Michel's was the Hôtel Lutétia. My last hope for news of Sonja was futile. I left Paris on July 10.

I returned to Villon briefly. I confided to my sister Violette that I had surrendered all hope of Sonja's survival. François should know without further delay that his mother was dead. But I was still a coward. I asked her to be kind to me and break the news to him after I left for Nice.

On July 12 I took the train from Auxerre to Nice and went directly from the station to Sonja's cousins. Maurice, Giselle, and their daughter were favorites of mine. Our reunion was a smaller scale version of the exuberance in Villon. We hugged and kissed and cried and talked on and on and on. Some of it was wildly joyful, some was very sad. I learned from them that Sonja's mother had died in Auschwitz. I had to tell them that Sonja and Paule had also died there. I was delighted when Maurice and Giselle asked me to stay with them.

I visited Monsieur Ramoine. He was speechless when I suddenly appeared in his office. The next moment he wrapped me in a great bear hug. It took time to calm our joy and excitement sufficiently so that we could sit down and talk about all that had happened since we had last seen each other. This scene was to be repeated every time I met one of the small number of our surviving comrades.

One of the things I learned from Ramoine was the chain of events that had landed my little son safely in Villon after my arrest. Jackie, my driver, had seen the Gestapo take us away from the headquarters of the organization. He immediately notified the people at the laundry near our hotel. One of them ran to the park where François and his governess always spent the mornings and warned them not to return to the hotel. The other one reached Ramoine, who arranged for one of his workers to meet them in the park and carry François safely to Mme. Billard.

Mme. Billard was my next stop. I walked unannounced into her raffia shop. She was overwhelmed to see me. My letters from Munich and Laurahutte had reached her, each one temporarily reversing her certainty that I must be dead. Then she had counted me among the unnumbered victims of Nazi destruction until Saint-Gast had given her the joyful news of my survival.

At eleven o'clock on the morning of July 14, the surviving Resistance fighters in Nice and the relatives of the dead began the solemn march down the Avenue de la Victoire to the ceremonial platform erected at the monument for those who had died in the First World War. The speech-making finally ended. The names of the dead were read off, one by one, a single haunting voice calling out "not present" after each name. The crowds remained silent. Taps were played for the dead. I felt cold inside. Tears rolled down my cheeks.

The presentations and reading of the individual citations began in the late afternoon. So many of us had died that the ceremony continued into the dawn of the next day. Sonja's decoration and citation were presented around midnight. After I accepted them I asked to be excused because I was very tired.

I arrived at Maurice and Giselle's wrapped in gloom. Light shone from under their front door. I was surprised that they had waited up so long past their bedtime for me. I had barely opened the door on a brightly lit living room when everyone began shouting at once. From an incomprehensible babble I finally caught, "Sonja and Paule . . . Sonja and Paule . . . Sonja and Paule!"

They thrust a telegram into my hands, a telegram from Saint-Gast with word that Sonja and Paule had been found in a prisoner-of-war camp in Russian-occupied Magdeburg, Germany.

They were alive! I crowed, I shouted, I laughed. I jumped around the room. A full five minutes passed before I could actually speak. Maurice brought out a bottle of brandy. Laughing and happy we all toasted *l'chaim*—to life! I took out Sonja's posthumous medal and cried and laughed.

The next morning I called Saint-Gast as early as politeness permitted. He was already waiting for my call. Sonja and her sister had escaped from the Auschwitz death march at Pless as Mr. Mandel had suspected. They were eventually picked up by the Polish Underground and taken in by a sympathetic Polish family. The arriving Russians took them to a French prisoner-of-war camp. Sifting out French collaborators and the disguised SS soldiers was slow work, turning the repatriation process into a laborious and time-consuming procedure. All I needed now was the patience to await their release.

I told Saint-Gast that the heads of the local Resistance groups in my network had decided to treat the returned deportees to a proper convalescence. He took the address at Les Abrets, where he would be able to find me for the next several weeks. What a coincidence that the Resistance committee should rent a hotel in Les Abrets, the village near Grenoble where Sonja and I had spent our memorable night in the mayor's cornfields.

Later that day I telegraphed my mother with the wonderful

news about Sonja and my plans to go to Les Abrets for the next several weeks. I wrote a note to Saint-Gast asking him to postpone my army decision until Sonja was back in France.

Pierre, still the mayor of Les Abrets, was intoxicated with happiness when I arrived. He took me home to a delicious dinner. He was still the same shrewd-eyed, intelligent peasant. His small brown eyes were as alive as ever. We drank too much *marc* after dinner and talked and laughed long into the night. The next day I slept. At last I was able to rest.

Pierre and I played chess daily. I lost every single game because one eye was constantly on the post office across the street from my hotel. It got so that everyone in town knew what I was waiting for. At 10:00 A.M., about ten days after I arrived at Les Abrets, the postmaster came tearing across the street waving a pale blue-gray telegram high over his head.

"Here it is!" he shouted. "Your wife is in Paris!"

"*Home with Paule. Meet you in Villon. Najon, najon.*" The signature was Sonja's fractured phonetic Hungarian for "much, much love."

The one train serving the tiny town of Les Abrets had already come and gone. I went to my friend Pierre for help. I waited in his office while he searched to find a way for me to get to Villon. He was all smiles when he returned. An antiques dealer in town from Nice was driving to Paris. He would drop me off at Villon in exchange for gasoline, which was still in very scarce supply. Pierre added: "And I have already arranged to give him gasoline from the town's reserve! After all, what are friends for?"

Half an hour later, the antiques dealer arrived to find me pacing the curb in front of my hotel. At last I was really on my way home!

Around midnight we pulled up to my mother's house. The lights were blazing. Through the front window I saw Sonja, dressed in a heavy Russian army uniform, suntanned and healthy looking, cocky as ever, wearing the same eyeglasses she had when we were arrested. She was sitting on the sofa with François asleep in her lap. I burst in and gathered them both in my arms, holding them close, feeling the warmth of their bodies, inhaling their smell. I couldn't get enough of them. This was the reunion I had dreamed of!

REPUBLIQUE FRANÇAISE

ORDRE NATIONAL DE LA LÉGION D'HONNEUR

HONNEUR PATRIE

LE PRÉSIDENT DE LA RÉPUBLIQUE FRANÇAISE
GRAND MAÎTRE DE L'ORDRE NATIONAL DE LA LÉGION D'HONNEUR

nomme par décret de ce jour, **Madame N A D E L Sophie épouse H A A S Z**
ancien Sous-Lieutenant des Forces Françaises Combattantes, Déportée-Résistante
née le **26 Décembre 1910** *à* **Anvers (Belgique)**

CHEVALIER DE LA LÉGION D'HONNEUR

pour prendre rang du **22 Mai 1980** *et jouir de tous les droits, honneurs et prerogatives attachés à cette qualité*

Fait à **Paris** le **30 Octobre 1979**

Scellé du sceau de l'Ordre sous le n° **2284 MU 79**
Le Secrétaire Général adjoint.

Par le Président de la République
LE GRAND CHANCELIER DE LA LÉGION D'HONNEUR.

Certificate naming Sonja Haas a Chevalier in the French Legion of Honor, which carries with it the Croix de Guerre with palm.

EXTRAIT

du DÉCRET en date du 27 MAI 1971

publié au J. O. du -5 JUIN 1971

portant nominations dans la LEGION D'HONNEUR

ARTICLE 1er.- Sont nommés dans l'Ordre National de la Légion d'Honneur :

AU GRADE DE CHEVALIER

DEPORTES - RESISTANTS

Armée de Terre

. . . .

HAASZ (Béla) 27.10.1911 - Sous-Lieutenant des Forces Françaises Combattantes - recrutement de Paris.

" S'est engagé dès 1942 dans un réseau de renseignements pour " défendre sa seconde patrie. Chargé de missions particulièrement délicates, " organisa un secteur qui donna de très beaux résultats. Arrêté par la " gestapo, ne dévoila, malgré les tortures, aucun des secrets de l'organisa- " tion à laquelle il appartenait.

" Paya par la déportation, sa fidélité à la France".

Cette citation annule la citation à l'ordre de la Division attribuée par Ordre Général n° 11 du 24 décembre 1946.

. . . .

CES NOMINATIONS COMPORTENT L'ATTRIBUTION DE LA CROIX DE GUERRE 1939-1945 AVEC PALME.

Par le Président de la République
Le Premier Ministre
signé : Jacques CHABAN-DELMAS

signé : Georges POMPIDOU

POUR AMPLIATION

L'Administrateur Civil de 1ère Classe BALAT
Chef du Bureau des Décorations
P.O. Le Chef de la Section "Invalidité"

Le Ministre d'Etat
Chargé de la Défense Nationale
signé : Michel DEBRE

Certificate naming Albert Haas a Chevalier in the French Legion of Honor and awarding him the Croix de Guerre with palm.

Haasz Albert

Répondant à l'appel de la France en péril de mort, vous avez rallié les Forces Françaises Libres. Vous avez été de l'équipe volontaire des bons Compagnons qui ont maintenu notre pays dans la guerre et dans l'honneur. Vous avez été de ceux qui, au premier rang, lui ont permis de remporter la Victoire! Au moment où le but est atteint, je tiens à vous remercier amicalement, simplement, au nom de la France!

1er Septembre 1945 C. de Gaulle

Commendation from Charles de Gaulle to Albert Haas, September 1, 1945.